Legume Crops for Food Security - Cultivation and Benefits

Edited by Jose C. Jimenez-Lopez and Julia Escudero-Feliu

Published in London, United Kingdom

Legume Crops for Food Security - Cultivation and Benefits
http://dx.doi.org/10.5772/intechopen.1001792
Edited by Jose C. Jimenez-Lopez and Julia Escudero-Feliu

Contributors
Akanksha Singh, Alane Farida, Ambrin Rajput, Anirban Roy, Chabaca Rabha, Chandrashekhar Biradar, Dhanesha Nanayakkara, Donna Harris, Eliezah M. Kamau, Esther N. Kimani, Hafeezullah Babar, Ioana Stanciu, Jacques Fils Pierre, Jeet Singh Sandhu, Jill F. Keith, Jim Heitholt, Mariem Bouhadida, Mohammad Zarif Sharifi, Nataliia Lavrenko, Payabrata Roy, Peter A. Dotray, Qurban Ali Panhwar, Rael Karimi, Samita Saha, Satyaki Nandan Maity, Senaratne Ranamukhaarachchi, Sergiy Lavrenko, Shiv Kumar, Smita Kaul, Sourav Garai, Sukamal Sarkar, Suman Dutta, Sushil Kumar Chaturvedi, Tania María Guzmán Armenteros, Tapash Dasgupta, Todd A. Baughman, William James Grichar

First published in London, United Kingdom, 2025 by IntechOpen
IntechOpen is the global imprint of INTECHOPEN LIMITED, registered in England and Wales, registration number: 11086078, 167-169 Great Portland Street, London, W1W 5PF, United Kingdom

For EU product safety concerns: IN TECH d.o.o., Prolaz Marije Krucifikse Kozulić 3, 51000 Rijeka, Croatia, info@intechopen.com or visit our website at intechopen.com.

British Library Cataloguing-in-Publication Data
A catalogue record for this book is available from the British Library

Legume Crops for Food Security - Cultivation and Benefits
Edited by Jose C. Jimenez-Lopez and Julia Escudero-Feliu
p. cm.

This title is part of the Agricultural Sciences Book Series, Volume 23
Topic: Agronomy and Horticulture
Series Editor: W. James Grichar
Topic Editor: Ibrahim Kahramanoglu
Associate Topic Editors: Murat Helvaci and Olga Panfilova

Print ISBN 978-0-85466-593-8
Online ISBN 978-0-85466-592-1
eBook (PDF) ISBN 978-0-85466-594-5
ISSN 3029-052X

If disposing of this product, please recycle the paper responsibly.

IntechOpen Book Series

Agricultural Sciences

Volume 23

Aims and Scope of the Series

The importance of agriculture cannot be overstated. It helps sustain life, as it gives us the food we need to survive and provides opportunities for economic well-being. Agriculture helps people prosper around the world and combines the creativity, imagination, and skill involved in planting crops and raising animals with modern production methods and new technologies. This series includes two main topics: Agronomy and Horticulture, and Animal Farming. This series will help readers better understand the intricacies of production agriculture and provide the new knowledge that is required to be successful. The success of a farmer in modern agriculture requires knowledge of events happening locally as well as globally that impact input decisions and ultimately determine net profit.

Meet the Series Editor

W. James Grichar has been employed with Texas A&M AgriLife Research for over 45 years with an emphasis on research in agronomy, plant pathology, and weed science. He obtained his BS from Texas A&M in 1972 and his Masters of Plant Protection in 1975. He has published 195 journal articles, over 330 research reports and briefs, 11 book chapters, and over 300 abstracts of profession meetings. He also directs research in many crops including corn, grain sorghum, peanuts, and sesame. He has held various positions in different professional societies including the American Peanut Research and Education Society, Southern Weed Science Society, and Texas Plant Protection Conference in addition to being Associate Editor for Peanut Science and Weed Technology. Significant accomplishments have included spearheading efforts to determine the optimum planting time for soybean production along the upper Texas Gulf Coast. These efforts have shown growers that soybean yields can be improved by 10 to 20% by following a late March to early April plant date. He also has been instrumental in developing a herbicide program for peanut production in the south Texas growing region. Through the development and use of herbicides that are effective against major weed problems in the south Texas region, peanut yields have increased by 25 to 30%.

Meet the Volume Editors

Dr. Jose C. Jimenez-Lopez holds a BS in Biochemistry and Molecular Biology (1998), a BS in Biological Sciences (2001), an MS in Agricultural Sciences (2004), and a PhD in Plant Cell Biology (2008) from the Spanish National Research Council (CSIC). He completed postdoctoral research at Purdue University (2008–2011) and was a Marie Curie Senior Research Fellow under the EU-FP7 program at the University of Western Australia (2012–2015). From 2016 to 2021, he was a Senior Ramón y Cajal Researcher at CSIC. Currently, he is a tenured scientist at CSIC, focusing on the functionality, health benefits, and molecular allergy aspects of seed proteins from agro-industrial crops like legumes. He has co-authored over 90 peer-reviewed articles, over 30 book chapters, and 3 patents and has presented at over 150 international conferences. He is an active member of several scientific societies, including the Spanish and International Legume Societies.

Dr. Julia Escudero Feliu is a postdoctoral researcher at the Estacion Experimental del Zaidin (EEZ), a research institute belonging to the Spanish National Research Council (CSIC), Granada, Spain. She holds a Ph.D. in Clinical Medicine and Public Health from the University of Granada and a degree in Biomedical Engineering. Her doctoral research focused on the potential of β-conglutin proteins from narrow-leafed lupin seeds as preventive, cytotoxic, and radio-sensitizing agents in breast cancer cells. This interdisciplinary work spanned oncology, cancer genetics, and nutraceuticals, leading to high-impact publications and an international patent. Currently, she integrates molecular biology and bioinformatics to develop pathogen detection tools for agricultural crops. She has participated in over 20 international conferences. She collaborates in scientific book editing, contributing to the dissemination of knowledge in nutraceuticals, legumes and their potential in translational biomedicine.

Contents

Preface

Legumes play a crucial role in global agriculture, not only as a rich source of plant-based proteins but also as key crops for soil sustainability and ecological health. The present book, *Legume Crops for Food Security – Cultivation and Benefits*, explores the vital importance of legumes in global food security, addressing their potential to improve human nutrition and promote resilient and sustainable agricultural practices.

This book also highlights the biodiversity of cultivated legumes, including those undervalued and underutilized species, and how their utilization can significantly contribute to a more inclusive and diverse agriculture. By considering aspects such as soil quality improvement, reducing the need for chemical inputs, and their ability to foster ecosystem health, legumes emerge as an indispensable resource in the fight against current global challenges, such as climate change and food insecurity.

The topics covered in this book also explore the crucial role of legumes in sustainable agricultural systems, particularly in extensive cropping systems. Legumes are vital for human nutrition and serve as valuable livestock feed and forage, contributing to food security across various production chains. Their ability to enhance soil fertility and their use in weed control make them a cornerstone of low-environmental-impact agriculture.

Through advancements in the genetics and cultivation of legumes, this book underscores how these crops can be key to sustainable agricultural intensification, both for food production and environmental conservation. Legumes offer not only nutritional and bioactive benefits for health but also play a fundamental role in enhancing the resilience of agroecosystems against the adverse effects of climate change.

In this context, using underutilized legumes is critical to diversifying our food sources and strengthening global food security. This book offers a valuable perspective on how legumes can provide solutions to both agricultural production challenges and the nutritional needs of the future.

Jose C. Jimenez-Lopez and Julia Escudero-Feliu
Spanish National Research Council (CSIC),
Granada, Spain

Section 1

Legume Uses and Benefits

Chapter 1

Crop Biodiversity: Utilization of Neglected and Underutilized Beneficial Legumes

Esther N. Kimani, Eliezah M. Kamau and Rael Karimi

Abstract

The current dependence on a limited variety of primary staple crops for food and nutritional security is not sustainable, particularly in light of the challenges posed by climate change. Numerous underutilized legumes are of regional importance and are predominantly produced by smallholder farmers. Despite being climate-smart crops with significant agronomic benefits that can promote sustainable livelihoods, these crops have garnered limited research attention compared to commercial crops. Underutilized legumes are a rich source of protein, minerals, and bioactive compounds. Their cultivation and use have persisted, albeit restricted to specific cultures, regions, or countries. These legumes are increasingly recognized for inclusion in modern diets, and in manufacturing protein-based packaging products. This review addresses the merits of underutilized legumes in nutrition, health, and possible industrial uses that can enhance their production, improve acceptance, and boost commercialization prospects. The challenges surrounding the cultivation and consumption of these lesser-known legumes are also explored.

Keywords: underutilized legumes, food security, nutritional security, soil conservation, distribution

1. Introduction

Food production faces numerous challenges stemming from emerging pests and diseases, changes in climatic conditions, an increase in population, and the degradation of soils. Several approaches have been identified to overcome these challenges and produce enough food for the rising population. These include the breeding of high-yielding and adapted crop varieties, the use of conservation agriculture and fertilizers to improve the soils, and the use of integrated approaches in the management of insect pests and diseases. These efforts have been mainly targeted towards improving a few selected major global crops like rice, maize, potato, wheat, and legumes such as soybeans and common beans [1]. However, numerous minor and underutilized crops are regionally important but have yet to receive minimal research and government attention for their development. Despite their minimal economic value in global markets, underutilized crops are predominantly farmed by traditional farmers, particularly women, to supply families with high-nutritious food security.

Underutilized plants are available for all types of crops, including cereals, vegetables, legumes, tubers, oilseeds, fruits, and forages [2, 3]. Legumes are widely cultivated and consumed globally, mainly because of their nutritional benefits to humans and animals. They are a great source of protein in human diets and animal feed, providing amino acids, minerals, and vitamins. The plant parts used are the leaves, immature green and mature dried seeds, pods, and, in some cases, tubers. While some legumes have been commercialized globally, others are cultivated and consumed only locally, and some still grow in the wild [4]. These neglected legumes' diverse cultural values and important agronomic, genetic, and biochemical traits have not been fully utilized. The regional and international markets have yet to fully harness these lesser-known legumes' economic potential [5].

The overlooked and underutilized legumes, among other crops, possess significant potential for enhancing food and nutritional security while also exhibiting resilience to the climatic changes currently being experienced. Due to their valuable untapped potential, these crops are considered the future's food. They also have the potential to provide income for smallholder farmers who predominantly produce them, especially in developing countries.

While colonialism disrupted the cultivation and consumption of certain indigenous crops in Africa and Asia, various regions within these continents have preserved these crops. They are now receiving acknowledgment through their incorporation into contemporary diets [6]. However, the consumption of indigenous legume crops is still very low compared to that of significant legumes. Initiatives have been undertaken to recognize the benefits of these legumes and encourage their cultivation, utilization, and market integration. These efforts would improve food security, promote biodiversity conservation, and provide sustainable productivity while leveraging their potential for economic benefits for farmers. This chapter discusses the origin and distribution, different uses, and challenges faced in the production and uptake for consumption of selected cultivated, neglected, and underutilized legumes.

2. Distribution and production of underutilized legumes globally

The global food system has changed over time, placing high importance on popular cereals and horticultural crops. Despite these, cultivation and utilization of neglected and underutilized legumes have continued, though limited to cultures, regions, or countries.

Studies on wild lima beans have pointed the origin to the northern Andes and Mesoamerica [7, 8]. Flat, medium-sized seeds characterize the domesticated event in the Andes gene pool, while the Mesoamerica cultivars had small globular and large flat seeds [8]. Lima bean, *Phaseolus lunatus*, is the second most important species of the genus Phaseolus. It is cultivated for human food and as an alternative income source in Brazil, where landraces and cultivars with high variability are available [8–11]. The bean is cultivated by both large- and small-scale farmers, and research activities in Brazil have been conducted on the nutrition, agronomic, phytochemical, breeding, pest, and disease aspects of lima bean [9, 12, 13]. Different accessions of lima beans are also available in African research institutions [14–16]. The accessions in Ethiopia, cultivated in the east Shoa zone and South Western area, have different levels of crude protein, crude fiber, iron, and total ash. It has been recommended that they be incorporated in addressing malnutrition challenges in the region [16, 17]. Approximately 80% of interviewed farmers in the Bono East region of Ghana were aware of lima bean [18]. The different varieties of lima beans available in this country are generally consumed because of their nutritional value [14, 18]. Similarly, the consumption of lima beans is presented among the rural inhabitants in the North Delta region of Egypt, South West and South East regions of Nigeria, and five administrative areas of Côte d'Ivoire [15, 19–21].

The Bambara groundnut (*Vigna subterranea L.*), a climate-resilient crop, originates in West Africa [22]. Nigeria is the largest producer of Bambara groundnut, with an average production of 0.1 million tonnes annually [22]. Zambia and Zimbabwe are the most prominent producers of legumes. The legume is characterized by significant variability morphologically and genetically in the landraces that are commonly grown by smallholder farmers [23, 24]. Variability has also been reported for studies carried out in accessions from Kenya, Uganda, southern Africa region (Namibia, Swaziland, Zimbabwe, Malawi), Singapore, Ghana, Burkina Faso, and Tanzania, among other countries in Africa [25–31]. The legume has been introduced from West Africa to the Asian continent, with reports from Indonesia and Thailand [32, 33]. The endearing characteristics of Bambara groundnut have earmarked it as a potential crop, especially in areas where malnutrition, food insecurity, and a growing population of people are present, for example, in China [34, 35].

Rice bean (*Vigna umbellata*), a legume with numerous potentials, is underutilized in several places. The landraces are available in Nepal, Myanmar, Thailand, Laos, India, southern China, and East Timor [36, 37]. The most extensive collection of accessions is held in gene banks in India, Japan, and Nepal [37]. The wild collections have shown genetic diversity, which is vital in crop improvement programs [38, 39]. Domestication of rice bean is believed to have spread its cultivation in other regions in the original countries and further spread to countries such as Indonesia, Japan, Korea Indies, USA, Australia, east Africa, Sri Lanka, and other countries [37, 39, 40], where production and utilization are low.

Canavalia ensiformis, commonly called Jack bean, is a crop native to South and Central America [41]. It is an annual or perennial crop found in the tropics and subtropics of Africa, Asia, Latin America, South America, India, and the South Pacific [42–44], with large-scale production in Congo. It is used as food for humans and livestock forage [45, 46].

The African yam bean (*Sphenostylis stenocarpa*) is another native legume crop to West Africa, believed to have originated in Nigeria. The plant is well adapted to the tropical climate in the region and is grown primarily for its edible tubers and seeds [47]. Accessions with high genetic variability are available in the International Institute of Tropical Agriculture gene bank in Ibadan, Nigeria [48, 49]. The distribution of the African yam bean extends beyond Nigeria to other West African countries such as Ghana, Benin, Togo, and Cameroon. It is also found in Central Africa, including the Democratic Republic of Congo and East Africa, Ethiopia, Malawi, and Zimbabwe [50]. In these regions, African yam beans are typically cultivated by small-scale farmers as a source of food and income.

The tropical legume *Psophocarpus tetragonolobus* (L.) DC. (winged bean) is a domesticated crop in Asia and the Western Pacific. Cultivation has also been reported in South America and West Africa [51, 52]. The original progenitor is undiscovered, though wild relatives of *Psophocarpus* have been identified in Africa [53]. Most plant parts are consumed as food, including mature and immature seeds, green young pods, tubers, leaves, and flowers [53, 54]. Yields of 5500 to 11,500 kg/ha tubers were reported in Papua New Guinea, with projections of 1.1 tonnes/ha and 800–1000 kg/ha of seed yields against a potential of 10 tonnes/ha [55]. There is evidence of variation of winged bean accessions using morphological attributes, molecular markers, and genetic mapping [56–60].

Another bean that originated in Africa is the Dolichos bean or Hyacinth bean (*Lablab purpureus*). A study on genetic diversity using nuclear simple sequence repeats on 493 accessions tagged the origin to Rwanda [61]. However, Tanzania holds the largest genetic resource in Africa at the Nelson Mandela African Institution of Science and Technology (NM-AIST) [62]. Other genetic resources holding accessions from different areas are maintained in Asia, America, Europe, and Oceania [62, 63]. The accessions have a range of variability from narrow to wide variability, depending on the source [62, 64–66]. The production and consumption in Africa have lost popularity in the last decades due to the introduction of the common bean [67], while the production in India in 2021/2022 was at 2.51 metric tonnes with an area harvested at 0.228 mha [68]. The crop is mainly cultivated in small acreage, either as monocrops or intercropped with cereals [69, 70].

Similar to the Dolichos bean, India is the largest producer of Horse gram (*Macrotyloma uniflorum* (Lam.) Verdc.), where it has its origin [71, 72]. It is also cultivated in Sri Lanka, the West Indies, and Malaysia. The wild and domesticated genotypes of horse gram are also available in Africa; however, cultivation of the legume is not well documented in this region [73]. The genotypes in India have shown variability in qualitative and quantitative traits and are a good source of breeding material for improving horse gram [74–76] for production.

Vigna aconitifolia (moth bean) is a native legume from India, where 85% of the production is from Rajasthan, the driest region [77, 78]. However, it is also cultivated in other parts of Asia, including Myanmar, Malaysia, Sri Lanka, Africa, and the USA, with different accessions in gene banks worldwide [79]. Variations in agro morphological traits of the moth bean accessions have been documented [79]. Although genetic diversity studies using molecular markers have been scarce, a preliminary study on 15 elite cultivars using random amplified polymorphic DNA (RAPD) markers has shown high polymorphism [80]. The agro morphological variation and the high polymorphism present a good background for further research and breeding opportunities to improve the moth bean, one of the identified foods for the future due to its valuable benefits.

Pigeon pea (*Cajanus cajan*), the most cultivated species of the Cajanus genus, is mainly produced in India. Its origin is attributed to Northern India, but it has been grown in East Africa for a long time [81]. India recorded 80% of all pigeon peas produced worldwide in 2022, while Eastern Africa produced 13% in the same year [82]. Malawi and Myanmar are also large producers of pigeon peas. Other regions where there is production include Southeast Asia, West Africa, and the Caribbean [82]. Morphological, agronomic, and molecular diversity has been revealed among accessions. However, the analysis has been on a limited number of accessions [83, 84]. Numerous other grain legumes are available in the regions but remain understudied and underutilized. The potential of the legumes remains unexploited due to the narrowing utilization of specific commercially available legumes. However, with the growing demand for legume protein, there are opportunities to tap into the beneficial legumes to advance food and nutrition security.

3. Benefits of cultivation and consumption of neglected and underutilized legumes

3.1 Increased biodiversity

The underutilized crops provide plant diversity in farming and agrosystems [85–87]. Plant diversification offers several environmental, animal feed, and human food security benefits. Neglected and underutilized legumes are represented in a diverse range of Genera in the family Fabaceae, many of which are indigenous to specific regions. The genus Vigna consists of the largest number of domesticated underutilized bean species. They include species *V. aconitifolia* (Jacq.) Maréchal (moth bean), *V. radiata* (mung bean), *V. umbellata* (rice bean), *Vigna subterranea (Bambara groundnut), and Vigna angularis (adzuki bean),* among others. Different accessions of the underutilized legumes are held in genetic resource banks globally. The accessions of the various underutilized legumes have genetic and morphological variability and can be evaluated to increase cultivated biodiversity [88].

3.2 Resistance to abiotic and biotic stresses

Declining soil fertility is a major abiotic stress factor affecting crop production. Nitrogen is one of the crucial nutrients in the soil, and its presence enhances crop yield. Application of fertilizers among smallholder farmers is limited as they have little option for external inputs due to high cost. Nitrogen fixation potential is embedded in underutilized legumes such as pigeon pea, Bambara groundnut, and lablab bean [89–91]. The symbiotic nitrogen fixation in legumes harnesses atmospheric oxygen and contributes to a positive nitrogen balance in the soil. Different accessions of the legumes fix varying amounts of nitrogen, with some known to fix up to 280 kg/ha depending on the production site [92]. The nitrogen then boosts the production of both the legume and companion crop when intercropped, and the subsequent crops, particularly cereals, when in rotation [91]. Different underutilized legumes can withstand varying soil properties such as salinity, varying pH, and waterlogging [2]. The underutilized legumes like pigeon pea, Bambara groundnut, African yam bean, and moth bean are known for their adaptability to poor soils and those with high salinity [79, 89, 92, 93], especially in marginal areas. The deep roots system that enables the legumes to withstand dry weather also helps bind the soil, reducing erosion even on steep slopes. Underutilized legumes such as pigeon peas and dolichos bean are also important in soil conservation due to the good ground cover, suppressing weeds, and reducing moisture loss [90, 91]. The legumes add considerable biomass to the soil, thus improving nutrient status, maintaining soil structure, promoting microbial activity, and improving soil health.

Reports have shown that underutilized legumes are grown in various environments [54]. Among them are those that exhibit tolerance to drought, frost, and various other stresses frequently encountered, mainly due to the fluctuating weather patterns induced by climate change [2, 4, 88]. Drought-tolerant underutilized legumes include Bambara groundnut, moth bean, horse gram, lablab, and pigeon pea. The drought tolerance is attributed to these legumes' deep tap root system, enabling them to utilize water from subterranean levels [74, 90, 94, 95]. The tolerance to harsh climatic conditions ensures food availability during these times [96]. Another benefit of incorporating neglected and underutilized legumes in crop rotation is breaking cycles of pests and diseases and managing weeds. In maize fields, Dolichos and faba beans are incorporated as trap crops in managing striga weed [97]. At the same time, the legumes are also a good option for crop rotation to reduce maize lethal disease by reducing virus inoculum levels [98].

3.3 Nutrition and health benefits

Underutilized crops provide diverse foods in diets [99–102] of different communities globally. They are a rich source of nutrients beneficial for healthy living and preventing and managing specific diseases. Despite the neglect of underutilized legumes, they are a rich source of affordable and diverse protein, one of the most essential nutrients for good health. Legumes constitute 33% of dietary protein requirements, and the inclusion of underutilized legumes increases the diversity of the source of the proteins. These crops are also a good source of minerals such as calcium, iron, and phosphorus. The protein levels in some of the underutilized legumes range from 8.6–43%, while calcium, iron, and phosphorus levels range from 28 to 850, 1.7–128, and 4.3–640 mg [88], depending on the variety. Dried winged bean seeds and other legumes are also good for manganese, potassium, sodium magnesium, and zinc [43, 103]. In particular, winged bean has protein levels comparable to soybean, one of the most produced and utilized legumes [104]. The bean is also rich in oil. Supplementing cereals with protein-rich legumes is considered one of the best solutions to protein-calorie malnutrition in the developing world [100] fodder. Broad adoption of diverse, resilient crops such as pigeon pea, Bambara groundnuts, and mung bean can provide synergy and short-term supply of dietary requirements and thus stand to play an important role in food security, providing balanced diets and alleviating poverty.

The legumes also have health benefits due to the presence of phytochemicals. Their consumption is associated with the management of conditions and diseases such as diabetes type II, obesity, hypercholesterolemia, cancer tumors, and heart conditions [19, 88, 105]. Crude fiber in the legumes maintains the digestive system by alleviating constipation, managing blood glucose levels, and minimizing cholesterol build-up [106]. The legumes are a good source of iron, with the red Bambara groundnut seeds showing higher iron levels. High iron levels are crucial in addressing iron deficiency [89]. Underutilized legumes are valuable as human food, and the foliage is used as fodder. Leaves from dolichos bean, winged bean, moth, and rice beans, among other legumes, have been incorporated in livestock feeds as a source of protein, minerals, and vitamins [45, 79, 81, 107].

Legumes have nutritional components that are beneficial for the health of humans. These components include proteins, minerals, vitamins, slowly digestible starch, dietary fibers, phytosterols, and other phytochemicals [108]. Legumes have a higher protein content than other cereals, constituting 33% of dietary protein requirements [88]. Underutilized legumes have protein content ranging from 8.6–43%. Winged bean, in particular, has protein levels of 27–43%, comparable to soybean, one of the most produced and utilized legumes [103, 104]. Other underutilized legumes with crude protein levels of above 30% include grass pea, horse gram, lablab, lima, and yard-long bean [88]. The proteins have nutritional and functional properties depending on their quality. However, the proteins have varying amino acids depending on the plant and variety [109]. Specifically, the legume proteins are a rich source of lysine, an amino acid lacking in cereals. Thus, protein from legumes can be used to fortify the cereals. Amino acids in legumes help enhance human genetic, reproductive, and digestive processes and boost immunity [110]. However, they may be deficient in some amino acids, especially the sulfur-containing amino acids methionine and cytosine, such as those found in Jack bean and pigeon pea, among other underutilized beans. Underutilized legumes also contribute to the range of amino acids, with some underutilized legumes having a higher content than commonly used beans. For example, total proteins from moth bean varieties were higher than those of cowpea, common bean, and chickpea [78].

Bioactive compounds from legumes are gaining attention as they benefit human health. Numerous bioactive compounds have been reported in underutilized legumes [52]. The compounds include polyphenols, alkaloids, saponins, carotenoids, and anthocyanins [45]. Different species and varieties provide varying nutraceutical properties [108], with the compounds addressing health issues such as cardiovascular diseases, diabetes, cancer, hypertension, and obesity and acting against inflammatory agents, among other health conditions [108, 109, 111]. Antioxidant compounds include polyphenols, which are abundant in legumes. The compounds act against the effects of microbial and viral infections, allergic reactions, cancer, mutations, and agents of inflammation [52]. Flavonoids are the most common polyphenols found in legumes. A study of 15 genotypes of horse gram bean revealed flavonoids with a mean value of 197.53 μgCE/g. However, this is lower than that recorded for pigeon peas and kidney beans [107]. Flavonoids were higher in brown beans as compared to black and cream lablab beans, with higher levels of 1-diphenyl-2-picrylhydrazyl (DPPH), a free radical [112]. The lablab bean is rich in isoflavones that induce an increase in the production of insulin by regulating gluconeogenic enzymes [113]. A mean of 47.48 μMTE/g of the scavenging activity of DPPH has also been reported from rice bean extracts [107]. Extracts of African yam bean, Bambara groundnuts, jack, and winged beans have an abundance of polyphenols [114, 115], with winged beans recording a range of antioxidant capacity of 1.3–1.8 mg AAE/g [115]. Traditional medicine has used winged beans as the compounds provide antibacterial, antifungal, anti-proliferative, and anti-inflammatory activities [115]. Biopeptides extracted from winged beans have been found to have high inhibitory activity angiotensin I-converting enzyme, thus managing hypertension [116]. Tannins, lectins, and saponins are beneficial in diabetes and cancer management. These have been reported in jack, mung, and adzuki bean extracts [111, 117], among other legumes. Tannins and flavanols also confer resistance to nematodes that infect the digestive system [118].

3.4 Industrial applications and socio-cultural values

Plant-based proteins have gained importance due to their low cost and availability compared to chemically synthesized ones. Proteins from underutilized legumes have been studied for their functionality and applicability for incorporation in food processing [52]. The protein isolate and concentrate are extracted and added to various foods during baking, preparation of beverages, and meat analogues [119]. The characteristics of the protein are determined by the amino acid profile, the structure of the protein, and its effects on water. The proteins obtained from the underutilized legumes are also being explored for use as food packaging and films [120]. The use of natural sources of packaging and films is due to the increasing awareness of the management of plastic and glass waste, which is the common material in the packaging of foods. Including protein-based packaging and films would impact the environment by reducing waste and improving health benefits since the packaging materials have been found to influence foods depending on the storage conditions and period.

Underutilized legumes are cultivated mainly by smallholder farmers, with males and females having different roles [81, 96]. The requirement of minimal farm inputs, harvesting, and processing is appealing to farmers due to the low cost of production. However, this is a cause for low yields that are achieved. Winged bean, which is nutritionally comparable to soybean, is one of the legumes that requires low inputs for production while growing in a wide range of environments [55]. The legumes diversify the households' diets and provide an alternative source of income when the surplus is sold. In many regions where the underutilized legumes are cultivated, the legumes are associated with a cultural or religious event. Bambara groundnuts are offered as gifts to circumcised boys in Cameroon to signify their initiation into manhood. It is consumed during funeral and wedding ceremonies [31]. Several underutilized legumes also have social, cultural, and religious values in Nepal, where the preparation methods are specific for the occasion [38].

4. Challenges in the production and uptake for consumption of underutilized legumes

Despite the numerous benefits associated with the neglected and underutilized legumes, the uptake for cultivation and consumption remains low over the decades. Their production and consumption have challenges that would need to be addressed to improve utilization. Most of the underutilized legumes have low grain yield [93, 121, 122], a factor hindering farmers' widespread adaptation and production. Research on minor legumes has been ignored and overlooked for a long time [45] and rarely receives funding for their systematic breeding and improvement. Modern breeding programs concentrate on enhancing critical agronomic traits of major legumes through controlled crosses and selection, which lead to significant yield improvements over time. Farmers, therefore, are more attracted to the high-yielding and widely accepted major legume varieties at the expense of the underutilized legumes [121].

Agricultural mechanization and intensification through high-yielding seeds, inorganic fertilizers, pesticides, irrigation, and monoculture are necessary to produce more food for the ever-growing population [123]. However, agricultural intensification, extensive monoculture, and preference for more high-yielding varieties have been cited as contributing to the underutilization of many traditional crops [124]. For instance, due to the expansion of groundnut (*Arachis hypogea*) production, Bambara groundnut has remained an underutilized crop in most parts of Sub-Saharan Africa [125].

Seed is a vital component in all crop production and is essential in developing an effective value chain for underutilized crops. More than 90% of the minor legumes grown in Sub-Saharan Africa are distributed through an informal seed system [121]. Farmers prefer planting farm-saved seeds due to their affordability and availability during the next planting season [126]. A lack of a regular supply of seed in predictable quantities and of high quality hinders the production of underutilized pulses outside the traditionally used communities [121, 127, 128].

Diseases and pests are also significant constraints to underutilized legume production. In dolichos bean production, for example, aphids and pod borers are the primary pests [69, 121, 129]. In contrast, pigeon peas, pod borer, and plant wilt disease are considered the major constraints of economic importance [130]. Insect pests can cause direct damage by feeding on crops or transmitting disease viruses. Generally, an integrated approach to managing insect pests and diseases, such as using certified seeds, chemicals, and resistance breeding, is usually recommended to reduce crop losses [131]. However, some of the solutions mentioned above, such as certified seeds, may be unavailable or prohibitively expensive for underutilized legume producers. Investment in breeding resistant varieties, training farmers in integrated pest and disease management, and enhancing seed systems of underutilized legumes can boost the productivity of these neglected legumes.

Another factor that has hindered the uptake of underutilized legumes is inadequate technical knowledge to support commercial cultivation [132]. The majority of the smallholder farmers in Sub-Saharan Africa base their agricultural decisions on locally developed indigenous knowledge. Nevertheless, there is a need to integrate it with scientific knowledge to increase the productivity and commercialization of these crops [133]. For instance, some traditional production systems disregard the application of soil amendments to legumes, reasoning that the pulses can generate their nutrients from the atmosphere. However, adding plant nutrients like phosphate fertilizers in depleted soils is necessary for adequate nitrogen fixation, hence higher yields [134] and profitability. The involvement of farmers while identifying and filling in particular technology gaps is important to increase the adoption of the new knowledge [132].

Legumes contain several antinutritional factors (ANFs) that negatively affect the digestion and bioavailability of essential nutrients [23]. These ANFs negatively affect the uptake of protein and carbohydrates from neglected and underutilized legumes despite some of the factors being bioactive compounds. The commonly reported antinutrients in pulses include condensed tannins, phytic acid, oligosaccharides, flatulence factors, haemagglutinin, and trypsin inhibitors [111]. High levels of these antinutrients, especially oligosaccharides, can cause consumers' flatulence, which makes them hesitant to eat the legumes [71, 110, 135, 136]. The antinutritional factors also inhibit the bioavailability of minerals [106]. It is, therefore, necessary to reduce the number of antinutritional factors in underutilized legumes to boost their effective human consumption and improve their nutritional quality [137]. The common processing methods for minimizing these factors include autoclaving, boiling, dehulling, roasting, soaking, and boiling in water. Careful selection of the processing method to reduce antinutrients is critical because some methods are more effective in reducing some particular antinutrients than others. For instance, dehulling was more effective in lowering tannin in cowpeas, chickpeas, green gram, and lentils, while the cooking method was more effective in reducing trypsin inhibitors [138], tannin, and oligosaccharide [137]. Proteins from some legumes causing allergic reactions include conglutins and vicilin. Proteins with IgE binding sites also cause allergic reactions [108]. It is necessary to raise the consumer's awareness of meal preparation techniques to improve the utilization of underutilized pulses.

The utilization of some minor pulses is challenged by the long cooking time caused by hard-to-cook traits [5]. The hard-to-cook characteristic is linked to the incapacity of the legume cotyledon to soften during cooking, necessitating a longer cooking duration to achieve the desired texture. Pulses with hardened cotyledon have low uptake among consumers because their prolonged cooking time requires a lot of energy and can result in diminished nutritional quality and bioavailability [138]. Even though hardening of testa has been reported in many legumes after prolonged storage under high temperatures and relative humidity [139], the problem is more pronounced in African yam and Bambara groundnut [138, 140]. Numerous processing methods for softening legume seeds during cooking have been suggested. These include soaking in water before cooking, [141] use of sodium bicarbonate and ash [142], and use of tenderizers [140].

5. Conclusion

Global food production faces numerous challenges arising from the emergence of pests and diseases, changes in climate, population growth, and the decline in soil health. Underutilized legumes, alongside other crops, hold considerable potential due to their resilience to the climatic changes presently impacting many regions. They possess high-nutritional content containing high protein levels and available minerals, vitamins, and phytochemicals and, therefore, provide affordable avenues to address the common malnutrition problem, especially in Asia and Sub-Saharan Africa. The large diversity of these grain legumes enables their adaptability to various environments, particularly in regions characterized by marginal agricultural production. Plant-based proteins from underutilized legumes also have enormous potential for incorporation in food processing, especially in baking, preparation of beverages, meat analogues, food packaging, and films. Despite the considerable promise indigenous legume crops hold, their production and utilization have been markedly limited compared to the leading legume species. The low grain yields of underutilized legumes render them less appealing to farmers. In addition, the need for a regular supply of high-quality seeds and adequate technical knowledge to support commercial cultivation hinders their uptake. Significant levels of antinutrients in these lesser-used legumes, particularly oligosaccharides that cause flatulence in individuals, impede their widespread use. Increasing the uptake of underutilized legumes would require research and breeding initiatives that specifically target the challenges presented by abiotic and biotic stresses affecting production, as well as the antinutritional factors that limit consumption. The preferences and needs of a population, which also determine the uptake of underutilized legumes, should also be documented to enhance the adoption of diets. Future research should also focus on the effects of the environment on the nutrients and antinutritional factors and the mitigation processes. Addressing these factors will enhance the utilization of beneficial underutilized legumes.

Conflict of interest

The authors declare no conflict of interest.

DOI: http://dx.doi.org/10.5772/intechopen.1007719

Author details

Esther N. Kimani[1*], Eliezah M. Kamau[2] and Rael Karimi[3]

1 Kabete Centre, Kenya Agricultural and Livestock Research Organization, Biotechnology Research Institute, Nairobi, Kenya

2 Kandara Centre, Kenya Agricultural and Livestock Research Organization, Horticultural Research Institute, Thika, Kenya

3 Katumani Centre, Kenya Agricultural and Livestock Research Organization, Agricultural Mechanization Research Institute, Machakos, Kenya

*Address all correspondence to: esther.kimani@gmail.com

References

[1] Dansi A, Vodouhè R, Azokpota P, Yedomonhan H, Assogba P, Adjatin A, et al. Diversity of the neglected and underutilized crop species of importance in Benin. The Scientific World Journal. 2012;**2012**:e932947

[2] Adelabu DB, Franke AC. Status of underutilized crop production: Its potentials for mitigating food insecurity. Agronomy Journal. 2023;**115**(5):2174-2193

[3] Ratnayake SS, Kumar L, Kariyawasam CS. Neglected and underutilized fruit species in Sri Lanka: Prioritisation and understanding the potential distribution under climate change. Agronomy. 2020;**10**(1):34

[4] Mabhaudhi T, Chimonyo VGP, Hlahla S, Massawe F, Mayes S, Nhamo L, et al. Prospects of orphan crops in climate change. Planta. 2019;**250**(3):695-708

[5] Popoola JO, Aworunse OS, Ojuederie OB, Adewale BD, Ajani OC, Oyatomi OA, et al. The exploitation of orphan legumes for food, income, and nutrition security in sub-Saharan Africa. Frontiers in Plant Science. 2022;**13**:782140. DOI: 10.3389/fpls.2022.782140

[6] van Ittersum MK, van Bussel LGJ, Wolf J, Grassini P, van Wart J, Guilpart N, et al. Can sub-Saharan Africa feed itself? National Academy of Sciences of the United States of America. 2016;**113**(52):14964-14969

[7] Martínez-Castillo J, Araujo ASF, Chacón-Sánchez MI, Santos LG, Lopes ACA, Gibson K, et al. International Lima bean network: From the origin of the species to modern plant breeding. Genetic Resources and Crop Evolution. 2023;**70**(6):1575-1583

[8] Oliveira-Silva RN, Lobo-Burle M, Gomes-Pádua J, de Almeida-Lopes AC, Ferreira-Gomes RL, Martínez-Castillo J, et al. Phenotypic diversity in lima bean landraces cultivated in Brazil, using the Ward-MLM strategy. Chilean Journal of Agricultural Research. 2017;**77**(1):35-40

[9] Alves AU, de Oliveira AP, Alves AU, Dornelas CS, Alves EU, Cardoso EA, et al. Lima beans production and economic revenue as a function of organic and mineral fertilization. Horticultura Brasileira. 2008;**26**:251-254

[10] Damas LR, Barroso PA, de Assunção Neto WV, de Lopes ACA, da Silva Junior JV, Gomes RLF, et al. Selection of lima bean accessions for fresh and dry production. Revista Brasileira de Engenharia Agrícola e Ambiental. 2023;**27**:625-631

[11] Silva RNO, Lopes ACA, Gomes RLF, Pádua JG, Burle ML. High diversity of cultivated lima beans (*Phaseolus lunatus*) in Brazil consisting of one Andean and two Mesoamerican groups with strong introgression between the gene pools. Genetics and Molecular Research. 2019;**18**(4):18441

[12] Gomes De Sousa G, Ferreira Goes G, De Melo MA, Pires De Souza V, De Sousa AM, Valdenor Pereira Filho J. Potassium fertilizer on productivity in the lima bean. Revista Agro@Mbiente On-Line. 2023;**17**:1-11

[13] Gomes RDSS, Martins JVDS, Silva ECD, Silva HAOD, Nascimento LCD. Reactions of lima bean (*Phaseolus lunatus* L.) accessions to *Colletotrichum truncatum*. Revista Caatinga. 2022;**35**:809-817

[14] Owusu KD, Newman CF, Gabriel AG. Evaluation of physical and

cooking characteristics of five improved Lima beans. World Journal of Food Science and Technology. 2018;**2**(2):38-43

[15] Soetan KO, Atanda TN. Proximate composition, phytochemical screening and some antinutritional factors of three accessions of lima beans (*Phaseolus lunatus*) an underutilized legume. Nigerian Journal of Animal Production. 2018;**45**(3):71-78

[16] Nasir LNL, Feyissa T, Asfaw Z. Genetic diversity analysis of Lima bean (*Phaseolus lunatus* L.) landrace from Ethiopia as revealed by ISSR marker. Ethiopian Journal of Science. 2021;**44**(1):81-90

[17] Gemede HF, Birhanu E. Nutritional, Antinutritional and phenolic properties of Lima bean (*Phaseolus lunatus*) accessions: Underutilized legume in Ethiopia. Acta Universitatis Cibiniensis Series E: Food Technology. 2020;**24**(2):195-204

[18] Agyekum PB, Dombrowski J, Lutterodt HE, Ofosu IW. Consumption patterns and usage of selected underutilized legumes in a Ghanaian community. Legume Science. 2023;**5**(4):e202

[19] Amoo IA, Balogun TM, Ibeto AU, Adeboye OO. Phytochemicals, antioxidants and Glycemic index assessment of *Lablab purpureus* (Lablab Bean) and *Phaseolus lunatus* (Lima Bean) seeds. Asian Journal of Applied Chemistry Research. 2023;**13**(1):54-63

[20] Badawy ASM, Ismail FS, Salem AK. Effect of sowing dates and potassium fertilizer rates on seed yield of Lima bean CV. (*Phaseolus vulgaris* L.). Journal of Plant Production. 2019;**10**(1):45-52

[21] Kouonon LC, Koffi KG, Adjoumani K, Goba KAE. Morphological variability of neglected and underutilized legumes (NULs) collection from Côte d'Ivoire. American Journal of Plant Sciences. 2020;**11**(2):226-244

[22] Paliwal R, Adegboyega TT, Abberton M, Faloye B, Oyatomi O. Potential of genomics for the improvement of underutilized legumes in sub-Saharan Africa. Legume Science. 2021;**3**:e69

[23] Tan XL, Azam-Ali S, Goh EV, Mustafa M, Chai HH, Ho WK, et al. Bambara groundnut: An underutilized leguminous crop for global food security and nutrition. Frontiers in Nutrition. 2020;**7**:601496

[24] Uba CU, Oselebe HO, Tesfaye AA, Abtew WG. Genetic diversity and population structure analysis of bambara groundnut (*Vigna subterrenea* L) landraces using DArT SNP markers. PLoS One. 2021;**16**(7):e0253600

[25] Asante B, Abadar R, Agyeman K, Ayeh S, Aidoo R, Afona D. Preference for improved varietal attributes of Bambara groundnut among smallholder farmers in Ghana. African Journal of Agricultural and Resource Economics. 2021;**16**(2):155-168

[26] Nandkangre H, Zongo KF, Kambou DJ, Ilboudo BY, Konate MN, Ouoba A, et al. Phenological, morphological and agronomic characterization of Bambara groundnut genotypes on plinthite soil in East-Centre area, Burkina Faso. Annual Research & Review in Biology. 2022;**37**(12):105-114

[27] Odongo F, Oyoo ME, Kimurto PK, Wasike VW. Morphological characterization of Bambara groundnut [*Vigna subterranea* (L.) Verdc.] landraces in Kenya. Agricultural Science Digest. 2023;**43**(3):282-288

[28] Kiryowa M, Ddamulira G, Alenoma G, Karwani G, Ifeyinwa MO, Umeugochukwu OP, et al. Morphological characterization and diversity of Bambara groundnuts in Uganda. Journal of Agricultural Science. 2022;**14**(9):p86

[29] Valombola JS, Awala SK, Hove K. Morphological characterization of Bambara groundnut (*Vigna subterranea* (L.) Verdc.) germplasm collections: A basis for crop improvement. Journal of Agriculture and Applied Biology. 2022;**3**(1):8-18

[30] Minnaar-Ontong A, Gerrano AS, Labuschagne MT. Assessment of genetic diversity and structure of Bambara groundnut [*Vigna subterranea* (L.) verdc.] landraces in South Africa. Scientific Reports. 2021;**11**(1):1-9

[31] Temegne NC, Dooh JPN, Nbendah P, Ntsomboh-Ntsefong G, Taffouo VD, Youmbi E. Cultivation and utilization of Bambara Groundnut (Vigna subterranea (L.) Verdc.), a neglected Plant in Cameroon. Asian Plant Research Journal. 2020;**4**(2):9-21

[32] Sari GN, Saptadi D, Kuswanto K. The yield stability and adaptability of Bambara Groundnut at three locations. Agrivita Journal of Agricultural Science. 2021;**44**(1):130-138

[33] Rungnoi OU, Suwanprasert J, Somta P, Srinives P. Molecular genetic diversity of Bambara Groundnut (*Vigna subterranea* L. Verdc.) revealed by RAPD and ISSR marker analysis. SABRAO Journal of Breeding and Genetics. 2012;**44**:87-101

[34] Khan MMH, Rafii MY, Ramlee SI, Jusoh M, Al-Mamun M. Bambara groundnut (*Vigna subterranea* L. Verdc): A crop for the new millennium, its genetic diversity, and improvements to mitigate future food and nutritional challenges. Sustainability. 2021;**13**(10):5530

[35] Jahanshiri E, Goh EV, Wimalasiri EM, Azam-Ali S, Mayes S, Tengku Mohd Suhairi TAS, et al. The potential of Bambara groundnut: An analysis for the People's republic of China. Food and Energy Security. 2022;**11**(2):e358

[36] Isemura T, Tomooka N, Kaga A, Vaughan DA. Comparison of the pattern of crop domestication between two Asian beans, Azuki bean (*Vigna angularis*) and Rice bean (*V. Umbellata*). Japan Agricultural Research Quarterly: JARQ. 2011;**45**(1):23-30

[37] Verma SK, Singh CK, Taunk J, Gayacharan, Chandra Joshi D, Kalia S, et al. Vignette of *Vigna* domestication: From archives to genomics. Frontiers in Genetics. 2022;**13**:960200

[38] Darai R, Neupane RK, Sarker A, Dhakal KH, Agrawal SK. Diversity, conservation and utilization of grain legumes: A potential source for food and nutritional security in Nepal. Journal of Food Legumes. 2023;**36**(4):229-243

[39] Raiger HL, Phogat BS, Dua RP, Sharma SK. Improved varieties and cultivation practices of Rice bean. Intensive Agriculture. 2009:16-20

[40] Guan J, Zhang J, Gong D, Zhang Z, Yu Y, Luo G, et al. Genomic analyses of rice bean landraces reveal adaptation and yield related loci to accelerate breeding. Nature Communications. 2022;**13**(1):5707

[41] Sheahan CM. Plant Guide for Jack Bean (*Canavalia ensiformis*). USDA-Natural Resources Conservation Service, Plant Materials Center: Cape May, NJ; 2012

[42] Darini MT, Susilaningsih SEP, Sunaryo Y. The potential of Jack bean (*Canavalia ensiformis* L.) developed in suboptimal soil to succeeding food sufficiency. International Journal of Current Science Research and Review. 2021;**4**(7):740-744

[43] Haq N. Underutilized food legumes: Potential for multipurpose uses. In: Pratap A, Kumar J, editors. Biology and Breeding of Food Legumes. 1st ed. UK: CABI; 2011. pp. 329-347

[44] de Sá Andrade ML, de Oliveira IA, Pires LL, Kitajima EW, da Silva GS, Beserra JEA. First report of yam bean mosaic virus in Brazil. Australasian Plant Disease Notes. 2019;**14**(1):22

[45] Popoola J, Ojuederie O, Omonhinmin C, Adegbite A. Neglected and underutilized legume crops: Improvement and future prospects. In: Shah F, Khan Z, Iqbal A, Turan M, Olgun M, editors. Recent Advances in Grain Crops Research. [Internet]. London, UK: IntechOpen; 2019. p. Ch. 8. Available from: https://www.intechopen.com/chapters/68218

[46] Soetan KO, Adeola AA. Comparative nutritional and functional properties of selected underutilized legumes. Nigerian Journal of Animal Production. 2020;**45**(3):96-106

[47] Olomitutu OE, Abe A, Oyatomi OA, Paliwal R, Abberton MT. Assessing intraspecific variability and diversity in African yam bean landraces using agronomic traits. Agronomy. 2022;**12**(4):884

[48] Usoroh JI, Eneobong EE, Abraham SO, Umoyen AJ. Evaluation of the genetic diversity of African yam bean (*Sphenostylis stenocarpa* (Hoechst. Ex. a rich.) harms.) using seed protein marker. Archives of Current Research International. 2019;**17**(4):1-10

[49] Abberton M, Paliwal R, Faloye B, Marimagne T, Moriam A, Oyatomi O. Indigenous African orphan legumes: Potential for food and nutrition security in SSA. Frontiers in Sustainable Food Systems. 2022;**6**:708124

[50] Ndubuisi PO, Micheal AT, Nwafor SC, Amadi CO. Factors affecting production and consumption of African yam bean (*Sphenostylisstenocarpa* Hochst. Ex. A. Rich) harms in southeast agro-ecological zone of Nigeria. Journal of Natural Sciences Research. 2022;**13**(10):43-49

[51] Csizinszky AA. Performance of two winged bean accessions in Florida. HortScience. 1981;**16**(1):62-63

[52] Popoola JO, Ojuederie OB, Aworunse OS, Adelekan A, Oyesola OL, Akinduti PA, et al. Nutritional, functional, and bioactive properties of African underutilized legumes. Frontiers in Plant Science [Internet]. 2023;**14**. Available from: https://www.frontiersin.org/journals/plant-science/articles/10.3389/fpls.2023.1105364/full

[53] Yang S, Grall A, Chapman MA. Origin and diversification of winged bean (*Psophocarpus tetragonolobus* (L.) DC.), a multipurpose underutilized legume. American Journal of Botany. 2018;**105**(5):888-897

[54] Sriwichai S, Monkham T, Sanitchon J, Jogloy S, Chankaew S. Dual-purpose of the winged bean (*Psophocarpus tetragonolobus* (L.) DC.), the neglected tropical legume, based on pod and tuber yields. Plants. 2021;**10**(8):1746

[55] Tanzi AS, Eagleton GE, Ho WK, Wong QN, Mayes S, Massawe F. Winged bean (*Psophocarpus tetragonolobus* (L.) DC.) for food and nutritional security:

Synthesis of past research and future direction. Planta. 2019;**250**(3):911-931

[56] Bhadmus A, Abberton M, Idehen E, Ekanem U, Paliwal R, Oyatomi O. Genetic diversity assessment of winged bean [*Psophocarpus tetragonolobus* (L.) DC.] accessions using agronomic and seed morphometric traits. Crops. 2023;**3**(2):170-183

[57] Shonde TEO, Adebayo MA, Bhadmus AA, Adejumobi II, Oyatomi OA, Faloye B, et al. Diversity assessment of winged bean [*Psophocarpus tetragonolobus* (L.) DC.] accessions from IITA Genebank. Agronomy. 2023;**13**(8):2150

[58] Eagleton G. Winged bean (*Psophocarpus tetragonolobus*) cropping systems. Biodiversitas [Internet]. 2020;**21**(12):5927-5946. Available from: https://smujo.id/biodiv/article/view/7146

[59] Thompson AE, Haryono SK. Winged bean: Unexploited tropical food crop. HortScience. 1980;**15**(3):233-238

[60] Ho WK, Tanzi AS, Sang F, Tsoutsoura N, Shah N, Moore C, et al. A genomic toolkit for winged bean *Psophocarpus tetragonolobus*. Nature Communications. 2024;**15**(1):1901

[61] Kongjaimun A, Takahashi Y, Yoshioka Y, Tomooka N, Mongkol R, Somta P. Molecular analysis of genetic diversity and structure of the lablab (*Lablab purpureus* (L.) sweet). gene pool reveals two independent routes of domestication. Plants. 2023;**12**(1):57

[62] Letting FK, Venkataramana PB, Ndakidemi PA. Breeding potential of lablab [*Lablab purpureus* (L.) sweet]: A review on characterization and bruchid studies towards improved production and utilization in Africa. Genetic Resources and Crop Evolution. 2021;**68**(8):3081-3101

[63] Letting FK, Venkataramana PB, Ndakidemi PA. Pre-breeding prospects of lablab (*Lablab purpureus* (L.) sweet) accessions in Tanzania: Morphological characterization and genetic diversity analysis. Agronomy. 2022;**12**(10):2272

[64] Kamau EM, Kinyua M, Waturu C, Kiplagat O, Wanjala B, Kariba R, et al. Diversity and Population Structure of Local and Exotic *Lablab purpureus* Accessions in Kenya as Revealed by Microsatellite Markers. Global Journal of Molecular Biology. 2021;**3**(8)

[65] Gayacharan PSK, Mondal N, Yadav R, Vishwakarma H, Rana JC. Mining legume germplasm for genetic gains: An Indian perspective. Frontiers in Genetics [Internet]. 2023;**14**:996828. Available from: https://www.frontiersin.org/journals/genetics/articles/10.3389/fgene.2023.996828/full

[66] Wang ML, Morris JB, Barkley NA, Dean RE, Jenkins TM, Pederson GA. Evaluation of genetic diversity of the USDA *Lablab purpureus* germplasm collection using simple sequence repeat markers. The Journal of Horticultural Science and Biotechnology. 2007;**82**(4):571-578

[67] Maass BL, Knox MR, Venkatesha SC, Angessa TT, Ramme S, Pengelly BC. *Lablab purpureus*—A crop lost for Africa? Tropical Plant Biology. 2010;**3**(3):123-135

[68] Shubha K, Choudhary AK, Dubey AK, Tripathi K, Kumar R, Kumar S, et al. Evaluation of lablab bean [*Lablab purpureus* (L.) sweet] genotypes: Unveiling superior pod yield, nutritional quality, and collar rot resistance. Frontiers in Nutrition. 2024;**10**:1243923

[69] Kamotho GN, Kinyua MG, Muasya RM. Assessing the contribution of Dolichos bean (*Lablab purpureus* L. (sweet)) to food security in a changing

climate in Kenya. International Journal of Agriculture, Environment and Bioresearch. 2022;**07**(02):12-21

[70] Missanga JS, Venkataramana PB, Ndakidemi PA. *Lablab purpureus*: Analysis of landraces cultivation and distribution, farming systems, and some climatic trends in production areas in Tanzania. Open Agriculture. 2023;**8**(1):20220156

[71] Bhartiya A, Aditya JP, Kant L. Nutritional and remedial potential of an underutilized food legume horsegram (*Macrotyloma uniflorum*): A review. The Journal of Animal and Plant Sciences. 2015;**25**(4):908-920

[72] Rana JC, Kumar NKK, Singh M, Yadav R, Tripathi K, Bhardwaj R, et al. Underutilized grain legumes: Rice bean, moth bean, adzuki bean, Faba bean and horse gram. In: Tyagi RK, Pandey A, Agrawal A, Varaprasad KS, Paroda RS, Khetarpal RK, editors. Regional Expert Consultation on Underutilized Crops for Food and Nutritional Security in Asia and the Pacific-Thematic, Strategic Papers and Country Status Reports. Bangkok, Thailand, November 13-15, 2017. Bangkok, Thailand: Asia-Pacific Association for Agricultural Research Institutions (APAARI); 2018. pp. 121-136

[73] Aditya JP, Bhartiya A, Chahota RK, Joshi D, Chandra N, Kant L, et al. Ancient orphan legume horse gram: A potential food and forage crop of future. Planta. 2019;**250**(3):891-909

[74] Sunil N, Kumar V, Natarajan S, Venkateswaran K, Pandravada S. Evaluation and diversity observed in Horsegram (*Macrotyloma uniflorum* (lam) Verdc.) germplasm from Andhra Pradesh, India. International Journal of Plant Research. 2014;**4**(1):17-22

[75] Geetha K, Divya S. Genetic divergence studies in horse gram (*Macrotyloma uniflorum* lam.Verdc) for quantitative traits using Mahalanobis D2 analysis. Madras Agricultural Journal. 2021;**108**:29321

[76] Pushpayazhini V, Sudhagar R, Vanniarajan C, Hepziba SJ, Souframanien J. Diversity analysis in horse gram [*Macrotyloma uniflorum* (lam.) Verdc.] mutants. Legume Research. 2022;**45**(7):834-838

[77] Gupta N, Shrivastava N, Singh PK, Bhagyawant SS. Phytochemical evaluation of moth bean (*Vigna aconitifolia* L.) seeds and their divergence. Biochemistry Research International. 2016;**2016**:3136043

[78] Bhadkaria A, Narvekar DT, Gupta N, Khare A, Bhagyawant SS. Moth bean (*Vigna aconitifolia* (Jacq.) Marechal) seeds: A review on nutritional properties and health benefits. Discover Food. 2022;**2**(1):18

[79] Kanishka RC, Gayacharan C, Basavaraja T, Chandora R, Rana JC. Moth bean (*Vigna aconitifolia*): A minor legume with major potential to address global agricultural challenges. Frontiers in Plant Science. 2023;**14**:1179547

[80] Choudhary KB, Sharma R, Solanki RK, Mahla HR, Jadon KS, Choudhary M, et al. Genetic relatedness in elite cultivars of moth bean using Morpho-agronomic and molecular markers. Legume Research. 2024;**47**(3):378-384

[81] Zavinon F, Adoukonou SH. A pigeon pea [*Cajanus cajan* (L.) Millsp] cultivation, its major constraints and ethnobotanical status in southern Benin. Journal of Agricultural and Crop Research. 2019;7(6):95-105

[82] FAOSTAT. FAOSTAT [Internet]. In: Food and Agriculture Organization of the United Nations. Rome, Italy; 2024.

Available from: https://www.fao.org/faostat/en/#data/QCL

[83] Kumar A, Chaudhary R, Kumar A, Sharma N. Assessment of genetic diversity in pigeon pea (*Cajanus cajan*) using microsatellite markers. Journal of Applied and Natural Science. 2023;**15**(2):530-537

[84] Ojo BT, Togun AO, Abberton M, Oyatomi O, Dada OA. Morphological and molecular characterization of pigeon pea (*Cajanus cajan* (L.) Millsp.) for improved utilization. Journal of Genetic Engineering and Biotechnology Research. 2023;**5**(3):186-197

[85] Isbell F, Adler PR, Eisenhauer N, Fornara D, Kimmel K, Kremen C, et al. Benefits of increasing plant diversity in sustainable agroecosystems. Journal of Ecology. 2017;**105**(4):871-879

[86] Oloyede FM, Ola DS, Iwalewa EA. Biodiversity conservation of the neglected and underutilized Nigerian horticultural crops. Acta Horticulturae et Regiotecturae. 2023;**26**(1):64-72

[87] Mabhaudhi T, Hlahla S, Chimonyo VGP, Henriksson R, Chibarabada TP, Murugani VG, et al. Diversity and diversification: Ecosystem services derived from underutilized crops and their co-benefits for sustainable agricultural landscapes and resilient food systems in Africa. Frontiers in Agronomy. 2022;**4**:859223

[88] Dwivedi SL, Chapman MA, Abberton MT, Akpojotor UL, Ortiz R. Exploiting genetic and genomic resources to enhance productivity and abiotic stress adaptation of underutilized pulses. Frontiers in Genetics. 2023;**14**: 1193780

[89] Quintieri L, Nitride C, De Angelis E, Lamonaca A, Pilolli R, Russo F, et al. Alternative protein sources and novel foods: Benefits, food applications and safety issues. Nutrients. 2023;**15**(6):1509

[90] Ekeleme F, Akobundu IO, Fadayomi RO, Chikoye D, Abayomi YA. Characterization of legume cover crops for weed suppression in the moist Savanna of Nigeria. Weed Technology. 2003;**17**(1):1-13

[91] Ooro PA, Birech RJ, Malinga JN, Thuranira E. Effect of legumes on nitrogen use efficiency of wheat in a short term crop rotation in Njoro Sub-County. Journal of Experimental Agriculture International. 2021;**43**(3):1-15

[92] Kwena K, Karuku G, Ayuke F, Esilaba A. Nitrogen deficiency in semi-arid Kenya: Can Pigeonpea fix it? East African Agricultural and Forestry Journal. 2019;**83**. DOI: 10.1080/00128325.2019.1658696

[93] Nnamani CV, Adewale DB, Oselebe HO, Atkinson CJ. African yam bean the choice for climate change resilience: Need for conservation and policy. In: Leal Filho W, Oguge N, Ayal D, Adeleke L, da Silva I, editors. African Handbook of Climate Change Adaptation [Internet]. Cham: Springer International Publishing; 2020. pp. 1-18. DOI: 10.1007/978-3-030-42091-8_203-1

[94] Chai HH, Massawe F, Mayes S. Effects of mild drought stress on the morpho-physiological characteristics of a bambara groundnut segregating population Euphytica. Euphytica. 2016;**208**:225-236

[95] Tiwari B, Kalim S, Bangar P, Kumari R, Kumar S, Gaikwad A, et al. Physiological, biochemical, and molecular responses of thermotolerance in moth bean (*Vigna aconitifolia* (Jacq.) Marechal). Turkish Journal of Agriculture and Forestry. 2018;**42**(3):176-184

[96] Chamara R, Wijerathna M, Herath I, Rajapaksha R. Knowledge, attitudes and practices (KAP) of rural women in the dry zone of Sri Lanka on underutilized vegetable crops. Ethnobotany Research and Applications. 2021;**21**:1-25

[97] Esilaba A. Options for Stiga Management in Kenya. Nairobi, Kenya: Kenya Agricultural Research Institute; 2006

[98] Prasanna BM, Suresh LM, Mwatuni F, Beyene Y, Makumbi D, Gowda M, et al. Maize lethal necrosis (MLN): Efforts toward containing the spread and impact of a devastating transboundary disease in sub-Saharan Africa. Virus Research. 2020;**282**:197943

[99] Knez M, Ranić M, Gurinović M. Underutilized plants increase biodiversity, improve food and nutrition security, reduce malnutrition, and enhance human health and well-being. Let's put them back on the plate! Nutrition Reviews. 2023;**82**(8):1111-1124

[100] Li X, Yadav R, Siddique KHM. Neglected and underutilized crop species: The key to improving dietary diversity and fighting hunger and malnutrition in Asia and the Pacific. Frontiers in Nutrition. 2020;**7**:593711

[101] Scarano A, Semeraro T, Chieppa M, Santino A. Neglected and underutilized plant species (NUS) from the Apulia region worthy of being rescued and Re-included in daily diet. Horticulturae. 2021;**7**(7):177

[102] Singh RK, Sreenivasulu N, Prasad M. Potential of underutilized crops to introduce the nutritional diversity and achieve zero hunger. Functional & Integrative Genomics. 2022;**22**(6):1459-1465

[103] Bepary RH, Roy A, Pathak K, Deka SC. Biochemical composition, bioactivity, processing, and food applications of winged bean (*Psophocarpus tetragonolobus*): A review. Legume Science. 2023;**5**(3):e187

[104] Adegboyega TT, Abberton MT, AbdelGadir AH, Dianda M, Maziya-Dixon B, Oyatomi OA, et al. Nutrient and Antinutrient composition of winged bean (*Psophocarpus tetragonolobus* (L.) DC.) seeds and tubers. Journal of Food Quality. 2019;**2019**:e3075208

[105] Saravanan M. Hypocholesterolemic effect of Indian medicinal plants - A review. Medicinal Chemistry[Internet]. 2015;**05**(01):40-49. Available from: https://www.omicsonline.org/open-access/hypocholesterolemic-effect-of-indian-medicinal-plants-a-review-2161-0444.1000040.php?aid=37082

[106] Adebo JA. A review on the potential food application of Lima beans (*Phaseolus lunatus* L.), an underutilized crop. Applied Sciences. 2023;**13**(3):1996

[107] Katoch R, Tripathi A, Kiran and. Rice bean (*Vigna umbellata*). In: Farooq M, Siddique KHM, editors. Neglected and Underutilized Crops [Internet]. Massachusetts, USA: Academic Press; 2023. pp. 647-680. Available from: https://www.sciencedirect.com/science/article/pii/B9780323905374000028

[108] Barman A, Marak CM, Barman RM, Sangma CS. Nutraceutical properties of legume seeds and their impact on human health. In: Legume Seed Nutraceutical Research [Internet]. London, UK: IntechOpen; 2019. Available from: https://www.intechopen.com/chapters/62638

[109] Hertzler SR, Lieblein-Boff JC, Weiler M, Allgeier C. Plant proteins: Assessing their nutritional quality and effects on health and physical function. Nutrients. 2020;**12**(12):3704

[110] Ayilara MS, Abberton M, Oyatomi OA, Odeyemi O, Babalola OO. Potentials of underutilized legumes in food security. Frontiers in Soil Science. 2022;**2**:1020193

[111] Thi HHP, Nguyen TL, Thi HHP, Nguyen TL. Nutraceutical properties of legume seeds: Phytochemical compounds. In: Legumes Research. Vol. 2 [Internet]. London, UK: IntechOpen; 2021. Available from: https://www.intechopen.com/chapters/79589

[112] Purwanti E, Hermanto FE, Prihanta W, Permana TI, Wiwekananda IGNA. Nutritional contents and bioactive compounds among several variants of Dolichos lablab: Fundamental facts for functional food development. Journal of Tropical Biodiversity and Biotechnology. 2023;**8**(2):81339

[113] Rai KK, Rai N, Pandey-Rai S, Rai KK, Rai N, Pandey-Rai S. Unlocking pharmacological and therapeutic potential of hyacinth bean (*Lablab purpureus* L.): Role of OMICS based biology, biotic and abiotic elicitors. In: Legumes Research. Vol. 2 [Internet]. London, UK: IntechOpen; 2021. Available from: https://www.intechopen.com/chapters/77832

[114] Odeku OA, Ogunniyi QA, Ogbole OO, Fettke J. Forgotten gems: Exploring the untapped benefits of underutilized legumes in agriculture, nutrition, and environmental sustainability. Plants. 2024;**13**(9):1208

[115] Bassal H, Merah O, Ali AM, Hijazi A, El Omar F. *Psophocarpus tetragonolobus*: An underused species with multiple potential uses. Plants. 2020;**9**(12):1730

[116] Wan Mohtar WAAQI, Hamid AA, Abd-Aziz S, Muhamad SKS, Saari N. Preparation of bioactive peptides with high angiotensin converting enzyme inhibitory activity from winged bean [*Psophocarpus tetragonolobus* (L.) DC.] seed. Journal of Food Science and Technology. 2014;**51**(12):3658-3668

[117] Akinyemi FA, Orishadipe AT, Ebun-Oluwa O, Aladesanmi OA, Akinyemi FA, Orishadipe AT, et al. Physico-chemical properties and functional characteristics of Jack beans (*Canavalia ensiformis)* starch. World Journal of Biology Pharmacy and Health Sciences. 2020;**3**(2):012-022

[118] Priego-Poyato S, Rodrigo-Garcia M, Escudero-Feliu J, Garcia-Costela M, Lima-Cabello E, Carazo-Gallego A, et al. Current advances research in nutraceutical compounds of legumes, Pseudocereals and cereals. In: Grain and Seed Proteins Functionality [Internet]. London, UK: IntechOpen; 2021. Available from: https://www.intechopen.com/chapters/76727

[119] Onyango E. Legume protein: Properties and extraction for food applications. In: Legumes Research. Vol. 2 [Internet]. London, UK: IntechOpen; 2022. Available from: https://www.intechopen.com/chapters/79220

[120] Mihalca V, Kerezsi AD, Weber A, Gruber-Traub C, Schmucker J, Vodnar DC, et al. Protein-based films and coatings for food industry applications. Polymers (Basel). 2021;**13**(5):769

[121] Letting FK, Venkataramana PB, Ndakidemi PA. Farmers' participatory plant selection of lablab (*Lablab purpureus* (L.) sweet) in Tanzania. Front Plant Science. 2022;**13**:784032

[122] Kamau E, Kinyua M, Waturu C, Karanja D. Genotypic variability, heritability and path analysis of yield

components of determinate lablab (*Lablab purpureus* (L.) sweet) inbred lines in Kenya. African Journal of Plant Science. 2021;**15**:266-276

[123] Imathiu S. Neglected and underutilized cultivated crops with respect to indigenous African leafy vegetables for food and nutrition security. Journal of Food Security. 2021;**9**(3):115-125

[124] Knez M, Ranic M, Gurinovic M, Glibetic M, Savic J, Mattas K, et al. Causes and conditions for reduced cultivation and consumption of underutilized crops: Is there a solution? Sustainability. 2023;**15**(4):3076

[125] Chivenge P, Mabhaudhi T, Modi AT, Mafongoya P. The potential role of neglected and underutilized crop species as future crops under water scarce conditions in sub-Saharan Africa. International Journal of Environmental Research and Public Health. 2015;**12**(6):5685-5711

[126] Padulosi S, Thompson J, Rudebjer P. Fighting Poverty, Hunger and Malnutrition with Neglected and Underutilized Species (NUS). Rome, Italy: Bioversity International; 2013. 1-56 p

[127] Ahmed H, Gregg BR, Louwaars NP. Seed systems for underutilized crops. Acta Horticulturae. 2009;**806**:459-464

[128] Oballim G, Obura M, Mutio J, Opile W, Ochuodho J. Production and utilization of Bambaranut (*Vigna subterranea* (L.) Verdc.) in northern and eastern Uganda. African Journal of Agricultural Research. 2022;**18**:977-990

[129] Rahman MM, Dash CK, Rahman MM, Hasan MM, Hannan A, Dev S, et al. Farmers' perceptions and knowledge of country bean (*Lablab purpureus* L.) insect pests, and diseases, and their management practices, in Bangladesh. Sustainability. 2022;**14**(20):13591

[130] Ayenan MAT, Ofori K, Ahoton LE, Danquah A. Pigeonpea [(*Cajanus cajan* (L.) Millsp.)] production system, farmers' preferred traits and implications for variety development and introduction in Benin. Agriculture & Food Security. 2017;**6**(1):48

[131] Graham PH, Vance CP. Legumes: Importance and constraints to greater use. Plant Physiology. 2003;**131**(3):872-877

[132] Zimmer S, Liebe U, Didier JP, Heß J. Luxembourgish farmers' lack of information about grain legume cultivation. Agronomy for Sustainable Development. 2016;**36**(1):2

[133] Muoni T, Barnes AP, Öborn I, Watson CA, Bergkvist G, Shiluli M, et al. Farmer perceptions of legumes and their functions in smallholder farming systems in East Africa. International Journal of Agricultural Sustainability. 2019;**17**(3):205-218

[134] Smith BM, Gathorne-Hardy A, Chatterjee S, Basu P. The last mile: Using local knowledge to identify barriers to sustainable grain legume production. Frontiers in Ecology and Evolution. 2018;**6**:102

[135] Cominelli E, Sparvoli F, Lisciani S, Forti C, Camilli E, Ferrari M, et al. Antinutritional factors, nutritional improvement, and future food use of common beans: A perspective. Frontiers in Plant Science [Internet]. 2022;**13**. Available from: https://www.frontiersin.org/journals/plant-science/articles/10.3389/fpls.2022.992169/full

[136] Chibarabada TP, Modi AT, Mabhaudhi T. Expounding the value

of grain legumes in the semi- and arid tropics. Sustainability. 2017;**9**(1):60

[137] Devi TB, Choudhary P, Tushir S, Tyagi SK. Processing interventions to reduce Antinutrients in legumes and pulses. In: Chakraborty I, Paul PK, Mani A, Tiwary AK, Prasad K, editors. Trends & Prospects in Processing of Horticultural Crops. New Delhi, India: Today & Tomorrow's Printers and Publishers; 2019. pp. 301-337

[138] Pal RS, Bhartiya A, Yadav P, Kant L, Mishra KK, Aditya JP, et al. Effect of dehulling, germination and cooking on nutrients, antinutrients, fatty acid composition and antioxidant properties in lentil (*Lens culinaris*). Journal of Food Science and Technology. 2017;**54**(4):909-920

[139] Perera D, Devkota L, Garnier G, Panozzo J, Dhital S. Hard-to-cook phenomenon in common legumes: Chemistry, mechanisms and utilization. Food Chemistry. 2023;**415**:135743

[140] Ojo M, Ade-Omowaye BIO. Some functional and physical properties of selected underutilized hard-to-cook legumes in Nigeria. American Journal of Food Science and Nutrition. 2015;**2**:73-81

[141] Ojo AM. Alleviation of hard-to-cook phenomenon in legumes: Effects of hydrothermal processing techniques on two species of *Vigna subterranean*. American Journal of Food Science and Technology. 2018;**6**(3):103-107

[142] Sathya A, Siddhuraju P. Effect of processing methods on compositional evaluation of underutilized legume, *Parkia roxburghii* G. Don (yongchak) seeds. Journal of Food Science and Technology. 2015;**52**(10):6157-6169

Chapter 2

Legumes: Cultivation, Uses, and Benefits

Senaratne Ranamukhaarachchi and Dhanesha Nanayakkara

Abstract

Legumes, also known as "pulses," belong to the Fabaceae plant family and the second most significant family of plant crops. Legumes are renowned as a sustainable source of protein, dietary fibre, and phytochemicals. Furthermore, legumes play diverse roles in fostering environmental and biodiversity stability, supporting animal husbandry, influencing economic factors, and making noteworthy contributions to human health. With concerns about environmental sustainability, animal welfare, and a growing desire for healthier protein alternatives, legumes have emerged as an ethical and sustainable primary choice of protein. Their protein content not only fulfills nutritional requirements but also aligns with the escalating trend of reducing dependence on animal-based proteins. With the rapidly growing global population, legumes contribute significantly to addressing food security challenges. Especially, the ability to fix nitrogen in the soil and incorporate it in different cropping systems enhances soil fertility, reducing the need for synthetic fertilizers. This intensifies the demand for legumes continuing to rise. However, out of several possible reasons, climatic changes, deterioration of soil fertility, inefficient farming practices, and reduction of arable lands has threatened the global food supply. In this chapter, legume cultivation and production, major benefits, and recommendations for future research are discussed.

Keywords: legumes, N fixation, protein source, pulses, conservation agriculture

1. Introduction

The Fabaceae plant family, which is commonly referred to as "legumes" or "pulses," consists of dicotyledonous, annual, or perennial crops and is one of the most extensive plant families that comprises approximately 800 genera and over 20,000 species, including grain, pasture, and agroforestry species [1, 2]. The Fabaceae family is the second most crucial family of plants, following the grass family (Poaceae) [3, 4]. Legumes stand out as a significant and sustainable source of protein, dietary fiber, and phytochemicals, as well as play diverse roles in the stability of the environment and biodiversity, economics, and the medical field [4–8]. Legumes perform to enhance human health [9], support the production of farm animals [10], enrich soil conditions [11], and mitigate the impact of greenhouse gases [12]. Grain legumes contribute to 27% of global crop production [13]. The key varieties of legumes include dry beans (*Phaseolus vulgaris* L.), chickpeas (*Cicer arietinum* L.), dry peas (*Pisum sativum* L.), cowpeas (*Vigna unguiculata* L. Walp), lentils (*Lens culinaris* Medik.), green gram

(*Vigna mungo* L.), and broad beans (*Vicia faba* L.), excluding those primarily used for oil extraction like soybeans and groundnuts [14]. These species are distributed across diverse habitats and exhibit a wide range of growth habits, spanning from herbaceous to arboreal, as outlined by Lewis et al. [1].

In addition to being a cost-effective nutrient source, legumes play a substantial role in generating income. The historical consumption of legumes dates goes back to the Roman era, where they utilized legumes in pastures and for soil improvement as early as 37 BC [15, 16]. Globally, grain legumes are grown approximately 81 million hectares and contributing to an annual global production exceeding 92 million tons [17]. Among the countries producing grain legumes, India stands out as the largest contributor, accounting for nearly a quarter of the global annual grain legume production, coupled with being the primary consumer. Other major grain legume-producing nations include China, Myanmar, Canada, Australia, Brazil, Argentina, the USA, and Russia [17]. With the ability to get well adapted to challenging environmental conditions, legumes possess a wide adaptability to different environmental conditions and high resistance to biotic and abiotic stress conditions [18, 19].

Among various uses of legumes, the increasing awareness on the importance of having healthy food towards better human health has enhanced consumers' interest in legumes, especially because of their nutritional and bioactive richness [20]. Genetic variations, varietal differences, geographical and climatic differences have resulted enormous variations in the nutritional and phytochemical compositions within the group of legumes [20]. The main features of legumes as significant sources of proteins, dietary fiber, vitamins, minerals, carbohydrates, and phytochemicals, all are essential for human health [21]. The importance of incorporating legumes into diets has gained special attention in human for disease prevention. Due to the health beneficial properties of high protein, fibre and antioxidant content, legumes are increasingly integrated into novel food product formulations. Especially, legumes hold the potential to be gluten-free and hence inclusion of legumes has proven to support in enhancing the nutritional profile of gluten-free products [22]. This has created a growing trend in novel product development with legumes. Further, legumes contribute to lowering the glycaemic index of various food products and enhancing functional properties such as water binding and fat absorption [22].

In addition to the above, legumes play a vital role in providing ecosystem services. Ecosystem services (ES) refer to the services that people derive from the functioning of ecosystems, which are essential for supporting the life on the Earth [23]. Ecosystem services produced by the production ecosystems, are known as agro-ecosystem services, encompass both biological and agronomic factors [23]. Integrating legumes into cropping systems at the field level, whether they are grown for food, feed/fodder, forage, or as service crops, offers agronomic benefits [24]. Legumes contribute nitrogen (N) to the soil through symbiotic N fixation, which reduces the reliance on nitrogen fertilizers for companion or subsequent crops [25]. This practice also enhances the utilization of soil nitrogen resources and, in certain instances, leads to increased yields in subsequent crops [5].

However, legumes continue to play a subordinate role compared to major cereal crops such as rice, wheat, and maize, primarily due to the escalating global demand for cereals driven by growing human populations. The cultivation and research priorities for food legumes typically prioritize after cereals in most farming systems. This is mainly because the susceptibility of legume crops to various biotic and abiotic stress conditions leading to higher cultivation risks, and lower yield potential compared to cereals [26]. The approach through genetic improvements would significantly help in agronomic enhancements to play a crucial role in narrowing the yield gap caused by diverse stress conditions [26]. Anyway, recent developments have been conducted towards the adoption of various approaches leading to enhance the uses of legumes [5]. Therefore, this chapter elaborates legume cultivation, production, major uses, and recommends future research work for legumes as a group of crops with invaluable importance.

2. Legume cultivation and production

Legumes are considered one of the first domesticated crops, claiming a hunter-gatherer lifestyle to agricultural techniques [8, 15]. Legume cultivation was prevalent in areas where agriculture was conducted, as indicated by archeological evidence of the coexistence of legumes and cereals [8]. Evidence suggests that lentils were cultivated in ancient Egyptian civilizations, and carbonized seeds dating back 7000 to 8000 years B.C. have been discovered in Turkey [15]. As reported by Ahmed and Hasan [15], peas and dwarf field beans were planted in Switzerland around 4000 to 5000 B.C. The cultivation of soybeans in China is thought to have started between 2000 and 3000 B.C. [15]. Archaeological excavations have provided evidence of the cultivation of bean crops dating back to 10,000 B.C. in Mexico and Peru [27]. The domestication of beans, soybeans, and other staple crops in America and Asia began more than 3000 years ago [15]. The Romans recognized the benefits of using legumes in pastures and for soil improvement as early as 37 B.C., demonstrating their understanding of legumes' ability to fix nitrogen [27]. During the 1960s, the primary cultivation of grain legumes in Europe, including chickpeas, cowpeas, ground nuts, lentils, and common beans, for human consumption accounted for 67% of the total production area. However, this percentage decreased to 27% by 2013 [28]. This fall in legume consumption among Mediterranean countries, which were the highest legume consumer populations in Europe, attributed mostly to increased competition from cheaper imports, particularly from Canada, and the substitution of legumes with meat products [29].

During the 1980s, peas and soybeans emerged as the predominant protein crops cultivated for animal feed [28]. According to Watson et al. [28], soybeans, field peas, and broad beans are the most prevalent grain legumes in Europe. The soybean crop has had a significant surge in recent decades, primarily due to the strong demand for protein-rich materials used in livestock feed in Europe [28]. According to the European Commission, soy production covered an area of 943,000 hectares in Europe during the 2018/2019 period [29]. It is expected that this area would increase by 44% to about 1.3 million hectares by 2030, making it the crop with the highest growth rate in Europe [29]. The total field production of peas and broad beans reported 4.4 million tons in 2018, and approximately 66% of the production is allocated for animal feed uses, where only about 20% is allocated for human consumption [29]. Lupins, lentils, and chickpeas are not widely available in the European Union (EU), and most lentils and chickpeas are grown for human use, while lupin is primarily used as animal feed [29]. In 2018, grain legumes accounted for merely 1.4% of the overall crop area in Europe, as reported by the European Commission [29]. This figure represents approximately 10% of their typical contribution to cropping systems globally, as indicated by Watson et al. [28]. In addition, European cropland produces just 43% of the dietary legumes consumed in Europe [28]. Europe's domestic production exhibits a shortfall of approximately 70% in terms of high-protein resources, with 87% of this deficit being met through imports of soybean and soymeal [28]. Research conducted by Stagnari et al. [30] indicates a persistent decrease in legume production in Europe, which can drive legumes to be less economically competitive than more profitable crops like cereals [28].

The global grain legume production has reported more than 92 million tons of production over an area of around 81 million hectares [17]. Of the legume growers, India is the largest contributor, facilitating nearly 1/4th of global annual grain legume production. Other than India, the USA, Russia, China, Canada, Australia, Brazil, Myanmar, and Argentina are the principal grain legume-producing countries [17]. Legumes exhibit adaptability to a broad spectrum of climatic circumstances, spanning from cool to warm seasons. India has a favorable environment for growing a variety of grain legumes, including chickpea (*Cicer arietinum* L.), pigeon pea (*Cajanus cajan* L.), lentils (*Lens culinaris* L.), field pea (*Pisum sativum* L.), cowpea (*Vigna unguiculata* L.), horse gram (*Macrotyloma uniflorum* L.), green gram (*Vigna radiata* L.), black gram (*Vigna mungo* L. Hepper), chickling pea (*Lathyrus sativus* L.), and dry bean (*Phaseolus vulgaris* L.). On a global scale, dry beans are the most extensively grown grain legume, covering 36.8 million hectares (**Figure 1**) and yielding 30.4 million tons (**Figure 1**). Chickpea comes next with a cultivation area of 14.81 million hectares (**Figure 1**), followed by cowpeas (dry) with 12.5 million hectares. Field pea demonstrates the highest productivity, yielding 1718 kg per hectare. Lentil follows with a yield of 1038 kg per hectare, and chickpea ranks third with a yield of 965 kg per hectare [31].

However, the recent decrease in grain legume production in Europe is in contrast with the gains observed in other locations, such as Australia and Canada [30]. Clear and coordinated policy actions are necessary to provide direction for the advancement of legume-based food and feed systems and to initiate change, particularly in the European setting. A study conducted by Balázs, et al. [32], reported that facilitation of the development of favorable policies and transition paths that can promote the production of legumes and the manufacture of legume-based products is recommended to ensure the legume use sustainability.

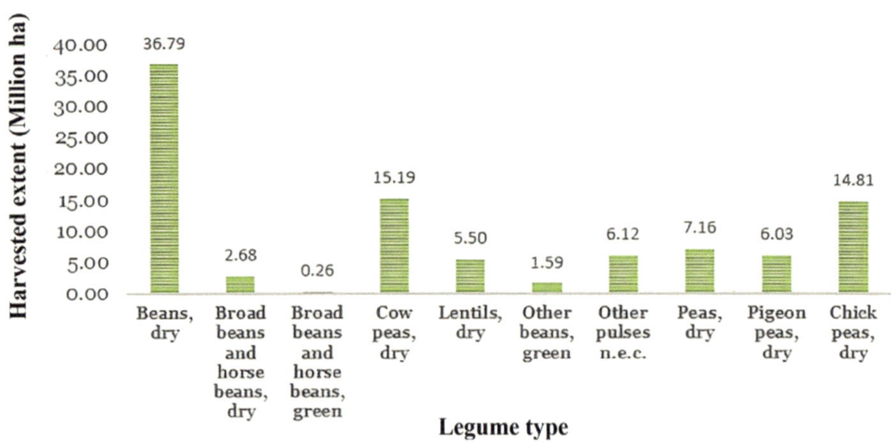

Figure 1.
Food legume area harvested in the world [31].

3. Uses of legumes

3.1 Legumes nutritional and nutraceutical benefits and uses

Legume seeds are considered one of the richest sources of proteins and amino acids for human nutrition [33]. Food legumes are often named "poor man's meat" as they are rich in protein (16–50%) [34]. The protein content of beans and grass peas ranges from 16.7 to 32.9 g/100 g and 22.4 to 31.7 g/100 g, respectively, and this value is higher than that of eggs (12.8 g/100 g), cereal crops (7.5–12 g/100 g), and meat (10–20 g/100 g) [35]. In addition to proteins, legumes are rich in dietary fibre (10–23%), vitamins, mono and polyunsaturated fatty acids, carbohydrates, and more than 15 essential mineral elements [36]. Among the essential mineral elements, legumes contain potassium (1244 mg), phosphorus (367 mg), calcium (113 mg), and magnesium (177 mg) per 100 g (**Table 1**) [37]. The main proteins found in legumes are salt-soluble globulins and water-soluble albumins [38]. Globulins contribute to the major protein fraction in legumes at around 72%, and albumins are the minor fraction, contributing to around 25% of the protein in legumes [38]. Legumin and vicilin are the primary types of globulins, while albumins consist of enzymatic proteins such as lectins, protease, and amylase inhibitors [39]. The nutritional quality of proteins in a particular food is determined by their essential amino acid composition and digestibility [39].

Furthermore, legumes are a source of folic acids and non-nutritional bioactive components like phytate, polyphenolics, lectins, and trypsin inhibitors [40]. Beans are rich sources of lysine and leucine, and other essential amino acids, such as methionine, are present in lower amounts [41]. As per the USDA Food Data Central [37] reports, legumes contain an average of 23.10 g proteins, 60.76 g carbohydrates,

Legume type	Ca	Fe	Mg	P	K	Na	Zn	Cu	Mn
Black bean	123	5.02	171	352	1483	5	3.65	0.84	1.06
Chickpeas	57	4.31	79	252	718	24	2.76	0.66	21.31
Cowpea	85	9.95	333	438	1375	58	6.11	1.06	1.54
Broad bean	22	1.9	38	95	250	50	0.58	0.07	0.32
Great northern bean	175	5.47	189	447	1390	14	2.31	0.84	1.42
Lentils	35	6.51	47	281	677	6	3.27	0.75	1.39
Lima bean (large)	81	7.51	224	385	1720	18	2.83	0.74	1.67
Lima bean (baby)	81	6.19	188	370	1400	13	2.6	0.67	1.69
Mung bean	132	6.74	189	367	1246	15	2.68	0.94	1.04
Navy bean	147	5.49	175	407	1185	5	3.65	0.83	1.15
Pigeon pea	130	5.23	183	367	1392	17	2.76	0.13	0.57
Pink bean	130	6.77	182	415	1460	8	2.55	0.81	1.38
Red kidney bean	83	6.69	138	406	1359	12	2.79	0.7	1.11
White bean	240	10.4	190	301	1800	16	3.67	0.98	1.8

Table 1.
Composition of major minerals in different legumes (mg/100 g) [37].

and 15.55 g dietary fibre per 100 g (**Table 2**). Legumes consist of complex carbohydrates, as well as resistant digestible starch (RDS) and slowly digestible starch (SDS) [38]. This helps the legumes to contribute to the component of diets high in fibre and low in glycaemic index (GI), rather than adding to the high sugar content of the human diet [42]. However, legumes contain mostly low amounts of fat, averaging around 1.86 g/100 g [38]. As per Maphosa and Jideani [34], legumes do not contain any cholesterol and are generally low in fat. Pigeon peas have 0.38 g of fat per 100 g, while lupin has 9.74 g (**Table 2**). The low-fat content of legumes notably gives them a marketing edge and makes them a desirable and healthful option for use in various culinary applications.

Legumes are generally reported to contain incomplete proteins because they contain relatively lower quantities of essential sulphur-containing amino acids, specifically cystine and methionine, found in higher concentrations in cereal grains. However, as reported by Curran [43] and FAO [44], cereal grains exhibit lower levels of lysine and tryptophan compared to legumes. Therefore, combining legumes with cereals in prepared foods can effectively enhance the overall amino acid profile of the diet [38].

Despite being rich in various nutrients, legume intake is limited due to the presence of several antinutritional factors (ANFs) such as trypsin, α-galactosidase, chymotrypsin inhibitors, phytates, lectins, and polyphenols [45, 46]. Many processing techniques, such as pressure cooking, heating, microwave cooking, sprouting, fermenting, and extrusion, have been shown to lower these ANFs and enhance the

Legume type	Water (g)	Energy (kcal/kJ)	Protein (g)	Total lipid/ fat (g)	Ash (g)	Carbohydrate (g)	Dietary fiber (g)
Black bean	11.02	341/1427	21.6	1.42	3.6	62.36	15.5
Chickpeas	7.68	378/1582	20.47	6.04	2.85	62.95	12.2
Cowpea	11.05	343/1435	23.85	2.07	3.39	59.64	10.7
Broad bean	10.98	341/1427	26.12	1.53	3.08	58.29	25
Great northern bean	10.7	339/1418	21.9	1.14	3.93	62.4	20.2
Lentils	8.26	352/1473	24.63	1.06	2.71	63.35	10.7
Lima bean (large)	10.2	338/1414	21.5	0.69	4.3	63.4	19
Lima bean (baby)	12.1	335/1402	20.6	0.93	3.55	62.8	20.6
Mung bean	9.05	347/1452	23.86	1.15	3.32	62.62	16.3
Navy bean	12.1	337/1410	22.33	1.5	3.32	60.75	15.3
Pigeon pea	10.59	343/1435	21.7	0.38	3.45	62.78	15
Pink bean	10.1	343/1435	21	1.13	3.66	64.2	12.7
Red kidney bean	11.75	337/1410	22.53	1.06	3.37	61.29	15.2
White bean	11.3	333/1393	23.4	0.85	4.2	60.3	15.2

Table 2.
Approximate nutrient composition of legumes (per 100 g) [37].

DOI: http://dx.doi.org/10.5772/intechopen.1009148

Legume type	Trypsin	Chymotrypsin inhibitor	α-Amylase inhibitor	Tannins	Saponins	Lectins	Phytic acid	Flavonoids
Black bean	+	+	+	+	+	+	+	+
Chickpeas	+	+	+	—	—	—	—	+
Cowpea	+	+	+	—	+	—	+	+
Broad bean	+	+	—	+	+	+	—	+
Lentils	+	+	—	—	—	+	—	+
Lima bean	+	+	—	—	—	+	+	—
Mung bean	+	+	+	+	—	—	+	+
Pigeon pea	—	+	+	—	—	—	+	+
Red kidney bean	+	+	+	+	—	+	+	+

Table 3.
Availability of anti-nutrients in major food legumes [51].

overall nutritional value and digestibility of legumes [47, 48]. ANFs are naturally occurring components of plant-based diets that have a negative impact on the consumption of traditional nutrients, as reported by Soetan and Oyewole [49]. The ANFs restrict nutrient availability in legumes, which leads to underutilization of nutrients [50]. These include of organic acids (like phytic acid), phenols (like tannins), proteins or peptides (such lectins and proteinase inhibitors—trypsin and chymotrypsin), and glycosides (like saponins). **Table 3** displays the often-detected ANFs in a selection of legumes. These ANFs can generally be classified in several ways, according to their physical features, specialized action, chemical structure, and/or origin in biosynthesis.

Ojokoh et al. [52] investigated the ANF content of breadfruit and fermented cowpea and found that, after fermentation, the amounts of phytic acids, oxalate, and hydrogen cyanide decreased. The fermentation of chickpeas also produced a 13% rise in protein content and a 45% decrease in phytates. Different strains of Lactobacillus were employed to start fermentation of sorghum (starting culture) in a different study by Adeyemo et al. [53] to examine their long-term effects on the antinutrient composition of the legume. After 5 days of fermentation, the decrease in the antinutrient content achieved by these strains was the most effective.

Furthermore, with respect to the nutritional composition, legumes usually contain a large amount of minerals; however, due to the presence of phytate, which is the primary inhibitor of iron and zinc absorption, they have relatively low bioavailability and digestibility [54]. The micronutrients included in legumes are essential for numerous cellular metabolic processes. By acting as vital cofactors for different antioxidant enzymes, they also help slow down the aging process naturally [55]. As a result of recent dietary recommendations to limit sodium, legumes are generally low in salt (**Table 1**). Furthermore, legumes are a good source of vitamins in the B-complex (i.e., folate, thiamin, and riboflavin). Still, they are not as good a source of vitamin C or fat-soluble vitamins [34].

In addition to the proteins, carbohydrates, dietary fibre, lipids, and minerals present in legumes, they include bioactive compounds that have several health benefits [34, 48, 56]. These chemicals include oligosaccharides, phenolics, peptides, low molecular weight proteins/peptides, carotenoids, saponins, phytic acids, and antioxidants. These bioactive chemicals were once considered to be antinutritional elements that hampered the digestion and absorption of important nutrients. However, in recent research, Pedrosa et al. [57] identified these chemicals as nutritionally active variables that have implications for health, including a decreased risk of type II diabetes, obesity, coronary heart disease (CHD), and body mass index (BMI). Furthermore, a thorough analysis of the bioactive substances present in common legumes and their corresponding characteristics was made by Conti et al. [58]. Peptides from the albumin and globulin fractions have been associated with anti-inflammatory, antioxidant, and antibacterial properties. It has been observed that phenolic chemicals, including phenolic acids, flavonoids, anthocyanins, and tannins, influence cell growth, homeostasis, and metabolism [58]. The major class of polyphenols known as flavonoids, which includes anthocyanins and anthoxanthins, is thought to have the greatest potential for antioxidant activity. They are a typical bioactive component of legumes, especially those having colorful seedcoats or testa [20]. The ability of legume flavonoids to scavenge free radicals, as well as their anti-inflammatory and anticancer characteristics, are some of their noteworthy functional traits. It is crucial to understand that although these chemicals are preserved in dry-milled materials (i.e., whole meals or flours), they are water-soluble and subject to leaching losses during aqueous processing [20, 59].

Despite being essential for nutrition because of their makeup of amino acids, legume proteins are a significant source of bioactive peptides. According to Duranti [60], prior research has validated the biological actions of legume proteins and peptides, indicating a noteworthy inhibitory effect on inflammatory indicators. Most bioactive peptides fall into the albumin or globulin fraction category. Lunasin is a 43-amino acid peptide of which the bioactivity has been shown by Jeong et al. [61] to specifically inhibit the division of malignant cells, but not healthy ones. Multivalent proteins called lectins bind carbohydrates and are found in a wide range of crops, including legumes. They can also limit the absorption of nutrients, making them an example of a bioactive substance with ambivalent tendencies. Numerous investigations have examined phytohemagglutinin in *P. vulgaris*, canavalin in jack bean, and other lectins from soybean, lentils, peas, and chickpeas; these have shown cytotoxicity against cancer cell lines in skin, bone, colon, breast, and liver [62, 63].

According to several studies, legume consumption and legume-based products have been linked to a decreased risk of diabetes, cardiovascular disease, cancer, and celiac disease (**Figure 2**) [33].

There is no controversy about the nutritional value of legumes, which are a great source of proteins, complex carbohydrates, and fiber. Legumes are important dietary components that promote satiety because they digest considerably more slowly than starchy cereals and tubers. Furthermore, lessening spikes following meals aids blood glucose regulation [58]. Legumes include dietary fibre, which is essential for healthy gut flora and may reduce the risk of several chronic illnesses, such as diabetes, heart disease, and several types of cancer.

Legumes possess the ability to manage body weight and combat obesity. Legume extracts have been discovered to include α-amylase inhibitors that decelerate the process of starch breakdown. This, in turn, facilitates weight and fat reduction [64]. According to Polak et al. [65], adult individuals who consume legumes have a BMI below 30, indicating a reduced likelihood of obesity. By analyzing data from eight trials, researchers found that over 50% of the 500 obese participants who had around one cup of legumes per day for a duration of 10 weeks experienced a decrease in both systolic and arterial blood pressure [65]. The results of a separate study, which included

Figure 2.
Human health benefits of food legumes.

113 individuals who were obese, showed a substantial reduction in weight, blood pressure, and waist measurements after consuming two portions of legumes and four portions of whole grains each day for a duration of 18 months. Notably, this improvement was achieved without the inclusion of any refined sugar in their diet [66].

The health benefits of legumes in humans are associated with their many bioactive components [67], which have been found to reduce the chance of acquiring noncommunicable diseases such as obesity and type 2 diabetes. Research has shown that obesity and a diet that is high in saturated and trans-fatty acids are all factors that contribute to the development of type II diabetes [68]. The glycemic index of legumes ranges from 28 to 52, which is significantly lower than the glycemic index of glucose, which is 100. This is because legumes include resistant starch, non-absorbable carbs, and oligosaccharides [69]. While they slow down the digestion of starch in the small intestine, these components slow down the rate at which the stomach empties. It is also possible that phytic acid, lectins, amylase inhibitors, or phenolic substances are responsible for the drop in blood glucose and insulin levels [70]. Bean-containing foods have been linked to higher insulin release and glucose absorption in the intestines [71]. Other research found that eating beans three or more times a week lowers the chance of getting type II diabetes [72]. People who are overweight and have diabetes can lose weight and lower the levels of inflammatory markers in their blood by eating legumes like chickpeas, peas, beans, and lentils [68].

Legumes have a high content of resistant starch (RS) and carbohydrates with a low glycaemic index (GI) [67]. Fiber, oligosaccharides (namely, raffinose, verbascose, and stachyose), RS traverse the stomach and small intestine without being digested, as there are no enzymes capable of breaking them down into simpler sugars [73]. Within the colon, they can serve as a reservoir of prebiotics that nourish resident probiotics. Bacterial fermentation leads to the production of short-chain fatty acids, such as butyrate, which have the potential to enhance colon health by stimulating a healthy population of microorganisms in the gut and reducing the likelihood of developing colon cancer [73]. Nevertheless, the process of fermentation can also result in the formation of gases, which can lead to symptoms such as bloating, cramps, and flatulence. Legumes contribute to weight loss by inducing a feeling of fullness [74] and regulating post-meal blood sugar levels, thus enhancing insulin sensitivity [59, 70].

Furthermore, the consumption of legumes can lower the risk of cardiovascular illnesses and reduce total cholesterol, low-density lipoprotein (LDL)-cholesterol, and triglyceride concentrations by approximately 10-15% [68]. Coronary heart disease ranks as the primary cause of mortality in most developed nations [43]. The incidence of cardiovascular disorders can be diminished with the intake of carbohydrates and essential vitamins such as B9, C, and E. Based on the findings of multiple epidemiological research, there has been an elevation in the recommendation for consuming legumes. Legumes are a highly abundant source of folic acid and consuming a higher amount of folate (0.8 mg folic acid per day) may decrease the likelihood of stroke by 24% and the likelihood of ischemic heart disease by 16% [43]. A recent study examining the correlation between legume consumption and cardiovascular disease found that senior individuals who consume legumes four or more times per week have a 34% lower chance of developing cardiovascular disease compared to those who consume legumes less than once a week [75]. Nevertheless, there was no substantial evidence to support the notion that regular consumption of legumes has a notable impact on cardiovascular disease in those under the age of 55 [75].

The low glycaemic index and slowly digestible starch content in legumes help prevent rapid spikes in blood sugar, promoting weight loss [72]. Additionally, flavonoids, phenolic acids, and tannins found in legumes have been linked to gastrointestinal benefits, antioxidant activity, and properties that mitigate inflammation and carcinogenic risks [76]. However, legumes contain compounds often classified as antinutritional, exhibiting dualistic functions with both positive and negative health effects [77]. For instance, digestive enzyme inhibitors such as proteinase inhibitors and amylase inhibitors are present in legumes and oligosaccharides cause flatulence yet are concurrently associated with prebiotic effects and potential protection for gastrointestinal health [78].

In addition to the above, consuming legumes directly or incorporating them into gluten-free goods might serve as a viable alternative for meeting the nutritional requirements for Celiac disease which is an auto-immune disorder caused by exposure to gluten [79]. Gluten induces damage in the small intestine of individuals with Celiac disease, leading to disruption in the assimilation of essential nutrients, thereby impacting all bodily systems [80]. Celiac illness can only be managed via adherence to a gluten-free diet [81]. However, there is apprehension regarding the gluten-free diet due to its tendency to result in a lower intake of proteins, vitamin B, iron, and dietary fiber compared to diets that include gluten [82]. Legumes are rich sources of proteins and vitamins B [79]. For example, a serving of 100 g of dry yellow pea flour can supply nearly 70% of the recommended daily intake of folate [83]. Patients with Celiac diseases are advised to consume a daily amount of 25–35 g of dietary fiber [84]. Nevertheless, gluten-free products frequently lack fiber due to the substitution of wheat flour with industrial starches, which typically have minimal impact on dietary fiber intake [22]. Hence, it is imperative to enhance the dietary fiber content of gluten-free foods. Yellow pea, lentil, and chickpea flours can enhance gluten-free products with a natural nutritional fiber that includes both soluble (33%) and insoluble fiber (26%) [85, 86].

The risk of colorectal cancer is reduced by consuming increased amounts of soybean and legume fiber, as they contain isoflavones [87]. There is a negative correlation between the consumption of legumes and cancer mortality. Papandreou et al. [88] reported that the intake of lentils, chickpeas, and beans had a positive impact on lowering cancer mortality rates in diabetic and obese males, compared to non-diabetic and non-obese females. Furthermore, an increased intake of legumes in general resulted in a 49% decrease in the risk of cancer, with 37% of this reduction specifically attributed to the consumption of lentils. Another study found that consuming a diet high in legumes, particularly chickpeas, can decrease the likelihood of developing colorectal cancer [89]. This is attributed to the high levels of protein and albumins found in legumes, which have been shown to effectively restrict the movement of colon carcinoma cells [90]. Consumption of legumes, namely, colorful beans, lentils, and black soybeans has been shown to suppress cell proliferation in prostate, ovarian, and breast cancer [91]. This observation pertained to the legumes' elevated concentration of phenolic compounds and their exceptional role as a dietary source of natural antioxidants [91]. Zhong et al. [92] reported a correlation between greater levels of isoflavones derived from legumes and a reduction in the incidence of endometrial cancer.

4. Nitrogen fixation of legumes

Biological nitrogen fixation (BNF) contributes to agricultural sustainability. It is revealed that up to 80% of the plant nitrogen requirement of field-grown legumes is fulfilled by the BNF [93]. The ability to fix atmospheric nitrogen (N) is obtained through the symbiotic relationship between legumes and *rhizobia*, which are bacteria found in the soil. In this symbiotic relationship, the legume transfers energy and carbon (C) to the *rhizobia* through the byproduct of photosynthesis, while the rhizobia reciprocate by providing the legume with nitrogen (N), primarily in the form of ammonium [94]. The symbiotic association commences when *Rhizobium* invades the roots of legumes, leading to the creation of root nodules where nitrogen fixation occurs. Nitrogen fixation is a biochemical process that involves the use of the nitrogenase enzyme as a catalyst to break down nitrogen molecules and transform them into ammonia. This process requires energy and an electron supply [95].

Tree legumes establish a mutually beneficial association with many types of *rhizobia*, including *Rhizobium* and *Bradyrhizobium*. Multiple strains of this *rhizobia* have been subjected to the phenotypic and genotypic investigation [96–98]. Nevertheless, the possibility of using them as symbiotic companions with tree legumes has not been explored. The nitrogen fixing ability of the symbiotic relationship between bacteria and plants varies among different legume species. The common bean is often considered to be a less efficient nitrogen fixer compared to other legume species [99–103]. According to Wilker et al. [104]. Mesoamerican indeterminate and climbing cultivars generally show better nodulation and higher N fixation compared to most Andean determinate (bush-type) cultivars. Sanyal et al. [105] found that pinto and navy beans exhibit a greater capacity for nitrogen fixation when compared to black and kidney beans.

The complex mechanism of legume biological nitrogen fixation (BNF) is affected by multiple environmental conditions, including temperature, water availability, nitrogen concentration, root zone pH, and the plant's nutritional state, including carbon and nitrogen substrates in the roots. Moreover, genetic variation is a factor in determining the inherent ability to carry out nitrogen fixation. The presence and functioning of nodules and nitrogenase activity in plants can be directly or indirectly affected by the concentrations of phosphorus (P) and potassium (K) in their nutritional status [106]. Further, legumes display differences in their susceptibility to nodulation in relation to varying nitrogen levels (**Figure 3a** and **b**) [108]. Several studies have found that the existence of soil mineral nitrogen in the area around the roots prevents the development of nodules in legumes [109, 110], the establishment of nodules [111], and the functioning of nitrogenase [112]. Legumes have a lower energy requirement for obtaining nitrogen from the soil compared to fixing nitrogen from the atmosphere through biological processes under high soil N concentrations [113–115]. The inhibition of nitrogen fixation by soil mineral nitrogen generally becomes stronger as the concentration of soil mineral nitrogen increases [116, 117].

Reports suggest that the presence of a particular quantity of mineral nitrogen (N) in the vicinity of the roots, referred to as "starter N," promotes the development of nodules and nitrogen fixation more effectively than non-mineral nitrogen in specific circumstances. The concentrations of "starter N" that enhance legume BNF vary depending on the cultivar and growth conditions. Typically, concentrations below 4 mM for ammonium (NH_4^+) and below 2 mM for nitrate (NO_3^-) are effective [114, 118–120]. The timing of exogenous nitrogen (N) application in relation to the developmental stage of legumes has a significant influence on the growth of nodules and nitrogen fixation. Applying N at a later stage result in reduced inhibition of nodule growth and nitrogen fixation in field peas (*Pisum sativum* L.) [121]. This is likely due to the fact that nodules are already fully developed in the early stages prior to

(a)

(b)

Figure 3.
a. Nodulated roots of a soybean plant. b. A detailed photograph of a divided nodule displaying its characteristic pink colour [107].

the application of nitrogen. Furthermore, studies have reported that the inhibition of nitrogen fixation by nitrate ions (NO_3^-) is more severe than that of ammonium ions (NH_4^+) in white clover, field pea, and soybean [120, 122–125]. Even recent studies revealed the N fixation inhibition by a substantial amount of N fertilizer [126]. Interestingly, Jiang et al. [123–125] discovered that nitrogen concentrations of 2.5 and 5 mM had a positive effect on the process of BNF in beans and Gan et al. [120] explained that low concentrations (1 mM and 3.75 mM) of inorganic nitrogen improve BNF in soybeans.

This type of symbiotic relationship can be observed in different classifications of legumes, including grain legumes, forage legumes, and specific leguminous trees. Grain legumes and forage legumes are used to improve soil nitrogen fertility in sustainable agriculture systems, especially in temperate regions [126–128]. The improvement of soil nitrogen fertility using grain/forage legumes is directly connected to the process of nitrogen fixation, as shown by Evans et al. [10], if the plant remains are incorporated into the soil after the crop is harvested. Therefore, it is essential to precisely quantify bean biological nitrogen fixation (BNF) to enhance our understanding of how to create sustainable agricultural systems. In these systems, subsequent crops can make use of the nitrogen fixation provided by legumes while simultaneously limiting nitrogen losses and so lowering environmental pollution.

5. Role of legumes in cropping systems

Legumes are known to contribute in numerous ways to cultivated crops. Jayakody et al. [129] reported that cassava intercropped with cowpea has reduced the weed occurrence and thus supported the growth of cassava. Further, Salgado et al. [130] revealed that legumes, namely, transferred nitrogen to cherry tomatoes when grown together. This promoted cherry tomatoes to undergo growth and maturation. The incorporation of legumes into the cropping system improves soil fertility and increases crop yield [30]. The advantages of integrating legumes into the cropping system can be attributed to various factors, such as the direct transfer of nitrogen, the existence of residual fixed nitrogen, enhanced nutrient availability and uptake, positive impacts on soil properties, disruption of pests' life cycles, and stimulation of other soil microbial activities (**Figure 4**) [131]. To fully maximize the beneficial impacts of legumes and the biological nitrogen fixation (BNF) system, it is advisable to integrate them into agricultural systems. Popular techniques for integrating legumes and their accompanying biological nitrogen fixation (BNF) into agricultural systems encompass crop rotation, intercropping, enhanced fallowing, green manuring, and alley cropping (**Figure 4**) [30].

However, the efficiency of nitrogen fixation can be increased by enhancing different components of the system. This entails the meticulous selection of appropriate legume genotypes, the inoculation with highly efficient rhizobia, the implementation of suitable agronomic practices, and the use of legumes in cropping systems [131]. Therefore, it is essential to maximize the utilization of legumes' complete genetic potential, utilize compatible rhizobia for inoculation of legumes, and implement appropriate agronomic practices and cropping systems to improve crop yields and food production efficiently. The integration of legumes into agricultural practices to improve agricultural productivity has gained significant momentum in meeting global food production demands. Hence, it is imperative to prioritize the assessment

Figure 4.
Maize-legume intercropping in the dry zone of Sri Lanka (Image by S. Ranamukhaarachchi, 1983).

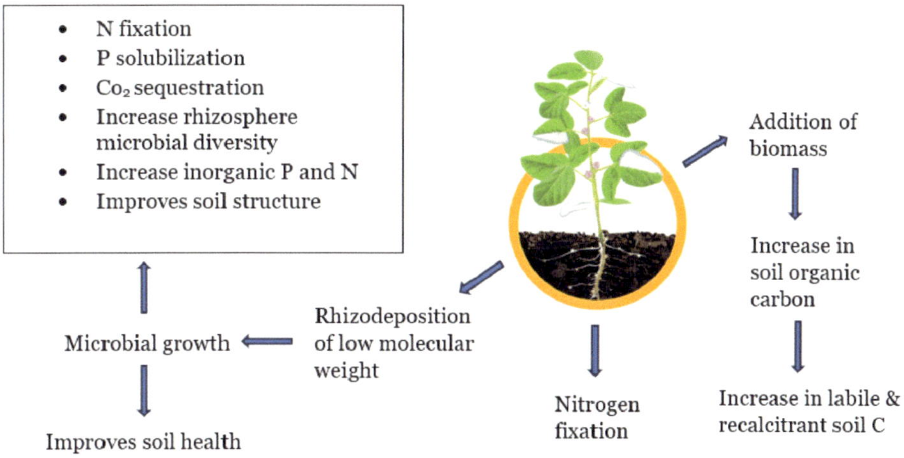

Figure 5.
Main functionalities of legume-incorporated innovative products.

of the BNF process by implementing available and novel sustainable technologies and enhancing our comprehension of the system [131]. In addition to BNF, the legume crops play a diverse role in cropping systems (**Figure 5**).

Legumes are used widely as a good component crop for double cropping to get a higher net return from arable lands. The potential of legumes for double cropping was evaluated through a study conducted in 1983 targeting the use in Central Pennsylvania, in the US [25]. This study revealed that soybean, being a short-day plant, could be planted late in the summer, and when planted no later than the first of July after harvesting winter wheat and winter barley, showed its feasibility for double cropping in Central Pennsylvania. Soybean as a component in double cropping system is common after winter wheat in the Central and Southern USA, particularly in Mississippi, Arkansas, Louisiana, and Alabama [132–136]. As reported by Kyei-Boahenand Zhang [137], the wheat-soybean double-crop system, utilizing soybeans from maturity group (MG), III or IV, and restricted irrigation, is more economically beneficial in comparison to the full-season soybean system.

Arshad et al. [24] reported the incorporation of legumes in intercropping and its effect on bioethanol production in sweet sorghum. According to its findings, the yield of the first-generation, second-generation, and overall bioethanol derived from biomass was much higher when sweet sorghum was intercropped with soybean, as opposed to when sweet sorghum was grown as a sole crop. The dry matter yields and composition of bioethanol-producing compounds obtained from intercropped sweet sorghum were reduced when intercropped with mung bean and simultaneously seeded, in contrast to intercropping with soybean and staggered planting. The sweet sorghum-soybean association established using an alternate single-row layout and staggered sowing demonstrated the highest bioethanol yield. Conversely, the sweet sorghum-mung bean intercropping, established with an alternate double-row design and simultaneous sowing, demonstrated the lowest bioethanol output. However, Ashard et al. [138] reported that simultaneous planting of sweet sorghum along with mung bean or soybean, under the application of nitrogen fertilizer at a rate of 60 kg ha^{-1} resulted in higher legume growth, increased crop yields, and improved bioethanol production. This was accomplished through the process of air pollution filtration.

6. Conservation agriculture

Legumes are well suited for two aspects of conservation agriculture: providing soil cover and serving as a rotational crop or residues. The advantages of incorporating legumes into cereal-based farming systems have been extensively examined by Siddique et al. [139] and can be summarized as follows. Legumes have a significant benefit in that they fixed nitrogen (N) and add to the soil and do not deplete the existing soil N stores. This is due to their ability to fulfill their own nitrogen requirements through biological nitrogen fixation. Legumes supply valuable organic matter to the soil due to their high nitrogen (N) to carbon (C) ratio, making them highly effective [26]. Additionally, legumes are beneficial in disrupting disease cycles in specific cereal crops as they have genetic differences from cereals, which means they host various pests and illnesses. Several legumes possess extensive root systems, which play a crucial role in the recycling of nutrients and the infiltration of water into the soil [140]. In addition, several legumes have the ability to obtain nutrients that are not easily accessible to other crops [106].

The soybean has a significant role in conservation agriculture systems in America, and Brazil, where this technique has been widely embraced for the past 30 years [141]. Brazilian instances of conservation agriculture have demonstrated consistent soil coverage, which minimizes erosion, and gradual enhancements in soil organic matter. These examples also involve the integration of livestock, the movement of surface-applied lime through the acidic soil profile, the disruption of compact soil layers, and a decreased dependence on agrochemicals. These practices have been successfully implemented across various climatic and edaphic conditions, farm sizes, and levels of mechanization [141]. The current target food legumes, which include lentils, chickpeas, peas, and faba beans, etc., are all a component of conservation agricultural systems in North America, Australia, and Turkey; however, the performance of these legumes in these systems has not been evaluated to the same extent as that of soybeans. It has been determined that there are several benefits associated with using minimum tilling for pulses in water-limited settings in Australia [141].

In the environments mentioned above, the retention of stubble in no-till systems, which involve less than 20% soil disturbance, has been found to considerably enhance soil water storage [142] and decrease soil compaction [143] in comparison to tilled soil. By reducing herbicide damage and by the stubble functioning as a trellis, combinations of wider row spacing (for example, 30 versus 20 cm) and stubble retention are proving to be beneficial for chickpea and lentil crops [144]. This results in improved harvestability and grain quality. According to Hawthorne [145], one of the most important agronomic adaptations for improving the yield of pulses cultivated in conservation agriculture systems in Australia has been the adjustment of row spacing to match the availability of water in the soil.

It has proven challenging to integrate all three conservation agriculture components—minimum tillage, soil cover, and rotation—in cropping systems, including target legumes, in developing agriculture, even though several systems in the Americas and Australia have done so successfully. The first stage is to implement conservation tillage, also known as minimum or zero tillage, in agricultural areas where full tillage, typically including animal traction, has been the norm. A significant limitation is the lack of availability of minimum tillage equipment specifically designed for use in large-scale agricultural operations. Nevertheless, in developing nations where four-wheeled tractors are employed, simplified, and inexpensive adaptations of minimum tillage techniques have been devised and embraced. Currently, there is an ongoing trial of low-cost Indian designs for minimum tillage planters in northern Iraq and Syria. These planters are locally made and used for some leguminous crops such as chickpeas, lentils, and others [146].

As discussed, food legumes that are well-grown can meet their own nitrogen (N) requirements and enhance the nitrogen economy of the entire cropping system. They achieve this by adding fixed nitrogen to the soil and minimizing or eliminating the need to extract nitrogen from soil reserves [147]. However, to achieve their maximum development and nitrogen-fixing capabilities in most agricultural settings, they typically need to be effectively inoculated with rhizobia bacteria [148, 149] and supplemented with additional elements like phosphorus (P), sulphur (S), boron (B), and molybdenum (Mo).

7. Ecosystem uses

Grain legumes have a significant impact on supplying ecological services. Grain legumes are favored by farmers in intensive cereal-based cropping systems due to their biological nitrogen fixation capacity, deep root systems, low input requirements, ability to live in poor soils, and resistance to abiotic stresses [150]. Cereal-based cropping systems that are practiced over a long period of time are intensive in terms of resource usage and are primarily responsible for the emission of greenhouse gases, pollution of groundwater, and other secondary concerns. Diversification within the cropping system had a crucial role in enhancing the ability of crops to withstand the impacts of environmental variations, climate change and various biological or non-biological pressures [151]. Grain legumes have a significant role in carbon sequestration due to their extensive leaf production and leaf shedding, extensive root system, symbiotic nitrogen fixation, and carbon-dense root exudates [152]. According to Bayer et al. [153], compared to cereals, legumes considerably increase soil carbon retention under decreased tillage conditions by increasing crop biomass and slowing the rate of C-mineralization.

Legumes transport a significant portion of their photosynthetic carbon as root exudates, which consist of amino acids, phenolic acids, organic acids, and lignin-rich compounds. This process plays a major role in carbon sequestration and helps reduce the carbon footprint of legumes [153]. According to Yue et al. [154], soybean cultivation in China is responsible for approximately 1% of greenhouse gas emissions. In comparison, cereals contribute 30%, while vegetables, fruit crops, and cash crops contribute 8, 6, and 4%, respectively to greenhouse gas emissions. Legumes, on the other hand, enrich the soil by providing nitrogen for future crops and reduce world carbon dioxide emissions by up to 50%, which are emitted by nitrogen-fertilizer businesses (urea) at a rate of around 300 Tg per year [127]. Legume-rich feeds are known to possess the ability to minimize CH_4 emissions from livestock businesses due to their low fiber content, faster rate of passage, and presence of condensed tannins and saponins. These compounds decrease cell wall digestion and alter rumen methanogenesis [155].

A study conducted by Jensen et al. (2012) revealed that legumes emit a lower amount of nitrous oxide (1.02 kg N_2O–N ha^{-1} $year^{-1}$) compared to cereals that are fertilized with nitrogen (2.71 kg N_2O–N ha^{-1} $year^{-1}$). The study conducted by Schwenke et al. [110] has documented the beneficial effects of decreased greenhouse gas emissions resulting from the cultivation of grain legumes. Watson et al. [28] reported that ecosystem services, such as the cycling of nutrients, possible reduction of greenhouse gas emissions, and the maintenance of biodiversity can be identified as ecosystem uses of legumes. However, there are instances where legumes have been shown to emit a greater amount of N_2O compared to cereals, as documented by Jeuffroy et al. [156] and Peyrard et al. [157]. The higher emission of N_2O by legumes can be attributed to two possible reasons: the faster decomposition rate of nitrogen-rich residues [156] and the denitrification of symbiotically fixed nitrogen within the nodule [157]. However, further research across multiple locations is necessary to obtain concrete evidence, as these emissions are highly influenced by location-specific climatic conditions and management practices [12, 158]. Nevertheless, the importance of legumes in mitigating greenhouse gases (GHGs) and sequestering deep soil carbon (C) is significant and necessitates thorough investigation.

Since grain legumes are superior to other plants due to adaptive strategies like higher rates of photosynthesis, a sharp increase in soluble sugar and starch content, and the rise of growth-promoting hormones like indole acetic acid and symbiotic nitrogen fixation (SNF), plants with C3 pathways are generally favored by elevated carbon dioxide (eCO_2) [159, 160]. Symbiotic nitrogen fixation is the primary process via which legumes adapt to elevated levels of atmospheric carbon dioxide (eCO_2). Legumes transport photosynthates from above ground to the nodules by symbiotic nitrogen fixation. This process provides a source of carbon and energy for the nitrogenase activity [161]. Legumes exhibit strong performance in environments of both moisture and high-temperature stress. Legumes are known to tolerate elevated temperature (eT) and moisture stress by generating antioxidants and reducing peroxide production [162]. Additional crucial aspects that significantly influence the adaptation to climate change include agronomic management approaches, nutritional supplementation, water availability, and the frequency of occurrence of abiotic and biotic stress conditions.

According to recent research work, the productivity of legumes is expected to significantly decrease in arid and semi-arid environments where water availability is a major issue [160, 163]. The combination of moisture stress, evapotranspiration (ET), and elevated carbon dioxide levels (eCO_2) has a devastating effect on the growth and production of legumes. The climate alteration has caused a shift in the geographic range of pests, leading to changes in pest outbreaks. In addition, there was a significant rise in secondary pest infestation, leading to the depletion of valued crops, according to Sharma et al. [164]. Hence, climate change has a diverse and complex effect on the growth and physiology of legumes. The next section comprehensively examines the influence of climate change drivers on legumes and their corresponding adaptation strategies. The growth stage and temperature magnitude influence the amount of damage caused by elevated temperature. Elevated temperatures are known to reduce the number of sinks, transfer carbohydrates away from grains, and hence impede seed development after experiencing stress [36]. The elevated temperature's detrimental effect of elevated temperature is seen in legumes, such as lentils. For every degree Celsius increase above 32°C, there is a reduction in the number of seeds by 0.10-0.20% [165]. However, legumes, being a climate-resilient crop, can withstand heat stress for up to 10 days without seeing a major decrease in output as described by Liu et al. The crop's indeterminate growth habit and high flexibility in phenology make them well-suited for future conditions [166]. Therefore, the process of choosing tolerant and durable genotypes in legumes should also consider all the characteristics listed. In addition to breeding, it is also essential to implement precision agronomic practices to reach the desired production targets and effectively manage natural resources in a sustainable manner.

Legume production in mixed crop and livestock systems has been found to lower the need for mineral nitrogen fertilizer usage and reduce nitrous oxide emissions [28]. On-farm production of legumes to feed and forage would enhance the self-sufficiency of farming systems and promote the nutrient circularity for nitrogen. From a consumer standpoint, the rise in the production and consumption of legumes is advantageous due to their provision of high-quality proteins, which are a crucial element of a nutritious diet [167]. Substituting meat with legumes to meet a significant amount of human protein requirements would promote more sustainable diets. This substitution would decrease the need for livestock production and its resulting environmental consequences [168].

Fulfilling human requirements with an ever-increasing population needs an undisturbed supply of food. This necessitates continuous land use for cultivations. The increase in deforestation has emerged as a challenge since many years ago deteriorating many ecosystem services [169]. Reforestation is often practiced as a solution for deforestation. Doi and Ranamukhaarachchi [170] reported following a study that used the legume, *Acacia* species to restore soil microbial activities in degraded lands in Thailand. The results revealed that *Acacia* has a greater impact on the rehabilitation process, including restoration of soil microbial activities compared to natural evergreen forest. Further, the incorporation of low-growing legumes as a live mulch for chili (*Capsicum annum* L.) has also shown improvement of soil conditions, reducing the high nitrogen application [171]. Furthermore, lowering weed competition and increasing chili yields were found compared to un-mulched plots. This study demonstrated that the use of low-growing legumes provides numerous benefits in short-term crops like chili.

8. Research work and recommendations for the future work in legumes

8.1 Food products that have been enhanced with legumes

Multiple studies have shown that legumes can be utilized as agents to increase the protein content of baked goods, namely in the form of flour [172–174]. For example, cereals have a low amount of lysine, but legumes do not have this limitation. On the other hand, methionine shows the opposite pattern. Hence, the amalgamation of grains and legumes enhances the protein quality and ensures a balanced distribution of important amino acids [41]. The addition of lentil flour to wheat bread resulted in higher levels of lysine, dietary fiber, and phenolic compounds, leading to an enhancement in antioxidant activity [174]. Another instance of adding legume supplementation is including hydrolysates of cowpea flour in the manufacturing process of pasta made from wheat flour. The cooked pasta with 10% cowpea exhibited a threefold increase in antioxidant capacity compared to pasta made solely from wheat [175]. Abou-Zaid et al. [176] produced a cake with a high protein content by combining raw and germinated broad bean flour with wheat flour. The blends containing a substantial amount of protein resulted in elevated amounts of amino acids. Millar et al. [177] created crackers by adding 40% broad bean and pea flour, resulting in a 55% increase in protein and total dietary fiber content. The phenolic content and antioxidant activity of the product were notably superior to that of the wheat crackers. Furthermore, consumer preference evaluations indicated that the product did not have any adverse impact on its appearance, hardness, or mouthfeel. Legume flour is a beneficial component utilized to enhance the nutritional value of gluten-free products (**Figure 6**).

Giuberti et al. [86] reported of a study that has found celiac patients may benefit from consuming gluten-free spaghetti made with bean flour. This is because the spaghetti has higher nutritional quality, with increased levels of protein, ash, dietary fiber, and resistant starch, compared to the control spaghetti made with 100% rice. Additionally, gluten-free spaghetti has a lower glycaemic index (**Figure 6**) [86].

Figure 6.
Main functionalities of legume-incorporated innovative products.

Separate research found that gluten-free cakes made using a combination of chickpea, pea, lentil, and bean flour along with rice flour (in a ratio of 50:50) included greater amounts of total protein, minerals, fat, and fiber compared to rice cake. The firmness and resilience of the gluten-free layer cake were enhanced by the addition of legumes, except for lentil flour. There was a reduction in the amount of starch that can be digested and in the glycaemic index, except for samples that included chickpeas. Overall, the addition of lentil flour produced the highest quality gluten-free cakes in terms of their physical and chemical qualities as well as their nutritional value [178].

Han et al. [179] examined the impact of using 100% legume flour (namely chickpea, green and red lentils, pinto beans, navy beans, and yellow peas) in the production of gluten-free cracker snacks. The composition and nutritional content of the chickpea crackers closely resembled those of the crackers already present in the market, except for the iron content. The chickpea crackers had three to six times more iron than the daily recommended intake per serving.

The development and enrichment of food with the incorporation of legumes have increased in the nutritional and functional value of foodstuffs.

8.2 Development of improved legume varieties for new habitats

Utilizing genetic manipulation to create high-yielding varieties is necessary to protect against newly emerging insect pests and to be able to withstand the harsh climate. The preservation of plant genetic resources, the identification of locally available hardy species that are genetically compatible, and the performance of gene manipulation to incorporate acceptable traits at the local level are all things that are desperately needed [180].

9. Advancement of climate-resilient management strategies for grain legumes and breeding for biotic and abiotic stress conditions

Implementing climate-wise management practices is essential for effectively tackling the difficulties posed by the increasing demand for food, fiber, and fuel. The implementation of diverse management practices, such as the selection of drought-resistant and high-yielding crop varieties, conservation agriculture techniques, crop rotations, soil and water conservation technologies, integrated tree-crop-livestock systems, utilization of non-conventional energy sources (such as solar power for drip irrigation), resource conservation technologies, and integrated nutrient and pest management practices, is essential for effective adaptation to changing climate conditions. The establishment of multi-stakeholder creative platforms is essential for capacity building and raising awareness about the adoption of climate-wise management methods. The adoption of such management approaches may face hurdles due to complexities in socio-cultural, political, economic, and ecological sectors. This issue can be effectively tackled by strategic interventions by both the government and private sector to promote practices that enhance resilience to climate change [180].

9.1 Utilization of crop models

Crop models are crucial instruments for comprehending the effects of certain crops in the context of climate change scenarios. Crop models encompass all facets of climate change and incorporate current knowledge to aid grain legumes in adapting to changing conditions. The biotic challenges that are aggravated by abiotic stress conditions require an integrated approach. Further, modern statistical methods, artificial intelligence, and machine learning software are utilized to forecast future climate conditions and adjust management operations accordingly. In recent years, there has been a growing interest in using modelling techniques to identify distinct limitations on crop yield in different regions, especially in the context of climate change. This method has been explored by researchers [181, 182].

10. Conclusions

Since being one of the first domesticated crops, legume crops have played a significant role in agriculture for thousands of years. Legumes have been cultivated in diverse geographical regions worldwide and served in numerous ways to sustain lives of all kinds and the environment. Legumes have a rich genetic diversity and provide heaps of uses to the consumers. Nutritional value and the biological nitrogen fixation ability of legume crops provide many benefits to consumers and growers, especially to ecosystem sustainability. With the expansion of the utilization of legume crops, new food products enhanced with legumes and advanced legume varieties to suit different traits are being developed with growing interest.

Conflict of interest

The authors declare no conflict of interest.

Author details

Senaratne Ranamukhaarachchi[1*] and Dhanesha Nanayakkara[1,2]

1 Sri Lanka Technology Campus, Meepe, Sri Lanka

2 School of Agriculture and Environment, College of Sciences, Massey University, Palmerston North, New Zealand

*Address all correspondence to: sranamukhaarachchi@sltc.ac.lk

References

[1] Lewis G, Schrire B, Mackinder B. Legumes of the World. Richmond, Surrey, UK: Royal Botanic Gardens. Kew; 2005

[2] Polhill RM, Raven PH, Stirton CH. Evolution and systematics of the Leguminosae. In: Polhill RM, Raven PH, editors. Advances in Legume Systematics Part 1. Royal Botanic Gardens, Kew: UK; 1981. pp. 1-26

[3] Gomes G, Silva G, Silva D, Oliveira R, Conceição G. Botanical composition of Fabaceae family in the Brazilian Northeast, Maranhão, Brazil. Asian Journal of Environment & Ecology. 2018;**6**(4):1

[4] FAO. Food and Agriculture Organization, FAOSTAT. 2017. Available from: http://www.fao.org/faostat/en/#home [Accessed: December 02, 2023]

[5] Ditzler L, van Apeldoorn DF, Pellegrini F, Antichi D, Bàrberi P, Rossing WA. Current research on the ecosystem service potential of legume inclusive cropping systems in Europe. A review. Agronomy for Sustainable Development. 2021;**41**:1-3

[6] Iannetta PP, Hawes C, Begg GS, Maaß H, Ntatsi G, Savvas D, et al. A multifunctional solution for wicked problems: Value-chain wide facilitation of legumes cultivated at bioregional scales is necessary to address the climate-biodiversity-nutrition nexus. Frontiers in Sustainable Food Systems. 2021;**5**

[7] Kouris-Blazos A, Belski R. Health benefits of legumes and pulses with a focus on Australian sweet lupins. Asia Pacific Journal of Clinical Nutrition. 2016;**25**(1):1-7

[8] Philips RD. Starchy legumes in human nutrition and culture. Plant Foods for Human Nutrition. 1993;**44**(3):195-211

[9] Tharanathan RN, Mahadevamma S. Grain legumes—A boon to human nutrition. Trends in Food Science & Technology. 2003;**14**(12):507-518

[10] Nichols PG, Loi A, Nutt BJ, Evans PM, Craig AD, Pengelly BC, et al. New annual and short-lived perennial pasture legumes for Australian agriculture—15 years of revolution. Field Crops Research. 2007;**104**(1-3):10-23

[11] Biederbeck VO, Zentner RP, Campbell CA. Soil microbial populations and activities as influenced by legume green fallow in a semiarid climate. Soil Biology and Biochemistry. 2005;**37**(10):1775-1784

[12] Bayer C, Gomes J, Zanatta JA, Vieira FC, Dieckow J. Mitigating greenhouse gas emissions from a subtropical Ultisol by using long-term no-tillage in combination with legume cover crops. Soil and Tillage Research. 2016;**161**:86-94

[13] Smýkal P, Coyne CJ, Ambrose MJ, Maxted N, Schaefer H, Blair MW, et al. Legume crops phylogeny and genetic diversity for science and breeding. Critical Reviews in Plant Sciences. 2015;**34**(1-3):43-104

[14] FAO. Food and Agriculture Organization, FAOSTAT. 2019. Available from: http://www.fao.org/faostat/en/#home [Accessed: December 05, 2023]

[15] Ahmed S, Hasan MM. Legumes: An overview. RADS Journal of Pharmacy and Pharmaceutical Sciences. 2014;**2**(1):34-38

[16] Sandlin C. The model system *C. elegans* demonstrates the health benefits of legumes and the potential benefits of legume consumption [thesis]. Louisiana State University and Agricultural & Mechanical College; 2010

[17] FAO. Food and Agriculture Organization, FAOSTAT. 2018. Available from: http://www.fao.org/faostat/en/#home [Accessed: December 10, 2023]

[18] Bravo L, Siddhuraju P, Saura-Calixto F. Composition of underexploited Indian pulses. Comparison with common legumes. Food Chemistry. 1999;**64**(2):185-192

[19] Rubiales D, Khazaei H. Advances in disease and pest resistance in faba bean. Theoretical and Applied Genetics. 2022;**135**(11):3735-3756

[20] Aguilera Y, Estrella I, Benitez V, Esteban RM, Martín-Cabrejas MA. Bioactive phenolic compounds and functional properties of dehydrated bean flours. Food Research International. 2011;**44**(3):774-780

[21] Parmar N, Singh N, Kaur A, Thakur S. Comparison of color, anti-nutritional factors, minerals, phenolic profile, and protein digestibility between hard-to-cook and easy-to-cook grains from different kidney bean (*Phaseolus vulgaris*) accessions. Journal of Food Science and Technology. 2017;**54**:1023-1034

[22] Fendri LB, Chaari F, Maaloul M, Kallel F, Abdelkafi L, Chaabouni SE, et al. Wheat bread enrichment by pea and broad bean pods fibers: Effect on dough rheology and bread quality. Lwt. 2016;**73**:584-591

[23] Zhang W, Ricketts TH, Kremen C, Carney K, Swinton SM. Ecosystem services and dis-services to agriculture. Ecological Economics. 2007;**64**(2):253-260

[24] Arshad M, Nawaz R, Shahnawaz M, Shahzad Sarfraz M, Ahmad S. First and second generation bioethanol production in sweet sorghum, mungbean and soybean in sole and intercropping systems. Tropical Agriculture. 2015;**92**(1)

[25] Ranamukhaarachchi SL. Potential of growing soybean (*Glycine max* L. Merrill) for double cropping in Central Pennsylvania [thesis]. Pennsylvania, USA: Department of Agronomy, The Pennsylvania State University, State College; 1983

[26] Siddique KH, Johansen C, Turner NC, Jeuffroy MH, Hashem A, Sakar D, et al. Innovations in agronomy for food legumes. A review. Agronomy for Sustainable Development. 2012;**32**:45-64

[27] Gomes AM, Vasconcelos MW. The legume grains: When tradition goes hand in hand with nutrition. In: Traditional Foods: General and Consumer Aspects. 2016. pp. 189-208

[28] Watson CA, Reckling M, Preissel S, Bachinger J, Bergkvist G, Kuhlman T, et al. Grain legume production and use in European agricultural systems. Advances in Agronomy. 2017;**144**:235-303

[29] European Commission. EU Agricultural Outlook for Markets and Income, 2018-2030. Brussels: European Commission; 2018

[30] Stagnari F, Maggio A, Galieni A, Pisante M. Multiple benefits of legumes for agriculture sustainability: An overview. Chemical and Biological Technologies in Agriculture. 2017;**4**:1-3

[31] FAO. Food and Agriculture Organization, FAOSTAT. 2022. Available

from: http://www.fao.org/faostat/en/#home [Accessed: December 10, 2023]

[32] Balázs B, Kelemen E, Szakál D. Transitions of legume-based agrifood systems-stakeholders' view from Hungary. The International Journal of Sociology of Agriculture and Food. 2021;**27**(1):119-140

[33] Carbas B, Machado N, Pathania S, Brites C, Rosa EA, Barros AI. Potential of legumes: Nutritional value, bioactive properties, innovative food products, and application of eco-friendly tools for their assessment. Food Reviews International. 2023;**39**(1):160-188

[34] Maphosa Y, Jideani VA. The role of legumes in human nutrition. In: Functional Food-Improve Health Through Adequate Food. Vol. 1. 2017. p. 13

[35] Kaur M, Singh N. Characterization of protein isolates from different Indian chickpea (*Cicer arietinum* L.) cultivars. Food Chemistry. 2007;**102**(1):366-374

[36] Wang N, Hatcher DW, Toews R, Gawalko EJ. Influence of cooking and dehulling on nutritional composition of several varieties of lentils (*Lens culinaris*). LWT-Food Science and Technology. 2009;**42**(4):842-848

[37] USDA. Food Data Central. US Department of Agriculture [Internet]; 2022. Available from: https://fdc.nal.usda.gov [Accessed: December 19, 2023]

[38] Affrifah NS, Uebersax MA, Amin S. Nutritional significance, value-added applications, and consumer perceptions of food legumes: A review. Legume Science. 2023;**5**(4):e192

[39] Carbonaro M, Maselli P, Nucara A. Structural aspects of legume proteins and nutraceutical properties. Food Research International. 2015;**76**:19-30

[40] Burstin J, Gallardo K, Mir RR, Varshney RK, Duc G. 20 improving protein content and nutrition quality. Biology and Breeding of Food Legumes. 2011:314

[41] Kan L, Nie S, Hu J, Wang S, Cui SW, Li Y, et al. Nutrients, phytochemicals and antioxidant activities of 26 kidney bean cultivars. Food and Chemical Toxicology. 2017;**108**:467-477

[42] Collins K. Carbohydrate quality part 2: Pulses and healthy eating. In: Nutrition Research in Perspective. [Internet]. 2020. Available from https://karencollinsnutrition.com/pulses-healthy-eating/ [Accessed: October 18, 2023]

[43] Curran J. The nutritional value and health benefits of pulses in relation to obesity, diabetes, heart disease, and cancer. British Journal of Nutrition. 2012;**108**(S1):S1-S2

[44] FAO. Research Approaches and Methods for Evaluating the Protein Content of Foods. Food and Agriculture Organization; 2014. Available from: https://www.fao.org/faostat/en/#data/QCL [Accessed: December 02, 2023]

[45] Fernández M, Aranda P, López-Jurado M, García-Fuentes MA, Urbano G. Bioavailability of phytic acid phosphorus in processed *Vicia faba* L. var. major. Journal of Agricultural and Food Chemistry. 1997;**45**(11):4367-4371

[46] Srivastava A, Rao LJ, Shivanandappa T. A novel cytoprotective antioxidant: 4-hydroxyisophthalic acid. Food Chemistry. 2012;**132**(4):1959-1965

[47] Popova A, Mihaylova D. Antinutrients in plant-based foods: A

review. The Open Biotechnology Journal. 2019;**13**(1)

[48] Wiesinger JA, Marsolais F, Glahn RP. Health implications and nutrient bioavailability of bioactive compounds in dry beans and other pulses. Dry Beans and Pulses: Production, Processing, and Nutrition. 2022:505-529

[49] Soetan KO, Oyewole OE. The need for adequate processing to reduce the anti-nutritional factors in plants used as human foods and animal feeds: A review. African Journal of Food Science. 2009;**3**(9):223-232

[50] Yacout MH. Anti-nutritional factors & its roles in animal nutrition. Journal of Dairy Veterinary & Animal Research. 2016;**4**(1):237-239

[51] Sathe SK. Chemistry and implications of antinutritional factors in dry beans and pulses. Dry Beans and Pulses Production, Processing and Nutrition. 2012;**5**:359-377

[52] Ojokoh AO, Daramola MK, Oluoti OJ. Effect of fermentation on nutrient and anti-nutrient composition of breadfruit (*Treculia africana*) and cowpea (*Vigna unguiculata*) blend flours. African Journal of Agricultural Research. 2013;**8**(27):3566-3570

[53] Adeyemo SM, Onilude AA, Olugbogi DO. Reduction of anti-nutritional factors of sorghum by lactic acid bacteria isolated from Abacha-an African fermented staple. Frontiers in Science. 2016;**6**(1):25-30

[54] Sandberg AS. Bioavailability of minerals in legumes. British Journal of Nutrition. 2002;**88**(S3):281-285

[55] Höhn A, Weber D, Jung T, Ott C, Hugo M, Kochlik B, et al. Happily (n) ever after: Aging in the context of oxidative stress, proteostasis loss and cellular senescence. Redox Biology. 2017;**11**:482-501

[56] Silva LR, Peix Á, Albuquerque C, Velàzquez E. Bioactive compounds of legumes as health promoters. Natural Bioactive Compounds from Fruits and Vegetables as Health Promoters, Part. 2016;**2**:3-27

[57] Pedrosa MM, Guillamón E, Arribas C. Autoclaved and extruded legumes as a source of bioactive phytochemicals: A review. Food. 2021;**10**(2):379

[58] Conti MV, Guzzetti L, Panzeri D, De Giuseppe R, Coccetti P, Labra M, et al. Bioactive compounds in legumes: Implications for sustainable nutrition and health in the elderly population. Trends in Food Science & Technology. 2021;**117**:139-147

[59] Didinger C, Thompson HJ. Defining nutritional and functional niches of legumes: A call for clarity to distinguish a future role for pulses in the dietary guidelines for Americans. Nutrients. 2021;**13**(4):1100

[60] Duranti M. Grain legume proteins and nutraceutical properties. Fitoterapia. 2006;**77**(2):67-82

[61] Jeong HJ, Lee JR, Jeong JB, Park JH, Cheong YK, de Lumen BO. The cancer-preventive seed peptide Lunasin from rye is bioavailable and bioactive. Nutrition and cancer. 2009;**61**(5):680-686

[62] Gautam AK, Shrivastava N, Sharma B, Bhagyawant SS. Current scenario of legume lectins and their practical applications. Journal of Crop Science and Biotechnology. 2018;**21**:217-227

[63] Shi Z, Chen J, Li CY, An N, Wang ZJ, Yang SL, et al. Antitumor effects of concanavalin A and *Sophora flavescens* lectin in vitro and in vivo. Acta Pharmacologica Sinica. 2014;**35**(2):248-256

[64] Cakir Ö, Ucarli C, Tarhan Ç, Pekmez M, Turgut-Kara N. Nutritional and health benefits of legumes and their distinctive genomic properties. Food Science and Technology. 2019;**39**:1-2

[65] Polak R, Phillips EM, Campbell A. Legumes: Health benefits and culinary approaches to increase intake. Clinical Diabetes. 2015;**33**(4):198-205

[66] Venn BJ, Perry T, Green TJ, Skeaff CM, Aitken W, Moore NJ, et al. The effect of increasing consumption of pulses and wholegrains in obese people: A randomized controlled trial. Journal of the American College of Nutrition. 2010;**29**(4):365-372

[67] Munro S. Legumes: Essentials of Human Nutrition. Oxford University Press; 2007. pp. 356-358

[68] Hosseinpour-Niazi S, Mirmiran P, Hedayati M, Azizi F. Substitution of red meat with legumes in the therapeutic lifestyle change diet based on dietary advice improves cardiometabolic risk factors in overweight type 2 diabetes patients: A cross-over randomized clinical trial. European Journal of Clinical Nutrition. 2015;**69**(5):592-597

[69] Jenkins DJ, Kendall CW, Augustin LS, Mitchell S, Sahye-Pudaruth S, Mejia SB, et al. Effect of legumes as part of a low glycemic index diet on glycemic control and cardiovascular risk factors in type 2 diabetes mellitus: A randomized controlled trial. Archives of Internal Medicine. 2012;**172**(21):1653-1660

[70] Mollard RC, Wong CL, Luhovyy BL, Anderson GH. First and second meal effects of pulses on blood glucose, appetite, and food intake at a later meal. Applied Physiology, Nutrition, and Metabolism. 2011;**36**(5):634-642

[71] Pérez-Ramírez IF, Becerril-Ocampo LJ, Reynoso-Camacho R, Herrera MD, Guzmán-Maldonado SH, Cruz-Bravo RK. Cookies elaborated with oat and common bean flours improved serum markers in diabetic rats. Journal of the Science of Food and Agriculture. 2018;**98**(3):998-1007

[72] Becerra-Tomás N, Díaz-López A, Rosique-Esteban N, Ros E, Buil-Cosiales P, Corella D, et al. Legume consumption is inversely associated with type 2 diabetes incidence in adults: A prospective assessment from the PREDIMED study. Clinical Nutrition. 2018;**37**(3):906-913

[73] Bird A, Conlon M, Christophersen C, Topping D. Resistant starch, large bowel fermentation and a broader perspective of prebiotics and probiotics. Beneficial Microbes. 2010;**1**(4):423-431

[74] Li SS, Kendall CW, de Souza RJ, Jayalath VH, Cozma AI, Ha V, et al. Dietary pulses, satiety, and food intake: A systematic review and meta-analysis of acute feeding trials. Obesity. 2014;**22**(8):1773-1780

[75] Nouri F, Sarrafzadegan N, Mohammadifard N, Sadeghi M, Mansourian M. Intake of legumes and the risk of cardiovascular disease: Frailty modeling of a prospective cohort study in the Iranian middle-aged and older population. European Journal of Clinical Nutrition. 2016;**70**(2):217-221

[76] García-Lafuente A, Moro C, Manchón N, Gonzalo-Ruiz A, Villares A, Guillamón E, et al. In vitro anti-inflammatory activity of phenolic rich extracts from white and red common beans. Food Chemistry. 2014;**161**:216-223

[77] Kalpanadevi V, Mohan VR. Effect of processing on antinutrients and in vitro protein digestibility of the underutilized legume, *Vigna unguiculata* (L.) Walp subsp. *unguiculata*. LWT-Food Science and Technology. 2013;**51**(2):455-461

[78] Sreerama YN, Sashikala VB, Pratape VM, Singh V. Nutrients and antinutrients in cowpea and horse gram flours in comparison to chickpea flour: Evaluation of their flour functionality. Food Chemistry. 2012;**131**(2):462-468

[79] Pastor-Cavada E, Juan R, Pastor JE, Alaiz M, Vioque J. Protein and amino acid composition of select wild legume species of tribe Fabeae. Food Chemistry. 2014;**163**:97-102

[80] Sapone A, Bai JC, Ciacci C, Dolinsek J, Green PH, Hadjivassiliou M, et al. Spectrum of gluten-related disorders: Consensus on new nomenclature and classification. BMC Medicine. 2012;**10**:1-2

[81] Mulder CJ, Wierdsma NJ, Berkenpas M, Jacobs MA, Bouma G. Preventing complications in celiac disease: Our experience with managing adult celiac disease. Best Practice & Research Clinical Gastroenterology. 2015;**29**(3):459-468

[82] Morreale F, Agnoli C, Roncoroni L, Sieri S, Lombardo V, Mazzeo T, et al. Are the dietary habits of treated individuals with celiac disease adherent to a Mediterranean diet?. Nutrition, Metabolism and Cardiovascular Diseases. 2018;**28**(11):1148-1154

[83] Muir JG, Varney JE, Ajamian M, Gibson PR. Gluten-free and low-FODMAP sourdoughs for patients with coeliac disease and irritable bowel syndrome: A clinical perspective. International Journal of Food Microbiology. 2019;**290**:237-246

[84] Safavi F, Rostami A. Role of serine proteases in inflammation: Bowman–Birk protease inhibitor (BBI) as a potential therapy for autoimmune diseases. Experimental and Molecular Pathology. 2012;**93**(3):428-433

[85] Boukid F, Vittadini E, Lusuardi F, Ganino T, Carini E, Morreale F, et al. Does cell wall integrity in legumes flours modulate physiochemical quality and in vitro starch hydrolysis of gluten-free bread?. Journal of Functional Foods. 2019;**59**:110-118

[86] Rocchetti G, Lucini L, Chiodelli G, Giuberti G, Gallo A, Masoero F, et al. Phenolic profile and fermentation patterns of different commercial gluten-free pasta during in vitro large intestine fermentation. Food Research International. 2017;**97**:78-86

[87] Zhu B, Sun Y, Qi L, Zhong R, Miao X. Dietary legume consumption reduces risk of colorectal cancer: Evidence from a meta-analysis of cohort studies. Scientific Reports. 2015;**5**(1):8797

[88] Papandreou C, Becerra-Tomás N, Bulló M, Martínez-González MÁ, Corella D, Estruch R, et al. Legume consumption and risk of all-cause, cardiovascular, and cancer mortality in the predimed study. Clinical Nutrition. 2019;**38**(1):348-356

[89] Perera HA. Why does regular legume consumption matter for colorectal cancer prevention?. [thesis]. Oregon State University; 2018

[90] Lima AI, Mota J, Monteiro SA, Ferreira RM. Legume seeds and colorectal cancer revisited: Protease inhibitors reduce MMP-9 activity and colon cancer cell migration. Food Chemistry. 2016;**197**:30-38

[91] Xu B, Chang SK. Comparative study on antiproliferation properties

and cellular antioxidant activities of commonly consumed food legumes against nine human cancer cell lines. Food Chemistry. 2012;**134**(3):1287-1296

[92] Zhong XS, Ge J, Chen SW, Xiong YQ, Ma SJ, Chen Q. Association between dietary isoflavones in soy and legumes and endometrial cancer: A systematic review and meta-analysis. Journal of the Academy of Nutrition and Dietetics. 2018;**118**(4):637-651

[93] Schulze J. How are nitrogen fixation rates regulated in legumes?. Journal of Plant Nutrition and Soil Science. 2004;**167**(2):125-137

[94] Howard JB, Rees DC. Structural basis of biological nitrogen fixation. Chemical Reviews. 1996;**96**:2965-2982

[95] Postgate JR. Biological nitrogen fixation: Fundamentals. Philosophical Transactions of the Royal Society of London. B, Biological Sciences. 1982;**296**(1082):375-385

[96] Zahran HH, Räsänen LA, Karsisto M, Lindström K. Alteration of lipopolysaccharide and protein profiles in SDS-PAGE of rhizobia by osmotic and heat stress. World Journal of Microbiology and Biotechnology. 1994;**10**:100-105

[97] Zahran HH. Diversity, adaptation, and activity of the bacterial flora in saline environments. Biology and Fertility of Soils. 1997;**25**:211-223

[98] Zhang X, Harper R, Karsisto M, Lindström K. Diversity of rhizobium bacteria isolated from the root nodules of leguminous trees. International Journal of Systematic and Evolutionary Microbiology. 1991;**41**(1):104-113

[99] Isoi T, Yoshida S. Low nitrogen fixation of common bean (*Phaseolus vulgaris* L.). Soil Science and Plant Nutrition. 1991;**37**(3):559-563

[100] Bliss FA. Breding common bean for improved biological nitrogen fixation. Plant and Soil. 1993;**152**:71-79

[101] Martinez-Romero E. Diversity of rhizobium-*Phaseolus vulgaris* symbiosis: Overview and perspectives. Plant and Soil. 2003;**252**:11-23

[102] Hardarson G, Serraj R. Enhancement of Symbiotic Nitrogen Fixation in Grain Legumes: Selected Results from the FAO/IAEA Program. Symbiotic Nitrogen Fixation: Prospects for Enhanced Application in Tropical Agriculture. 2004. pp. 163-171

[103] Peoples MB, Hauggaard-Nielsen H, Jensen ES. The potential environmental benefits and risks derived from legumes in rotations. Nitrogen Fixation in Crop Production. 2009;**52**:349-385

[104] Wilker J, Navabi A, Rajcan I, Marsolais F, Torkamaneh D, Pauls KP. Agronomic performance and nitrogen fixation of heirloom and conventional dry bean varieties under low-nitrogen field conditions. Frontiers in Plant Science. 2019;**10**:462964

[105] Sanyal D, Goos RJ, Chatterjee A. Determining symbiotic nitrogen fixation in dry bean cultivars using ureide method and isotope dilution techniques. Communications in Soil Science and Plant Analysis. 2018;**49**(16):2042-2052

[106] Liu Y, Wu L, Baddeley JA, Watson CA. Models of biological nitrogen fixation of legumes. Sustainable Agriculture. 2011;**2**:883-905

[107] Legume Hub [Internet]. 2021. Available from: https://www.legumehub.eu/ [Accessed: December 02, 2022]

[108] Harper JE. Uptake of organic nitrogen forms by roots and leaves. Nitrogen in Crop Production. 1984:165-170

[109] Abdel Wahab AM, Zahran HH, Abd-Alla MH. Root-hair infection and nodulation of four grain legumes as affected by the form and the application time of nitrogen fertilizer. Folia Microbiologica. 1996;**41**:303-308

[110] Schwenke GD, Herridge DF, Scheer C, Rowlings DW, Haigh BM, McMullen KG. Soil N_2O emissions under N_2-fixing legumes and N-fertilised canola: A reappraisal of emissions factor calculations. Agriculture, Ecosystems & Environment. 2015;**202**:232-242

[111] Imsande J. Inhibition of nodule development in soybean by nitrate or reduced nitrogen. Journal of Experimental Botany. 1986;**37**(3):348-355

[112] Purcell LC, Sinclair TR. Nitrogenase activity and nodule gas permeability response to rhizospheric NH_3 in soybean. Plant Physiology. 1990;**92**(1):268-272

[113] Thornley JH, Cannell MG. Dynamics of mineral N availability in grassland ecosystems under increased [CO_2]: Hypotheses evaluated using the Hurley pasture model. Plant and Soil. 2000;**224**(1):153-170

[114] Bethlenfalvay GJ, Abu-Shakra SS, Phillips DA. Interdependence of nitrogen nutrition and photosynthesis in *Pisum sativum* L. Plant Physiology. 1978;**62**(1):127-130

[115] Wood M. Nitrogen Fixation: How much and at what cost? In: Legumes in Sustainable Farming Systems: Occasional Symposium. Reading, UK: British Grassland Society; 1996. pp. 26-35

[116] Macduff JH, Jarvis SC, Davidson IA. Inhibition of N_2 fixation by white clover (*Trifolium repens* L.) at low concentrations of NO_3- in flowing solution culture. Plant and Soil. 1996;**180**:287-295

[117] Lamb JF, Barnes DK, Russelle MP, Vance CP, Heichel GH, Henjum KI. Ineffectively and effectively nodulated alfalfas demonstrate biology nitrogen fixation continues with high nitrogen fertilization. Crop Science. 1995;**35**(1):153-157

[118] Schomberg HH, Weaver RW. Nodulation, nitrogen fixation, and early growth of arrow leaf clover in response to root temperature and starter nitrogen. Agronomy Journal. 1992;**84**(6):1046-1050

[119] Gulden RH, Vessey JK. The stimulating effect of ammonium on nodulation in *Pisum sativum* L. is not long lived once ammonium supply is discontinued. Plant and Soil. 1997;**195**:195-205

[120] Gan Y, Stulen I, van Keulen H, Kuiper PJ. Low concentrations of nitrate and ammonium stimulate nodulation and N_2 fixation while inhibiting specific nodulation (nodule DW g^{-1} root dry weight) and specific N_2 fixation (N_2 fixed g^{-1} root dry weight) in soybean. Plant and Soil. 2004;**258**:281-292

[121] Waterer JG, Vessey JK. Effect of low static nitrate concentrations on mineral nitrogen uptake, nodulation, and nitrogen fixation in field pea. Journal of Plant Nutrition. 1993;**16**(9):1775-1789

[122] Svenning MM, Junttila O, Macduff JH. Differential rates of inhibition of N_2 fixation by sustained low concentrations of NH_4+ and NO_3-in northern ecotypes of white clover (*Trifolium repens* L.). Journal of Experimental Botany. 1996;**47**(6):729-738

[123] Bollman MI, Vessey JK. Differential effects of nitrate and ammonium supply on nodule initiation, development, and distribution on roots of pea (*Pisum sativum*). Botany. 2006;**84**(6):893-903

[124] Jiang Y, MacLean DE, Perry GE, Marsolais F, Hill B, Pauls KP. Evaluation of beneficial and inhibitory effects of nitrate on nodulation and nitrogen fixation in common bean (*Phaseolus vulgaris*). Legume Science. 2020;**2**(3):e45

[125] Hossain SA, Strong WM, Waring SA, Dalal RC, Weston EJ. Comparison of legume-based cropping systems at Warra, Queensland. 2. Mineral nitrogen accumulation and availability to the subsequent wheat crop. Soil Research. 1996;**34**(2):289-297

[126] Akter Z, Pageni BB, Lupwayi NZ, Balasubramanian PM. Biological nitrogen fixation by irrigated dry bean (*Phaseolus vulgaris* L.) genotypes. Canadian Journal of Plant Science. 2018;**98**(5):1159-1167

[127] Jensen ES, Peoples MB, Boddey RM, Gresshoff PM, Hauggaard-Nielsen H, Alves JR, et al. Legumes for mitigation of climate change and the provision of feedstock for biofuels and biorefineries. A review. Agronomy for Sustainable Development. 2012;**32**:329-364

[128] McCallum MH, Peoples MB, Connor DJ. Contributions of nitrogen by field pea (*Pisum sativum* L.) in a continuous cropping sequence compared with a lucerne (*Medicago sativa* L.)-based pasture ley in the Victorian Wimmera. Australian Journal of Agricultural Research. 2000;**51**(1):13-22

[129] Jayakody DPP, Ranamukhaarach chi SL, Gunasena HPM. Weed dynamics and crop yields in cassava-based intercropping under coconut. Tropical Agricultural Research. 1991;**3**:195-215

[130] Salgado GC, Ambrosano EJ, Rossi F, Otsuk IP, Ambrosano GM, Santana CA, et al. Biological n fixation and n transfer in an intercropping system between legumes and organic cherry tomatoes in succession to green corn. Agriculture. 2021;**11**(8):690

[131] Kebede E. Contribution, utilization, and improvement of legumes-driven biological nitrogen fixation in agricultural systems. Frontiers in Sustainable Food Systems. 2021;**5**:767998

[132] Anon. Market Developments and Policy Evaluation Aspects of the Plant Protein Sector in the EU—Final Report. Vol. 41. Brussels: European Commission; 2005. pp. 347-359

[133] Caviness CE, Collins FC, Sullivan M. Effect of wheat residue on early growth of soybean. Arkansas Farm Research. Arkansas Agric. Exp Stn. 1986;**35**:8

[134] Board JE, Hall W. Premature flowering in soybean yield reductions at nonoptimal planting dates as influenced by temperature and photoperiod. Agronomy Journal. 1984;**76**(4):700-704

[135] Hovermale CH, Davis JD, St. Louis DG. Double cropping systems for soybean and beef production. In: Technical Bulletin. Mississippi Agricultural and Forestry Experiment Station; 1992. p. 181

[136] Touchton JT, Johnson JW. Soybean tillage and planting method effects on yield of double-cropped wheat and soybeans 1. Agronomy Journal. 1982;**74**(1):57-59

[137] Kyei-Boahen S, Zhang L. Early-maturing soybean in a wheat–soybean double-crop system yield and net returns. Agronomy Journal. 2006;**98**(2):295-301

[138] Arshad M, Nawaz R, Shahnawaz M, Sarfaraz MS, Ahmad S, Ranamukhaarachchi SL. Effect of legume type, nitrogen dose and air quality on biomass and bioethanol production in sweet sorghum-legume intercropping. Agriculture & Forestry/Poljoprivreda i Šumarstv. 2014;**60**(3)

[139] Siddique KHM, Johansen C, Kumar Rao JVDK, Ali M. Legumes in sustainable cropping systems. In: Proceedings of the Fourth International Food Legumes Research Conference (IFLRC-IV), New Delhi, India, 2005. Vol. 1. 2008. pp. 787-819

[140] Kumar S, Meena RS, Datta R, Verma SK, Yadav GS, Pradhan G, et al. Legumes for carbon and nitrogen cycling: An organic approach. Carbon and Nitrogen Cycling in Soil. 2020:337-375

[141] Bolliger A, Magid J, Amado JC, Neto FS, dos Santos Ribeiro MD, Calegari A, et al. Taking stock of the Brazilian "zero-till revolution": A review of landmark research and farmers' practice. Advances in Agronomy. 2006;**91**:47-110

[142] O'Leary GJ, Connor DJ. A simulation study of wheat crop response to water supply, nitrogen nutrition, stubble retention, and tillage. Australian Journal of Agricultural Research. 1998;**49**(1):11-20

[143] Li Y, Tullberg JN, Freebairn DM. Traffic and residue cover effects on infiltration. Soil Research. 2001;**39**(2):239-247

[144] Crowe T, Brand J. The role of pulses in the no-till system. Pulse Update Annual. 2010;**9**:10

[145] Hawthorne W. Wider row pulses in stubble systems. Pulse Update Annual. 2010;**9**:24

[146] Piggin C, Haddad A, Khalil Y. Development and promotion of zero tillage in Iraq and Syria. In: Proceedings of the Fifth World Congress on Conservation Agriculture, Brisbane, Australia. 2011. pp. 26-29

[147] Walley FL, Clayton GW, Miller PR, Carr PM, Lafond GP. Nitrogen economy of pulse crop production in the Northern Great Plains. Agronomy Journal. 2007;**99**(6):1710-1718

[148] Clayton GW, Rice WA, Lupwayi NZ, Johnston AM, Lafond GP, Grant CA, et al. Inoculant formulation and fertilizer nitrogen effects on field pea: Nodulation, N_2 fixation and nitrogen partitioning. Canadian Journal of Plant Science. 2004;**84**(1):79-88

[149] Gan Y, Hanson KG, Zentner RP, Selles F, McDonald CL. Response of lentil to microbial inoculation and low rates of fertilization in the semiarid Canadian prairies. Canadian Journal of Plant Science. 2005;**85**(4):847-855

[150] Kumar N, Hazra KK, Nath CP, Praharaj CS, Singh U. Grain legumes for resource conservation and agricultural sustainability in South Asia. Legumes for Soil Health and Sustainable Management. 2018:77-107

[151] Newton AC. Exploitation of diversity within crops—The key to disease tolerance?. Frontiers in Plant Science. 2016;**7**:191877

[152] Ghosh PK, Venkatesh MS, Hazra KK, Kumar N. Long-term effect of pulses and nutrient management on soil organic carbon dynamics and sustainability on an inceptisol of Indo-Gangetic plains of India. Experimental Agriculture. 2012;**48**(4):473-487

[153] Hazra KK, Ghosh PK, Venkatesh MS, Nath CP, Kumar N,

Singh M, et al. Improving soil organic carbon pools through inclusion of summer mungbean in cereal-cereal cropping systems in Indo-Gangetic plain. Archives of Agronomy and Soil Science. 2018;**64**(12):1690-1704

[154] Yue Q, Xu X, Hillier J, Cheng K, Pan G. Mitigating greenhouse gas emissions in agriculture: From farm production to food consumption. Journal of Cleaner Production. 2017;**149**:1011-1019

[155] Peoples MB, Hauggaard-Nielsen H, Huguenin-Elie O, Jensen ES, Justes E, Williams M. The contributions of legumes to reducing the environmental risk of agricultural production. In: Agroecosystem Diversity Academic Press; 1 Jan 2019 pp. 123-143

[156] Jeuffroy MH, Baranger E, Carrouée B, de Chezelles E, Gosme M, Hénault C, et al. Nitrous oxide emissions from crop rotations including wheat, oilseed rape, and dry peas. Bio Biogeosciences. 2013;**10**(3):1787-1797

[157] Peyrard C, Mary B, Perrin P, Véricel G, Gréhan E, Justes E, et al. N_2O emissions of low input cropping systems as affected by legume and cover crops use. Agriculture, Ecosystems & Environment. 2016;**224**:145-156

[158] Hauggaard-Nielsen H, Lachouani P, Knudsen MT, Ambus P, Boelt B, Gislum R. Productivity and carbon footprint of perennial grass–forage legume intercropping strategies with high or low nitrogen fertilizer input. Science of the Total Environment. 2016;**541**:1339-1347

[159] Lin XY, Ye YQ, Fan SK, Jin CW, Zheng SJ. Increased sucrose accumulation regulates iron-deficiency responses by promoting auxin signaling in Arabidopsis plants. Plant Physiology. 2016;**170**(2):907-920

[160] Palit P, Kudapa H, Zougmore R, Kholova J, Whitbread A, Sharma M, et al. An integrated research framework combining genomics, systems biology, physiology, modelling, and breeding for legume improvement in response to elevated CO under climate change scenario. Current Plant Biology. 2020;**22**:100149

[161] Irigoyen JJ, Goicoechea N, Antolín MC, Pascual I, Sánchez-Díaz M, Aguirreolea J, et al. Growth, photosynthetic acclimation and yield quality in legumes under climate change simulations: An updated survey. Plant Science. 2014;**226**:22-29

[162] AbdElgawad H, Peshev D, Zinta G, Van den Ende W, Janssens IA, Asard H. Climate extreme effects on the chemical composition of temperate grassland species under ambient and elevated CO_2: A comparison of fructan and non-fructan accumulators. PLoS One. 2014;**9**(3):e92044

[163] Jagadish SV, Bahuguna RN, Djanaguiraman M, Gamuyao R, Prasad PV. Implications of high temperature and elevated CO_2 on flowering time in plants. Frontiers in Plant Science. 2016;**7**:166037

[164] Sharma HC, War AR, Pathania M, Sharma SP, Akbar SM, Munghate RS. Elevated CO_2 influences host plant defense response in chickpea against *Helicoverpa armigera*. Arthropod-Plant Interactions. 2016;**10**:171-181

[165] Delahunty A, Nuttall J, Nicolas M, Brand J. Response of lentil to high temperature under variable water supply and carbon dioxide enrichment. Crop and Pasture Science. 2018;**69**:1103-1112

[166] Liu F, Andersen MN, Jensen CR. Loss of pod set caused by drought stress

is associated with water status and ABA content of reproductive structures in soybean. Functional Plant Biology. 2003;**30**(3):271-280

[167] Semba RD, Ramsing R, Rahman N, Kraemer K, Bloem MW. Legumes as a sustainable source of protein in human diets. Global Food Security. 2021;**28**:100520

[168] Ferreira H, Pinto E, Vasconcelos MW. Legumes as a cornerstone of the transition toward more sustainable Agri-food systems and diets in Europe. Frontiers in Sustainable Food Systems. 2021;**5**:694121

[169] Carr DL, Suter L, Barbieri A. Population dynamics and tropical deforestation: State of the debate and conceptual challenges. Population and Environment. 2005;**27**:89-113

[170] Doi R, Ranamukhaarachchi SL. Slow restoration of soil microbial functions in an *Acacia* plantation established on degraded land in Thailand. International Journal of Environmental Science and Technology. 2013;**10**:623-634

[171] Ranamukhaarachchi SL, Surasena JM. Weed dynamics and green pod yield of chili (*Capsicum annum* L.) under chili-live mulch associations. Sri Lankan The Journal of Agricultural Science. 1998;**35**:124-136

[172] Angioloni A, Collar C. High legume-wheat matrices: An alternative to promote bread nutritional value meeting dough viscoelastic restrictions. European Food Research and Technology. 2012;**234**:273-284

[173] Mohammed I, Ahmed AR, Senge B. Dough rheology and bread quality of wheat–chickpea flour blends. Industrial Crops and Products. 2012;**36**(1):196-202

[174] Turfani V, Narducci V, Durazzo A, Galli V, Carcea M. Technological, nutritional and functional properties of wheat bread enriched with lentil or carob flours. LWT. 2017;**78**:361-366

[175] Drago SR, Franco-Miranda H, Cian RE, Betancur-Ancona D, Chel-Guerrero L. Bioactive properties of *Phaseolus lunatus* (Lima Bean) and *Vigna unguiculata* (Cowpea) hydrolyzates incorporated into pasta. Residual activity after pasta cooking. Plant Foods for Human Nutrition. 2016;**71**:339-345

[176] Abou-Zaid A, Ramadan M, Al-Asklany S. Utilization of faba bean and cowpea flours in gluten-free cake production. Australian Journal of Basic and Applied Sciences. 2011;**5**:2665-2672

[177] Millar KA, Barry-Ryan C, Burke R, Hussey K, McCarthy S, Gallagher E. Effect of pulse flours on the physiochemical characteristics and sensory acceptance of baked crackers. International Journal of Food Science & Technology. 2017;**52**(5):1155-1163

[178] Gularte MA, Gómez M, Rosell CM. Impact of legume flours on quality and in vitro digestibility of starch and protein from gluten-free cakes. Food and Bioprocess Technology. 2012;**5**:3142-3150

[179] Han JJ, Janz JA, Gerlat M. Development of gluten-free cracker snacks using pulse flours and fractions. Food Research International. 2010;**43**(2):627-633

[180] Dutta A, Trivedi A, Nath CP, Gupta DS, Hazra KK. A comprehensive review on grain legumes as climate-smart crops: Challenges and prospects. Environmental Challenges. 2022;**7**:100479

[181] Azadeh A, Tarverdian S. Integration of genetic algorithm, computer

simulation and design of experiments for forecasting electrical energy consumption. Energy Policy. 2007;**35**(10):5229-5241

[182] Feng YY, Zhang LX. Scenario analysis of urban energy saving and carbon abatement policies: A case study of Beijing city, China. Procedia Environmental Sciences. 2012;**13**:632-644

Section 2

Legumes for Food Security

Chapter 3

Legumes: Cornerstones of Global Food Security and Sustainable Agriculture

Jacques Fils Pierre

Abstract

Amidst global challenges to food security and sustainability, this chapter explores the pivotal role of legumes. It addresses how the increasing population and climate change impact agriculture, advocating for sustainable practices aligned with international goals, such as those of the FAO. Legumes are highlighted as vital to achieving food security and sustainability through their positive effects on soil health, nitrogen fixation, and erosion control. The chapter emphasizes legumes' contributions to soil fertility, water retention, and resilience in diverse environments, *including their ability to sequester carbon and combat climate change*. Additionally, legumes play a strategic role in pest resistance through biocontrol and genetic diversity. The chapter also acknowledges challenges, *such as precision agriculture technologies and genetically modified varieties*. Ultimately, integrating legumes into agricultural systems promises improved productivity, rural development, and solutions to environmental concerns. *Embracing these benefits is essential for cultivating a sustainable, nourished world*.

Keywords: nitrogen fixation, intercropping, biodiversity, crop rotation, carbon sequestration

1. Introduction

In today's world, we encounter significant challenges regarding global food security and sustainable agriculture practices. The potential for widespread food shortages and the depletion of crucial natural resources essential for sustaining food production highlight the urgent need for action. This chapter aims to provide a clear understanding of how legumes play a crucial role in addressing these pressing challenges. As we face a growing world population and the impacts of climate change on agricultural systems, the importance of adopting sustainable practices becomes increasingly evident. Esteemed organizations, such as the Food and Agriculture Organization [1], and researchers like Godfray et al. [2] stress the global significance of shifting toward sustainability. The United Nations' Sustainable Development Goal 2 urges nations to collectively end hunger, achieve food security, and promote sustainable agriculture by 2030 [3, 4]. Challenges exacerbated by climate change bring about unpredictable weather patterns and resource scarcity [5], extending the impact beyond immediate concerns for our food systems to affect global well-being. In this vast and interconnected context, the chapter focuses on legumes as indispensable contributors to the global pursuit of food security and sustainability. Legumes, with their diverse attributes, offer compelling and versatile alternative solutions for creating resilient and resource-efficient farming systems. By addressing current agricultural challenges, legumes contribute to shaping a future that is secure in terms of food and environmentally balanced. This holistic approach aims to create a future where we can confidently meet our nutritional needs while safeguarding the health and sustainability of our planet. Throughout this exploration, legumes take center stage as the central theme, providing solutions that extend beyond individual farms to impact the broader global landscape of agriculture.

2. Legume intercropping: enhancing soil health and structure

In our modern world, we face significant challenges concerning global food security and sustainable agricultural practices. This section explores the innovative approach of legume intercropping, cultivating leguminous plants alongside other crops, particularly cereals, to emphasize the significant benefits for soil health and structure. As a sustainable agricultural strategy, legume intercropping offers profound advantages. The process begins with nitrogen fixation, which is crucial for both legumes and main crops like cereals. Legumes enhance the soil with bioavailable nitrogen through a symbiotic relationship with nitrogen-fixing bacteria [6]. This nutrient enrichment extends beyond individual crops, contributing to the overall soil fertility of the entire agroecosystem. Research by Pierre et al. [7] underscores the significance of legume intercropping in maintaining soil nutrient levels and reducing nitrogen loss.

Specific legume varieties, such as Sunn hemp (*Crotalaria juncea* L.) and Cowpea, remarkably decrease nitrogen loss compared to monocropping systems with maize (*Zea mays* L.), supporting sustainable agricultural practices and reducing reliance on external fertilizers (Unpublished data). Furthermore, a global-scale analysis conducted by Jensen et al. in 2020 indicates that cereal-legume intercropping has been shown to improve the use of soil nitrogen resources and reduce the requirement for synthetic fertilizer nitrogen. Relay intercropping soybean with maize maintains soil fertility and increases nitrogen recovery efficiency by reducing nitrogen input [8]. Additionally, legume intercropping benefits organic matter accumulation, enhancing soil health as a tool for erosion control and soil structure improvement. The extensive root systems of legumes play a crucial role in preventing soil erosion and enhancing overall soil stability [9]. The previous study clearly analyzes the role of legumes in soil protection against erosion and its implications for the conservation of ecosystem services. This study clearly shows that the usage of leguminous species for soil fertility-building is widespread and included mainly in cover-cropping systems for their ability to fix atmospheric nitrogen (N) besides the soil protection. Therefore, including legumes in intercropping systems increases vegetation cover, which can lead to erosion control mainly by preserving the soil surface against the strength of wind and/or raindrops, consequently reducing erosion by improving soil quality properties.

In this holistic approach, legume intercropping amplifies agricultural productivity and fosters sustainable soil management [10]. Addressing soil health, nutrient conservation, and erosion control contributes to building a resilient and sustainable agricultural system, which has positive implications for farmers and the broader environment. In this context, this section emphasizes the multifaceted advantages of legume intercropping, showcasing its pivotal role in promoting sustainable agricultural practices. Legumes, from nitrogen fixation to erosion control, emerge as key players in fostering a resilient and environmentally conscious farming system. The subsequent sections will further explore the diverse contributions of legumes to sustainable agricultural practices.

3. Legumes and sustainable agricultural practices

In the ongoing effort to combat unsustainable agricultural practices, legumes emerge as vital allies, seamlessly integrating into key strategies such as intercropping and crop rotation, thereby establishing sustainable farming systems worldwide. Through a dynamic partnership with nitrogen-fixing bacteria, legumes, such as soybeans and lentils, play a transformative role in sustainable agriculture by establishing a dynamic partnership with nitrogen-fixing bacteria [6]. Integrated into crop rotations, these legumes act as generous contributors to nitrogen for subsequent crops, reducing reliance on synthetic fertilizers. This aligns with the global pursuit of sustainable agriculture and directly addresses the environmental impact of fertilizers, as emphasized in the previous section. This agricultural interaction involving legumes takes center stage, creating a resilient and efficient system. Wang et al. [9] highlight how legumes contribute to soil conservation, establish deep-reaching root systems, and prevent erosion. The organic matter they leave behind enriches the soil, creating a nurturing environment that promotes health, retains water, and unlocks nutrients. It also acts as a shield against soil degradation and erosion, directly countering the threats outlined in the introduction.

In a world marked by water scarcity and unpredictable climates, legumes, especially deep-rooted varieties such as chickpeas (*Cicer arietinum* L.) and pigeon peas (*Cajanus cajan* L. Millsp.), play a pivotal role in retaining water within intercropping systems [11]. The maize-cowpea (*Vigna unguiculata* L.) intercropping system is presented as a more efficient method in reducing water loss compared to sole maize [12], directly addressing concerns raised in the introduction section about climate change impacts on farming systems. In this regard, it is clearly showed that the integration of legumes such as cowpea into maize-based cropping systems with deeper roots promotes water use efficiency in associated maize plants by enhancing soil moisture retention and nutrient availability, facilitated by the symbiotic relationship between the legume and maize, ultimately contributing to improved overall crop productivity. Beyond serving as a reliable food source, legumes showcase remarkable adaptability to diverse environmental conditions. Additionally, intercropping Proso Millet (*Panicum miliaceum* L.) enhances water use efficiency by promoting extensive root growth and distribution, leading to increased soil water absorption and improved water utilization efficiency, while shading effects on shorter Mung Beans (*Vigna radiata* L.) results in decreased biomass and water use efficiency [13]. Pierre et al. [12] demonstrate how legumes alter their physiological characteristics when intercropped with crops like maize (**Table 1**), positioning them as climate-resilient crops capable of thriving in varying climates.

This adaptability aligns directly with the challenges posed by climate change, reinforcing legumes as essential players in mitigating its impacts. Furthermore, legumes shine as carbon sequestration champions in the battle against climate change.

Treatment	A_N (µmol m^{-2} s^{-1})	g_s (mol m^{-2} s^{-1})	C_i (µmol mol^{-1})	E (mmol m^{-2} s^{-1})	WUE (µmol CO_2/ mmol H_2O)
Maize/crotalaria	42.53 ± 6.61^b	0.44 ± 0.1^b	154 ± 29^a	9.35 ± 1.48^b	4.56 ± 0.47^b
Maize/cowpea	48.83 ± 5.68^a	0.49 ± 0.1^a	139 ± 24^b	10 ± 1.57^a	4.76 ± 0.29^a
Sole maize	44.03 ± 5.53^b	0.45 ± 0.01^{ab}	153 ± 24^a	9.15 ± 1.39^b	4.84 ± 0.31^a
Significance	$p < 0.001$	$p = 0.014$	$p < 0.002$	$p < 0.001$	$p < 0.001$

A_N (CO_2 assimilation rate or photosynthesis); g_s (stomatal conductance); C_i (intercellular carbon); E (transpiration rate); and WUE (water use efficiency).
^{a,b}Means ± standard deviation with different letters are significantly different, $p \leq 0.05$, Tukey test.
Source: [12].

Table 1.
Physiological characteristics of maize in the maize monocrop and in two intercropping systems (maize/crotalaria and maize/cowpea) under rainfed conditions. Becal, Campeche, 2021.

Through symbiotic relationships with nitrogen-fixing bacteria, legumes enhance soil fertility and increase organic matter incorporation, transforming the soil into a natural carbon sink [14]. This not only secures food production but actively contributes to a sustainable future by capturing and storing carbon in the very soil we cultivate, addressing the global imperative highlighted in the introduction. In essence, this section underscores how legumes, through their multifaceted contributions, directly tackle the challenges outlined in the introduction, positioning them as indispensable components in the pursuit of a sustainable and secure global food system.

4. Enhancing resistance to pests and diseases

In the context of global food security, legumes play a central role, serving not only as nutritional powerhouses but also as crucial guardians, preserving the health of crops. Recent research has shown that legumes, like peas and beans, can fight pests and diseases by working with helpful microorganisms in ways like allelopathy and symbiosis [15]. This sophisticated defense system acts as a sentinel, contributing to the suppression of pests and diseases in the agricultural landscape. Pierre et al. [12] conducted a study in the Yucatan Peninsula, which demonstrates that legumes such as cowpea and crotalaria can attract more beneficial insects compared to sole maize (**Table 2**). These beneficial insects play a vital role in reducing phytophagous insects, those that feed on plants that can harm maize plants, highlighting the pivotal role of legumes as natural deterrents against specific pests and diseases threatening crops like maize.

However, it is important to acknowledge the potential drawbacks and challenges associated with relying solely on legumes for pest control. Factors such as environmental conditions, specific pests, and the type of crop being protected can influence the effectiveness of legumes' biocontrol capabilities. In some cases, depending exclusively on legumes for pest management may not provide comprehensive protection, necessitating integrated pest management approaches that consider a range of strategies. Additional research by Pierre et al. [16] explores the effects of intercropping maize and legumes on insect diversity and abundance, indicating positive influences on both aspects (**Table 2**). The maize-legume intercropping system becomes an attractive habitat for beneficial insects, altering population dynamics and providing essential resources. For a deeper understanding of how cereal-legume intercropping

DOI: http://dx.doi.org/10.5772/intechopen.1006180

Family	Maize/crotalaria	Maize/cowpea	Sole maize
Predators			
Coccinellidae	–	–	+
Formicidae	+	+	+
Araneidae	+	+	+
Thomisidae	+	+	+
Syrphidae	+	+	+
Chrysomelidae	+	+	+
Oxypidae	+	+	+
Vespidae	+	+	+
Reduviidae	+	+	+
Carabidae	–	+	–
Asilidae	–	+	–
Salthicidae	+	–	–
Dolichopodidae	+	–	–
Parasitoids			
Eurytomydae	–	+	–
Braconidae	+	–	–
Tachinidae	–	+	–
Pteromalidae	–	+	–
Scelionidae	+	+	–
Figitidae	+	–	–

Source: [12].

Table 2.
Diversity of beneficial insects (predators and parasitoids) observed in both monoculture and intercropping systems (+: Presence of beneficial insect; –: Absence of beneficial insect) under rainfed conditions. Becal, Campeche, 2021.

systems manage crop pests, we recommend reviewing Pierre et al.'s [16] publication, "A review of the impact of maize-legume intercrops on the diversity and abundance of entomophagous and phytophagous insects." Despite these benefits, it is crucial to acknowledge that complete reliance on legumes for pest control may have limitations, and a balanced approach considering various pest management strategies is essential.

At the core of legumes' biocontrol capabilities lies a complex interplay of genetic factors conferring resistance against pests and diseases. Studies by Varshney et al. [17] emphasize the critical role of genetic diversity within legume species in their ability to repel attacks from pathogens and pests. While genetic diversity offers promising avenues for breeding programs focused on developing varieties with biocontrol capabilities, challenges such as potential genetic vulnerabilities or limitations should also be considered. In essence, legumes emerge not only as nutritional allies but as strategic defenders in our ongoing efforts to secure the world's food supply. However, recognizing and addressing potential challenges associated with exclusive reliance on legumes for pest control is vital to ensuring a balanced and effective approach to sustainable agriculture.

5. Future prospects and navigating challenges in legume-based sustainable agriculture

5.1 Refining precision farming technologies

As precision farming technologies continue to evolve, there is a need for focused research to refine these tools for optimized legume cultivation. Tailoring precision farming technologies to specific legume varieties and local conditions is paramount. The study by Kihara et al. [18] in sub-Saharan Africa, where smallholder farmers predominate, exemplifies the positive impact of customized precision farming tools on yields and resource-use efficiency for legumes like cowpeas. Future research should prioritize user-friendly, cost-effective precision tools that account for varying regional needs and ensure their widespread adoption and impact.

5.2 Advancing genetically modified (GM) legume varieties

The transformative potential of GM legume varieties demands ongoing research efforts to address safety concerns and ethical considerations [17]. Future studies should delve into the long-term environmental impact, allergenic potential, and unintended consequences on biodiversity associated with GM legumes. Transparent communication with farmers, rigorous safety assessments, and the involvement of diverse stakeholders, as demonstrated in the case study by Rajkhowa and Qaim [19] in India, should be integral to future research initiatives. Emphasizing responsible adoption and building trust through open dialogs are key aspects of navigating the complex landscape of GM legumes.

5.3 Scaling up agroecological approaches

Agroforestry systems with leguminous trees offer promising benefits, but challenges in scalability persist [7, 20, 21]. Future research should focus on evidence-based field trials and carefully implemented strategies to assess the feasibility of scaling up these agroecological practices. Drawing inspiration from the case study in Southeast Asia by Mbow et al. [22], collaborative studies across diverse regions can provide insights into the universal applicability of such approaches. By addressing critics' concerns and ensuring effective enhancement of biodiversity and soil health, these systems can be promoted for sustainable land management.

5.4 Overcoming seed accessibility challenges

Limited access to high-quality legume seeds remains a barrier to widespread adoption [23]. Future research interventions should explore innovative solutions, including alternative seed sources, efficient seed distribution models, and community-based seed-saving initiatives. The Ethiopian Seed Enterprise (ESE) case study [24] demonstrates the potential of community-centric approaches in enhancing seed distribution and ensuring equitable access to high-quality legume seeds. Collaborative research involving local communities and seed banks can contribute to sustainable seed access globally.

5.5 Addressing socioeconomic influences

The influence of socioeconomic factors on the success of legume-centric agricultural systems necessitates in-depth research into strategies that create supportive market structures, provide financial incentives, and foster partnerships within the legume value chain [25]. Future research should explore the dynamics of legume markets and understand economic factors influencing farmers' decisions. Several studies have illustrated the positive impact of creating supportive market structures and implementing financial incentives [26, 27]. Tailoring interventions to diverse economic realities will be essential for the large-scale adoption of economically viable legume-based practices.

5.6 Leveraging nitrogen-fixing abilities

While leveraging legumes for nitrogen fixation is crucial, potential disruptions in existing agricultural practices must be considered [28]. Future research efforts should emphasize a balanced approach, harmoniously integrating innovation with the preservation of traditional knowledge. The case study in China demonstrates successful nitrogen-fixing soybean (*Glycine max*)—maize relay intercropping, reducing synthetic fertilizer dependence. Research should continue to explore similar practices globally, addressing concerns related to socioeconomic implications and potential disruptions in existing agricultural practices.

6. Conclusion

Legumes offer a multifaceted solution to global challenges, pivotal in enhancing soil health, managing pests, and reducing environmental impact in agricultural systems worldwide. Their incorporation holds significant promise for addressing food security concerns on a global scale. However, barriers such as seed accessibility, difficulties in scaling up through mechanized intercropping systems, and socioeconomic factors impede widespread adoption. Overcoming these challenges demands strategic interventions and supportive policies. By embracing legumes, we not only enhance productivity but also foster rural development and mitigate environmental issues. Embracing the full spectrum of legume benefits is essential for cultivating a sustainable and interconnected world. Looking ahead, the knowledge gained from studying legumes can pave the way for innovative agricultural practices and policies that promote food security and environmental sustainability globally.

Acknowledgements

We would like to express our sincere gratitude to our esteemed co-authors, including Luis Latournerie-Moreno, René Garruña-Hernández, Krista L. Jacobsen, Carrie A. M. Laboski, Francisco Guevara-Hernández, and Esaú Ruiz-Sánchez, for their invaluable contributions to previous scientific publications throughout 2022–2023. Their insights and expertise have greatly enriched the content of this current publication. The author acknowledges the use of ChatGPT for language polishing of the manuscript.

Conflict of interest

The authors declare no conflict of interest.

Author details

Jacques Fils Pierre
International Fertilizer Development Center (IFDC), Muscle Shoals, AL, USA

*Address all correspondence to: jpierre@ifdc.org

References

[1] Mekouar MA. Food and Agriculture Organization of the United Nations (FAO). In: Yearbook of International Environmental Law. Vol. 31, No. 1. 2020. pp. 326-340. DOI: 10.1093/yiel/yvab061

[2] Godfray HCJ, Beddington JR, Crute IR, Haddad L, Lawrence D, Muir JF, et al. Food security: The challenge of feeding 9 billion people. Science. 2010;**327**(5967):812-818. DOI: 10.1126/science.1185383

[3] Lee BX, Kjaerulf F, Turner S, et al. Transforming our world: Implementing the 2030 agenda through sustainable development goal indicators. Journal of Public Health Policy. 2016;**37**(Suppl. 1):13-31. DOI: 10.1057/s41271-016-0002-7

[4] United Nations. General Assembly Resolution A/RES/70/1. Transforming our World, the 2030 Agenda for Sustainable Development. 2015. Available from: http://www.un.org/ga/search/view_doc.asp?symbol=A/RES/70/1&Lang=E

[5] Wheeler T, Von Braun J. Climate change impacts on global food security. Science. 2013;**341**(6145):508-513. DOI: 10.1126/science.1239402

[6] Giller KE, Andersson JA, Corbeels M, Kirkegaard J, Mortensen D, Erenstein O, et al. Beyond conservation agriculture. Frontiers in Plant Science. 2015;**6**:870. DOI: 10.3389/fpls.2015.00870

[7] Pierre JF, Latournerie-Moreno L, Garruña-Hernández R, Jacobsen KL, Guevara-Hernández F, Laboski CA, et al. Maize legume intercropping systems in southern Mexico: A review of benefits and challenges. Ciência Rural. 2022a;**52**(11):e20210409

[8] Qing D, Zhou L, Chen P, Liu X, Song C, Yang F, et al. Relay-intercropping soybean with maize maintains soil fertility and increases nitrogen recovery efficiency by reducing nitrogen input. The Crop Journal. 2020;**8**(1):140-152. DOI: 10.1016/j.cj.2019.06.010

[9] Wang J, Sun P, Li Y, Liu Y, Yu J, Ma X, et al. Hierarchically aligning 10 legume genomes establishes a family-level genomics platform. Plant Physiology. 2017;**174**(1):284-300. DOI: 10.1104/pp.16.01981

[10] Layek J, Das A, Mitran T, Nath C, Meena RS, Yadav GS, et al. Cereal+legume intercropping: An option for improving productivity and sustaining soil health. In: Madhava Rao K, Srivastava SL, Meena NK, editors. Legumes for Soil Health and Sustainable Management. 2018. pp. 347-386. DOI: 10.1007/978-981-13-0253-4_11

[11] Peoples MB, Hauggaard-Nielsen H, Jensen ES. The potential environmental benefits and risks derived from legumes in rotations. In: Emerich DW, Krishnan HB, editors. Nitrogen Fixation in Crop Production. American Society of Agronomy; 2009. pp. 349-385. Available from: https://orgprints.org/id/eprint/18989/

[12] Pierre JF, Latournerie-Moreno L, Garruña R, Jacobsen KL, Laboski CAM, Us-Santamaría R, et al. Effect of maize–legume intercropping on maize physio-agronomic parameters and beneficial insect abundance. Sustainability. 2022b;**14**(19):12385. DOI: 10.3390/su141912385

[13] Gong X, Dang K, Lv S, Zhao G, Tian L, Luo Y, et al. Interspecific root interactions and water-use efficiency

of intercropped proso millet and mung bean. European Journal of Agronomy. 2020;**115**:126034. DOI: 10.1016/j.eja.2020.126034

[14] Jensen ES, Peoples MB, Boddey RM, Gresshoff PM, Hauggaard-Nielsen H, Alves BJ, et al. Legumes for mitigation of climate change and feedstock in a bio-based economy-A review. Agronomy for Sustainable Development. 2011. DOI: 10.1007/s13593-011-0056-7. hal00930531

[15] Mishra J, Fatima T, Arora NK. Role of secondary metabolites from plant growth-promoting rhizobacteria in combating salinity stress. In: Arora NK, editor. Plant Microbiome: Stress Response. 2018. pp. 127-163. DOI: 10.1007/978-981-10-5514-0_6

[16] Pierre JF, Jacobsen KL, Latournerie-Moreno L, Torres-Cab WJ, Chan-Canché R, Ruiz-Sánchez E. A review of the impact of maize-legume intercrops on the diversity and abundance of entomophagous and phytophagous insects. PeerJ. 2023;**11**:e15640. DOI: 10.7717/peerj.15640

[17] Varshney RK, Pandey MK, Bohra A, Singh VK, Thudi M, Saxena RK. Toward the sequence-based breeding in legumes in the post-genome sequencing era. Theoretical and Applied Genetics. 2019;**132**(3):797-816. DOI: 10.1007/s00122-018-3252-x

[18] Kihara J, Sileshi GW, Nziguheba G, Kinyua M, Zingore S, Sommer R. Application of secondary nutrients and micronutrients increases crop yields in sub-Saharan Africa. Agronomy for Sustainable Development. 2017;**37**:25

[19] Rajkhowa P, Qaim M. Personalized digital extension services and agricultural performance: Evidence from smallholder farmers in India. PLoS One. 2021;**16**(10):e0259319

[20] Lebrazi S, Fikri-Benbrahim K. Potential of tree legumes in agroforestry systems and soil conservation. In: Advances in Legumes for Sustainable Intensification. Academic Press; 2022. pp. 461-482

[21] Raj A, Jhariya MK, Banerjee A, Yadav DK. Multifunctional role of legumes in agroforestry. In: Environmental and Sustainable Development through Forestry and Other Resources. Apple Academic Press; 2020. pp. 51-74

[22] Mbow C, Smith P, Skole D, Duguma L, Bustamante M. Achieving mitigation and adaptation to climate change through sustainable agroforestry practices in Africa. Current Opinion in Environmental Sustainability. 2014;**6**:8-14

[23] Gowda SJM, Radhika P, Kadoo NY, et al. Molecular mapping of wilt resistance genes in chickpea. Molecular Breeding. 2009;**24**(2):177-183. DOI: 10.1007/s11032-009-9282-y

[24] Alemu D, Bishaw Z, Zeray T, Ayana A, Hassana M, Getnet M, et al. National Seed Sector Coordination in Ethiopia: Status, Challenges, and Way Forward. Ethiopian Institute of Agricultural Research: Addis Ababa, Ethiopia; 2023

[25] Garforth CJ, Malla YB, Neopane RP, Pandit BH. Socioeconomic factors and agro-forestry improvements in the hills of Nepal. Mountain Research and Development. 1999;**19**(3):273-278. Available from: http://www.jstor.org/stable/25164033

[26] Van Hecken G, Merlet P, Lindtner M, Bastiaensen J. Can financial incentives

change farmers' motivations? An agrarian system approach to development pathways at the Nicaraguan agricultural frontier. Ecological Economics. 2019;**156**:519-529

[27] Tang L, Sun S. Fiscal incentives, financial support for agriculture, and urban-rural inequality. International Review of Financial Analysis. 2022;**80**:102057

[28] Zhang S, Shen T, Yang Y, Li YC, Wan Y, Zhang M, et al. Controlled-release urea reduced nitrogen leaching and improved nitrogen use efficiency and yield of direct-seeded rice. Journal of Environmental Management. 2018;**220**:191-197

Chapter 4

Legumes Crops Cultivation for Food, Feed and Soil Health

Jeet Singh Sandhu and Sushil Kumar Chaturvedi

Abstract

Legume crops taxonomically belong to the family Leguminosae (Fabaceae) which comprises four subfamilies: Papilionoidae, Caeslpinoidae, Mimosoideae and swartzinoidaea. These subfamilies have more than 800 genera and over 20,000 species of trees, shrubs, vines and herbs. Among them, herbs and vines include legume crops, which are grown worldwide in a range of agro-climatic zones under semi-arid, arid and dry temperate ecology. These crops are primarily annuals and cultivated as a solo crop or an intercrop. Legume crops have multiple uses in food, feed and soil amendment. Legume crops like soybean and groundnut are important sources of edible oil, while alfalfa is used as animal feed and grazing. Meanwhile, legumes like chickpeas, lentils and peas are known as pulse crops and are important sources of protein, dietary fibres, minerals and vitamins. The legumes are well known to fix atmospheric nitrogen, and many of them are used as green manuring to improve soil fertility and the physical properties of soil. Some important areas like seed storage studies are lacking for the storage of legume seed for the short and long term including mitigation of alfalfa toxin in groundnut seed, which needs to be initiated to avoid losses.

Keywords: legumes, cultivation, protein, rainfed, biological nitrogen fixation

1. Introduction

Legumes belong to the third largest family of angiosperms, known as Fabaceae/ Leguminosae [1]. The family Fabaceae is further classified into four subfamilies, viz., Mimosoideae, Faboideae (Papilionoideae), Caesalpinioideae and swartzinoidaea. These subfamilies possess 800 genera that comprise over 20,000 species [2, 3]. The legumes are trees, shrubs, vines and herbs. However, in this article, the information is restricted to cultivated legumes that are herbs and vines. The important legumes covered in this chapter in detail are soybean (*Glycine max*), peanuts, also called groundnut (*Arachis hypogaea*), alfalfa (*Medicago sativa*), chickpea or gram (*Cicer arietinum*), lentil (*Lens culinaris*) and pea (*Pisum sativum*), and other locally important legume crops. Legumes are cultivated under rainfed/conserved moisture conditions and in equatorial zones known for dry winters and frequent moisture stress conditions. The legumes are grown as solo crops or, more frequently, in combination with various others as intercrops due to fragile conditions uncertainty of moisture availability. Legumes are known for protein-rich food, edible oils, feed and fibre.

Legumes enrich soil fertility by adding nitrogen and soil organic carbon through biological nitrogen fixation (BNF) and green manuring. Geographically, legumes are cultivated in between the latitudes 50^0 north and 50^0 south and are confined to Tropics of Capricorn (23^0 26′ 16″) and Tropics of Cancer (23^0 27′ 30″) [4].

2. Soybean

The soybean (*Glycine max*) is one of the leading crops of the world having a share of one-fourth of global edible oil production and a valuable source of protein-rich livestock feed [5]. It is also called a miracle bean while it possesses a very high protein content of ~40 per cent, fairly good content of edible oil ~20 per cent and other important nutrients and vitamins.

2.1 Distribution and climate

Soybeans are adapted well to diverse agro-climatic regions like tropics, subtropics and temperate regions. The medical records of China illustrate that soybeans were being utilised for human consumption, dating back 6000 years [6]. It is evidenced that soybeans were being cultivated in China, and it is believed that from China, soybeans spread to nearby countries such as Myanmar, Japan, India, Indonesia, Malaysia, Nepal, the Philippines, Thailand and Vietnam. Among these countries in China, soybean cultivation has taken the lead while in other countries, its cultivation was restricted to limited acreage [7]. Soybean was introduced in the United States in the eighteenth century and strategic planning was drawn to initiate a sound improvement programme. It aimed to bring overall improvement, specifically to develop photo-insensitive, determinate growth habits and be amenable to combining harvesting plant types. These initiatives led to the development of high-yielding varieties from initially an inefficient fodder-type crop. New plant types of soybean were more productive, which led the United States to become the world leader in soybean production [8, 9]. The soybean is being cultivated today in more than 37 countries of the world and among them, major production comes from Brazil, the United States, Argentina, China, India, Paraguay and Canada. The area, production and yield of major soybean-growing countries are given in **Table 1**.

2.2 Agronomic practices

Soybean grows well in all kinds of soil except waterlogged, saline and poorly aerated soils. It is generally sown from mid-June to the first week of July to get the best yield in South Asia and in some regions; it is also sown in the spring season from the end of February to mid-March. It is mainly grown as a solo crop but also does well as intercrop with pigeon pea, cotton, sorghum and pearl millet. Soybean sowing is done with a seed drill on well-prepared flat or raised beds in water-logged or heavy rain situations, it is also grown on broad furrow beds. Row-to-row spacing varies from 30 cm to 45 cm and plant-to-plant distance is maintained at 10 cm. The seed rate is 25–35 kg/ha, depending on the seed size of the improved varieties used. For the soil test base, a fertiliser doze of 80 kg P2O5 and 40 kg K2O per hectare is recommended, and a whole dose of fertiliser is drilled at the time of sowing. The soybean needs very little nitrogen fertiliser of 20 kg/ha for growth initiation. The seed treated with Rhizobium culture initiates early nodulation and enhances germination. The weed

Country	Area (mha)	Production (mt)	Productivity (kg/ha)
Brazil	40.89	120.70	2952
United States of America	34.94	116.38	3331
Argentina	15.87	43.86	2763
China	10.24	20.29	1980
India	1.21	12.99	1069
Canada	2.12	6.54	3089
Russian Federation	3.36	6.00	1789
Paraguay	3.52	4.53	1288
Bolivia (Plurinational State of)	1.52	3.46	2267
Ukraine	1.53	3.44	2255
Total of Top 10 countries	115.20	338.19	
Other Countries	28.83	30.95	1073
Total	144.03	369.14	2563

Source: https://www.fao.org/faostat/en/#data/QCL (Accessed on 3 January 2024).

Table 1.
Area, production and yield of soybean of top 10 countries of world during 2022.

management is done using recommended herbicides, namely TokE 25 or Lasso at 4 litres per ha. It also advocated the use of inter-cultures/manual hoeing at different intervals. Soybean crops are generally grown rainfed, but for successful crop harvest, they need 600 to 700 mm of water [10].

2.3 Improved varieties

The improved varieties are not only high-yielding but also mature uniformly to facilitate the combined harvesting; grains are uniform in shape and colour and look attractive to fetch better prices in the market. Generally, the improved varieties take 130–160 days to mature, and some varieties are very early and mature in 80–90 days. The varietal composition varies with the agro-climatic zone. The details of improved varieties with special traits are given in **Table 2**.

2.4 Plant protection

The improved varieties recommended for cultivation possessed resistance against some diseases and were tolerant to insect pests. Plant protection starts with seed treatment with fungicides like thiram (3 g per kg seed) or carbendazim (2 g per kg seed) to save seed-borne diseases. Leaf spot disease can managed with foliar application of mancozeb (0.3%), carbendazim (0.1%) or benomyl (0.1%). Seed treatment with thiamethoxam 30 FS at 10 ml/kg seed or imidacloprid 48 FS at 1.25 ml/kg is recommended to control yellow mosaic virus and stem fly. Pod blight damage can be reduced with Zineb at 2–5 kg/ha in 1000 litre of water as a foliar application. Thrips are minor insects and are managed well by spraying Metasystox 25 EC/rogur 30EC at 600–700 ml per ha, which has been recommended in soybean-growing states of India.

Character	Varieties
Resistance to pod shattering	PK 472, PK 416, JS 335, NRC 7, NRC 37, Bragg, JS71–05
Tolerance to drought	NRC 7, JS 71–05, Hardee
Tolerance to excessive soil moisture	JS 97–52
Good Seed longevity	JS 95–05, JS 95–60, JS 355, NRC 2, JS 80–21, NRC 37, Punjab 1, NRC 2, JS 335, Kalitur
Suitable machine harvesting	MACS 58, NRC 37, Type 49, Durga, Punjab 1
Suitable for summer season	MACS 57, Punjab 1, JS 335, Pusa 16, PS 564, JS 335, Ahilya-1, PK 472, Punjab 1 etc.
Resistance to lodging	JS 71–05, Pusa-16, Ahilya-1
Low in trypsin inhibitor	Hardee, Punjab 1
Low lipoxygenase	Shilajeet, KHSb 2, Punjab 1
High Protiein (40%)	ADT 1, MACS 58
High Oil (20%)	NRC 7, VLS 1, PK 416
Low linolenic acid	VLS 59
High Oleic acid	LSb 1
High Isoflavines	Hardee, ADT 1
Low Oligosaccharide	SL 525
Tofu quality	Punjab 1, Hardee, PK 472
Collar rot or sclerotial blight	NRC 37, PS 1225
Sclerotium rolfsii	PK 1024, PK 1029, JS 80–21, Indira
Rust	Ankur, PK 1024, PK 1029, JS 80–21, MAUS 61–2, Indira soya 9, MAUS 61
Myrothecium leaf spot	Bragg, JS 71–05, JS 335, MACS 13, MACS 124, MAUS 47, NRC 7, PK 564
Yellow mosaic	PK 416, PK 472, PK 1024, PK 1029, PK 1042, SL 525, SL 688,
Mungbean Yellow mosaic virus	PS 1347, Pusa 9712, JS 97–52
Girdle beetle	NRC 12, NRC 7, Bragg, Indira Soya 9, JS 93–05, RAUS 5, PK 262, Punjab 1
Defoliators	NRC 12, NRC 7, JS 80–21, JS 90–41, MACS 450, MAUS 47, Monetta
Blue Beetle (*Cerotoma* sp.)	RAUS 5, Pusa 16, JS 95–60
Stem Fly	JS 335, NRC 12, NRC 7, MACS 124, JS 90–41, Indira Soya 9, JS 93–05, MACS 124, MAUS 47, PS 564, JS 95–60

Source: Agarwal et al. [5].

Table 2.
Improved soybean varieties with specific characters.

2.5 Harvesting

Harvesting of the crop is to be done when 75% of pods turn brown, and plants drop the leaves and plants are dried up. Harvesting is done with a combine harvester and sometimes also harvested with a small machine like a reaper or manually where the farm size is small, or the area is in an isolated place. The manually or reaper harvested crop mechanically thrashed. It is well known that the seed coat of soybean is very delicate, so due care should be taken, and there should be no damage to the seed coat. Seed coat damage attracts diseases and insect pests, leading to loss of seed viability and vigour.

2.6 Storage of seed

Freshly harvested produce is to be sun dried to reduce the moisture level to 10%. Seeds are properly cleaned before storage. Seed can be stored in silos or gunny bags in a cool place. Very high bag stacking is to be avoided, which puts pressure on lower bags and the seed to be damaged. Damaged seeds generally lose germination.

2.7 Usage and composition

Soybean possesses high nutritional value, and its grains contain ~20% high-quality edible oil and more than 40% protein (**Table 3**). Besides, soybean grains are rich in minerals, fibres, calcium, phosphorous and vitamins like thiamine, nicotinic acid and riboflavin. It is a rare crop plant that is rich in protein as well as fairly rich in edible oil. Soybean protein is rich in essential amino acids like lysine (5%) while cereals like wheat and rice are deficient. In Japan, soybean unfermented milk is used to make *tofu* or *tofu skin* while fermented soybean is used to make soy sauce, fermented bean paste, natto and other foodstuffs and feed. Soybean meal is rich in protein and is mostly used as animal feed (79%) [11]. Soybean grains also contain anti-nutritional elements like phytic acid and trypsin inhibitors. Therefore, it is advised that its grains are to be roasted or soaked in water for some time before consumption which helps to destroy/minimise the inhibitor and improves protein quality.

Serial No.	Component	Amounts (g)
1.	Moisture	0.1 g
2.	Protein	43.2 g
3.	Fat	19.5 g
4.	Carbohydrate	20.9 g
5.	Energy	432 Cal
6.	Minerals	4.6 g
7.	Fibres	3.7 g
8.	Calcium	240 mg
9.	Phosphorus	690 mg
10.	Thiamine	9.0 mg
11.	Nicotinic acid	23.0 mg
12.	Riboflavin	2.3 mg

Table 3.
Nutritive value of soybean (Per 100 g edible portion).

3. Groundnut

Groundnut or peanut (*Arachis hypogea)* is an annual allotetraploid (2n = 4x = 40, AABB) species. It is an important vegetable oilseed and food crop. Groundnuts are known by different names like monkey nut, goober or earthnut, manilla nut and pinda. It has oil content ranging from 42 to 55% and has a share of 25% in the total edible oil of the world. It is also a good source of vegetable protein ranging from 22 to 30% and a source of animal feed. Two types of groundnut varieties, viz., Virginia or spreading type and Valencia or Spanish type or bunchy type, are grown across the world.

3.1 Distribution and climate

Groundnut or peanut occupies a larger area in the Tropics and subtropics in 40^0 north and 40^0 south latitudes [12], covering more than 100 countries. Groundnuts probably originated in Northern Argentina and South Bolivia, where maximum species diversity exists. The countries like Bolivia, Paraguay, Argentina and Uruguay are also important for species diversity [13]. China ranks first in groundnut production, followed by India, the United States, Nigeria, etc. [14]. The area, production and yield of major groundnut-growing countries are given in **Table 4**.

3.2 Agronomic practices

Groundnut requires light soils, warm and moist conditions. Sandy-loam soils with good aeration and rich in organic matter are ideal for groundnut cultivation. It is grown mainly in the rainy season (Kharif) and small acreage in the spring season in India. The ideal temperature is $25C^0$ to $35C^0$ for better growth and yield of groundnut. Groundnut sowing starts from mid-June to July in India and line sowing is recommended as it

Country	Area (m ha)	Production (mt)	Productivity (kg/ha)
China	4.46	18.38	4122
India	5.70	10.13	1777
Nigeria	3.40	4.28	1260
United States of America	0.56	2.53	4505
Sudan	3.00	2.50	833
Myanmar	1.19	1.74	1458
Senegal	1.23	1.50	1226
Argentina	0.41	1.35	3312
Guinea	0.85	1.03	1200
Brazil	0.22	0.85	3848
Total of Top 10 countries	21.00	44.29	—
Other Countries	13.99	28.32	2024
Total	34.99	72.61	2075

Source: https://www.fao.org/faostat/en/#data/QCL (Accessed on 3 January 2024).

Table 4.
Area, production and yield of groundnut of top 10 countries of world during 2022.

facilitates weed management and other inter-culture practices. In some regions, in the spring/summer season, groundnut cultivation is also popular as it gives better yield due to less weed infestation and diseases and is insect pest-free due to off-season. Seed rate of 100–110 kg pods/ha for bunch type and 95–100 kg pods/ha spreading type is recommended. It is generally grown as a rainfed crop, as a rainy season crop. However, the area under spring/summer crops is showing increasing trends. In India, rainy season groundnut crop requires 20 kg N, 40 kg P2O5 and 60 K2 O kg/ha as a basal dose. It is advised that phosphorus be given through single superphosphate, since it contains sulphur, which is important for edible oilseed crops. Spring or summer crops do well in heavy soils under better management as they require more nutrients than Kharif season crops [15]. At the pod initiation stage, 20 kg/ha of nitrogen as a top dressing helps in better pod development. Micronutrients like zinc, manganese and copper are to be also applied in case the soil is deficient. Groundnut can withstand without water for 25 days. However, irrigation at flower initiation, pod initiation and pod filling stages is critical to have a good harvest. Sprinkler or drip irrigation is best suited for undulating sandy soils. Weed management is generally done through inter-culture, which also helps to earthen the plants. Alternatively, two-hand hoeing at 20 days and 35 days after sowing is recommended in case of weed intensity. Groundnuts are also grown as intercrops with pigeon pea, cotton, pearl millet, sesame, sorghum, etc.

3.3 Improved varieties

There are two types of improved varieties, one Virginia or spreading types, best suited for rainfed conditions and second Valencia or Spanish types or bunchy type/ semi-prostrate in habit and do well under irrigated conditions. A good number of improved varieties of both types are recommended for different ecologies. Rainy season varieties mature in 90–100 days, while crops sown in the spring/summer season take 110–120 days to mature. The Details of improved varieties are given in **Table 5**.

3.4 Plant protection

There are number of diseases and insect pests attack the groundnut crop and some of them cause heavy loss up to 90%. The prominent diseases are Tikka, collar rot, stem rot, early leaf spot, rust and peanut bud necrosis. The diseases like tikka and rust are effectively managed by spraying carbendazim 0.05% + mancozeb 0.2%(0.5 g/l + 2 g/l, respectively) at 2- to 3-week intervals and two to three times. No effective control of bud necrosis is available to date. Groundnut pests like aphids, jassids, thrips and white flies appear when the plant canopy is thick and there is high humidity. Insecticides like phosphamidon 0.03% or dimethoate 0.03% or methyl-O-demeton 0.025% can be applied to control these sucking pests. Leaf miners can be also controlled with the application of dichlorvos 0.05% or monocrotophos 0.04% or quinalphos 0.05% or endosulphan 0.07% or carbaryl 0.2% dust at an interval of 15 days. In areas where groundnut is grown continuously, white grub becomes a very severe pest, and the pupae of this insect reside deep in the soil; phorate 10% granules at 25 to 30 kg/ha to be drilled in about 10 cm deep when fields are vacant. Alternatively, trim the trees/shrubs in and around the field as male adults reside on trees. Insecticides like imidacloprid 17.8SL at 1.5 ml/litre or monocrotophos 36SL at 1.6 ml/litre of water to be sprayed on the trees/shrubs around the field to control adult insects [16]. Integrated pest management (IPM) is the best strategy to minimise crop losses and the recommended IPM is as follows:

Name of cultivars /germplasm accession	Country of origin or Affiliation	Characteristics and remarks
CG 7	ICRISAT	High yielding confectionery type, longer self- life
ICGV86325	India	Released jointly by ICAR and ICRISAT for rainy season cultivation
BARD92[ICGS(E)56]	Pakistan	High yielding released by the Pakistan Agriculture Research Council.
Stella (ICGV-SM85048), Veronica (ICGV-SM 86715)	Mauritius	High yielding released by Mauritius Sugar Industry Research Institute.
ICGV91004	ICRISAT	Higher and balanced pod yields in relation to fodder yield across the locations.
ICGS76,ICGS44	ICRISAT	Drought tolerant and high yielding cultivars
Kadiri3,J11,TAG 24,	India	High yielding and usually taken as National checks.
ICG7625and5856	ICRISAT	High oil content
ICG5369and5856	ICRISAT	High O/L ratio
ICGV87123	ICRISAT	High P/S ratio
ICGV86552	ICRISAT	Resistant to insects and pests, bud necrosis disease, tolerant of end-season drought.
ICGV86606	ICRISAT	Foliar disease resistant
ICGV86398 and86393	ICRISAT	Pest resistant
ICGV87480	ICRISAT	High yielding and short-duration
ICGV91278	ICRISAT	Resistant to seed infection by *Aspergillus flavus*
ICGV87350	ICRISAT	Foliar disease resistant
ICGS(E)56(ICGV 86015)	ICRISAT	A short duration variety
NRCG1800(ICG2530)	India	Resistance to ash-weevil and root knot nematode
NRCG8440(ICG3563)	India	Tolerant to Fe-deficiency chlorosis
NRCG664 (NCAc 17,090)	Peru	High peg strength
NRCG609(ICG3404)	Argentina	Salinity and drought tolerance
NRCG2615(ICG1587)	India	Cold tolerance and resistance to *Spodoptera litura*
Chico (ICG476)	USA	Short-duration
TAG24	India	High harvest index (HI)
ICG86031	—	Higher water use efficiency (WUE)
TKG19 A	India	High shelling percentage (68to 70%)
CSMG84–1	India	High yielding and rust resistant Virginia runner
GG20	India	High yielding semi-spreading variety
Luhia15	China	High-yielding, small-seeded with O/L ratio 1.31

Name of cultivars /germplasm accession	Country of origin or Affiliation	Characteristics and remarks
BirsaBold1	India	Promising new confectionery variety
BR1	Brazil	High yielding cultivar, seed yield1.3 t ha^{-1}
NCAc343	ICRISAT	Multiple insect pest resistance

Table 5.
Groundnut cultivars and elite germplasm being used in various developing countries for their useful traits.

- In the months of April to May, deep ploughing is used to open well the soil profile and soil-borne pupae of white grub come in direct contact with high day temperatures and predatory birds.
- Recommended improved and having tolerance to insect pests varieties to be sown
- Timely follow the intercultural practices to manage weeds and volunteer plants.
- Take measures like pheromone trap installation, putting perches for birds and spraying botanicals.
- Use recommended bio-agents like *Trichogramma chilonis* and *Bracon hebetor* at 7–10 days intervals.

3.5 Harvesting

Groundnut crop are to be harvested when the foliage turns yellow and plants start dropping leaves. The Spinish or bunch type varieties are to be harvested by pulling of plants while spreading types varieties are to be harvested by local plough or blade harrow. The harvested crop is allowed to dry up for 2–3 days in the field. Then, pods are detached from plants either manually or by using a thrasher. Pods are cleaned and dried further to maintain the moisture content at around 9% and kernel moisture at around 8%. Moisture content maintenance during storage is also important, as higher moisture content invites aflatoxin production. Dried leaves and plants left over after thrashing are used for fodder to feed the animals, particularly small ruminants.

3.6 Usage and composition

Groundnut kernels contain 42–55% oil and 22–30% protein. They provide 564 Kcal of energy from 100 g of kernels. Groundnut oil is free from cholesterol, contains less than 20% saturated fatty acids and is also low in sodium. Hence, its oil is considered heart-friendly. Its kernel is also a rich source of minerals like phosphorus, calcium, magnesium and potassium and vitamins E, K and B group. Groundnut oil contains some palmitic acid, arachidonic acid, behenic acid, lignoic acids and other fatty acids [17]. Oil cakes left after oil extraction are rich in protein and used to feed the milch animals. Large seeded groundnut is also used for table purposes as raw or roasted. Groundnut kernels are used in various dishes as well. Groundnut Butter is prepared by grinding and blanched kernels. Aflatoxin produced by the fungus *Aspergillus flavus* in pod/kernels is a serious problem in its trading.

4. Alfalfa

Alfalfa (*Medicago sativa*), also called "Queen of Forages", is an herbaceous perennial legume plant. The name "alfalfa" comes from the Arabic word "Al-Fasfasa", which means "father of all plants" [18]. The name alfalfa is used in North America, while Lucerne is commonly used in the United Kingdom, South Africa, Australia, New Zealand and India. The alfalfa is a unique plant where the whole plant is used in a variety of food, feed and non-food items. Alfalfa is the number one forage crop and is commonly used for grazing, hay and silage-making purposes. It is also used as green manure and as a cover crop. A perennial crop with a deep root system holds the soil together to check soil erosion.

4.1 Distribution and climate

Alfalfa is cultivated in more than 80 countries of the world, from 36^0 S to 58^0 N latitude and from the sea level up to 2400 m [19]. It originated in Iran, Afghanistan and Southwest Asia of the Mediterranean region, and it is an ancient crop used as a fodder crop [20]. It is mainly grown in warm, temperate and cool sub-tropical regions of the world. The United States and Argentina are the main producers of alfalfa and other prominent alfalfa-growing countries are the Middle East, Africa, South America and Australia.

4.2 Agronomic practices

Alfalfa is widely grown as a forage crop for animal feeding across the world. Alfalfa has a diverse genetic base and hence has good adaptability to different environmental conditions [21]. Alfalfa is a long-day plant and bright sunlight helps to grow well. It can be seeded in variable periods from mid-April to the whole of May, and even though, in some parts of the world, in early June, sowing is also preferred. In India, it is sown in October, and the broadcasted method is generally followed. Deep, well-drained, sandy-loam soils with 6.5–7.5 soil pH are considered best for the growth of alfalfa. It is tolerant to drought due to its tap root system, which allows it to utilise the moisture from lower profiles of soil. When the crop is sown by broadcasting method, 20–25 kg/ha seed is required, while for line sowing, 12–15 kg/ha seed is good enough. Alfalfa seed is sown at the 1.50 cm depth, and line-to-line distance is maintained at 25–30 cm. The Farm Yard Manure (FYM) gives better crops in light textured soils, and 20 ton/ha FYM is recommended every year before sowing; the seed should be treated with Rhizobium, particularly where the crop is sown for the first time. Chemical fertilisers at a rate of 20 kg/ha, 60–75 P2O5 and 40 k2O kg/ha are required as basal doses to have a good crop. The micronutrients like sulphur, zinc, boron and molybdenum may be sprayed depending on the deficiency symptoms. Initially, the alfalfa crop requires irrigation at intervals of 7–10 days, and when the crop is well established, it irrigates crops at 25- to 30-day intervals. The crop needs 15–20 irrigations in a year [22].

4.3 Plant protection

There are a number of diseases and insect pests attack the alfalfa crop. Among the diseases, rust, leaf spot, downy mildew and phytophthora root rot are prominent diseases that appear on alfalfa crop when conditions are conducive. Apply dithane

M-45 (0.25%) to control rust and leaf spot diseases and mancozeb is effective against downy mildew. Phytophthora root rot appears when either water stagnates for a period or the soil remains wet for a longer time. Lucerne weevil and aphids are insects that cause damage to the crop and are effectively managed with the spray of neem oil at 30 ml/litre of water.

4.4 Harvesting

Alfalfa crop is a multi-cut perennial legume crop, and a well-managed crop can be retained for more than three years. The first cut can be taken when the crop attains a height of about 60 cm, and generally, it takes 50–55 days after sowing. Further, a well-managed crop can be harvested on a regular basis at an interval of 25–30 days for green fodder. Approximately 8–10 harvests can be taken in a year. Green fodder about 80-100 t/ha and 18-20 t/ha dry fodder can be harvested in a year. Alfalfa crop produces enough forage, so animals are to be allowed to graze in open field. Grazing affects plant stand, and that reduces the life span of crops in comparison to mechanically harvested crops.

4.5 Usage and composition

Alfalfa is primarily used as green fodder, hay or silage making for ruminants, non-ruminants and poultry. Rabbits fed on green fodder of alfalfa showed better growth. As a catch crop, it fits well in the cropping system and improves soil fertility, being it a leguminous crop. Alfalfa is a source of high-quality protein and crude fibre and has better digestibility than most other forage crops. It also contains other vitamins [23]. Besides, lucerne comprises other valuable bioactive materials such as polysaccharides, amino acids, flavonoids, plant sterols, saponins and minerals like Ca, Fe and K [24–26]. Nutrient content of tropical alfalfa dry matter (DM) 17.7 to 21.2%, organic matter (OM) 87.72 + 3.55%, crude protein 15.3–32.27%, crude fibre 25.47 + 4.01%, extract ether 8.21 + 1.54%, nitrogen-free extract 39.19 + 1.43%, total digestible nutrients 56.27 + 1.90%, DM (*in vitro* digestibility 68.78 + 3.97% and DM *in vitro* HCl) digestibility 21.61+ 0.54% and contains a number essential amino acids.

5. Pulses

Many legume crops are categorised as pulse crops, where split grains or cotyledons, immature pods and immature green grains are directly consumed as food called pulses. More than a dozen and half such grain legumes, commonly called pulses, are grown worldwide in more than 55 countries. Pulse grains contain protein 20–22% and other important minerals. Among various pulses, chickpeas, lentils and peas/field peas are prominent. Besides, there are a number of minor pulse crops grown locally in different parts of the world that are also for discussion in this chapter. Therefore, it is pertinent to discuss these crops in detail.

5.1 Chickpea

Chickpea (*Cicer arietinum*) is an annual herbaceous plant grown for its nutritious seeds for food purposes. The *Cicer reticulatum* is a wild species which is considered to be the progenitor of the cultivated chickpea species *Cicer arietinum*. Chickpea has two distinct types;

desi type has small seed size, is angular in shape, and has a dark and rough seed coat, while kabuli types have large beaked (ram's head shape) seeds with smooth white in colour seed coat in appearance. Chickpea is known by various names in different parts of the world like garbanzo (Spanish), poischiche (French), kichar or chicher (German), chana (Hindi) and gram or Bengal gram (English). In Turkey, Romania, Bulgaria, Afghanistan and adjacent parts of Russia, chickpea is called "nakhut" or "nohut" [27].

5.1.1 Distribution and climate

Chickpea is grown in more than 55 countries of the world. Asia is a major producer of chickpea and has a share of 89.7% in area, followed by countries like Africa at 4.3%, Oceania at 2.6% and America and other countries at around 0.4% area of the world (**Table 6**). Among Asian countries, India is the largest producer of chickpea and contributes more than 68% to the total chickpea production of the world. Chickpea is grown as a winter crop in South Asia and spring/summer season in temperate/Mediterranean regions. It is grown under wide agro-climatic conditions ranging from 10^0 N to $40^{0\ N}$ latitude in the northern region [28] and 27^0 S to 38^0 S latitude in the Southern region. Chickpea originated from Southeastern Anatolia in modern Turkey [29]. There are two types of chickpeas: the kabuli type is mostly grown in temperate climates, while the Desi type is cultivated in tropical environments [30, 31]. However, recently, through breeding efforts, short-duration Kabuli varieties have been developed and are now successfully grown in tropical regions [32].

5.1.2 Agronomic practices

In India, it is grown in the post-rainy season, called the rabi season, while in the Mediterranean region, chickpea is grown in the spring season. Chickpea is mainly

Country	Area (m ha)	Production (mt)	Productivity (kg/ha)
India	10.74	13.54	1261
Australia	0.62	1.06	1725
Türkiye	0.46	0.58	1271
Ethiopia	0.23	0.49	2170
Russian Federation	0.34	0.47	1358
Myanmar	0.34	0.45	1315
Pakistan	0.86	0.32	367
Mexico	0.11	0.19	1693
Iran (Islamic Republic of)	0.43	0.18	410
United States of America	0.14	0.17	1199
Total of Top 10 countries	14.27	17.45	-
Other Countries	0.54	0.66	1222
Total	14.81	18.11	1223

Source: https://www.fao.org/faostat/en/#data/QCL (Accessed on 3 January 2024).

Table 6.
Area, production and yield of chickpea of top 10 countries of world during 2022.

grown as rainfed under conserved moisture conditions and sown from mid-October to the end of November/early December, while in other areas, it is sown in the month of February as a spring season crop. In Australia, chickpea is mainly sown in June/July, where crop season coincides with the rainy season. Sowing the chickpea crop in winter with cultivars tolerant to cold and to *Ascochyta* blight in the Mediterranean region minimises the adverse effect of high temperature at maturity and moisture stress, which ensures higher yield and provides much-needed stability in productivity [33]. Chickpea is a long day plant that need 8–9 hours' sunshine hours per day to achieve optimum plant biomass and yield. Chickpea grows well under medium to heavy black soils, and alluvial soils. Line sowing gives a better yield than a broad casted crop. Line sowing allows for the intercultural operations and management of optimum plant population. Row-to-row spacing is usually maintained at 30 cm for rainfed crop and 40–45 cm for irrigated crop. Plant to plant distance of 5–7 cm is considered optimum. Desi-type chickpea requires 65–80 kg/ha seed for sowing depending upon seed size, whereas in general white or large/extra-large kabuli types require higher seed rates ranging from 90 to 100 kg/ha. The seed size (100-seed weight) usually ranges from 15 to 28 g for varieties of Desi types and 25 g to 50 g for kabuli types. There is not much of a problem with weeds as a crop is mostly sown under rainfed conditions on conserved moisture conditions. For effective weed management, removal of weeds manually at 30–35 days after sowing or use of pre-emergence herbicide is in practice. Although, where soil moisture is deficient, pre-sown irrigation is provided for optimum germination and plant stand. One irrigation at the pod development stage is recommended to obtain a good yield. The duration of crop varies from 150 to 160 days in temperate regions, while in tropical region, crop matures in 90–100 days. The chickpea crop needs 15–20 kg N and 50–60 kg P2O5 per ha fertiliser as a basal dose for optimum growth. In rainfed or delayed seeded crop, foliar application of 2% Urea at the flowering stage helps in better pod and grain development [34]. Chickpea can be grown as a intercrop with linseed, mustard, safflower, barley, sugarcane, etc.

5.1.3 Improved varieties

A wide spectrum of improved varieties of both Desi and Kabuli types are recommended for varying growing conditions like winter/spring season and irrigated or rainfed situations. Most varieties also possess multiple resistances to major diseases, insect-pests and tolerance to drought and terminal heat stress. A number of early maturing varieties of chickpeas have also been developed, and these have opened doors for crop diversification and ensuring higher cropping intensity. Some of the improved varieties are Digvijay, KWR 108, Alok, HC 1, HC 5, Pusa 391, Pusa 372, BGM 547, Pusa 362, JG 16, JG 322, JG 130, RSG 963, Anvita, JG 14, JSC 55 and JSC 56, Pant G 186, Rajas, JAKI 9218, Vaibhav, BGD 78, Gujarat Gram 1, Gujarat Gram 2, GNG 1581 having resistance against wilt and root rot; GNG 469, Himachal Chana 1 and PBG 5 against Ascochyta blight; lodging tolerant DCP 92-3 for high input management condition; RSG 888, JG 74 and SCS 3 (IPC97-67) for drought-prone areas; salt tolerant CSG 8962 (Karnal Chana 1), heat tolerant JG 14; large seeded kabuli varieties Pusa 1003, Pusa 1053, L551, L555, Vihar, JGK 1, KAK 2, Shubhra, Ujjawal, HK 05-169. In the recent past, extra-large seeded (>50 g per 100 seed weight) kabuli chickpea varieties such as JGK 5, MNK 1, PKV kabuli 4-1 and Phule G0517 (Kripa) have been developed in India. As a result, impressive growth in the area has been observed. The other prominent varieties developed in India and other countries are Nandyal

Gram 857 (NBeG 857), CG Akshay Chana (RG 2016-134), IPCK 2013-163 (Madhav), Gujarat Kabuli Gram 1 (GJGK 1617), Kota Kabuli Chana 3 (RKGK 13-414), Karan Kabuli 4 (CSJK 174), Samriddhi (IPCMB 19-3), IPC L 4-14 (IPC MAS-1) and Pusa Chickpea 4005 (BG 4005) [32]. DZ 10-4, DZ 10-11, Dubie, Mariye, Worku (DZ 10-16-2), Akaki (DZ 10-9-2), Arerti (FLIP 89-84C), Shasho (ICCV 93512), Habru (FLIP 88-42C) and Chefe (ICCV 92318) [35, 36].

5.1.4 Plant protection

The important diseases affecting chickpea crops are fusarium wilt, collar rot, dry root rot, *Ascochyta* blight and botrytis grey mould (BGM). Management of soil-borne diseases, viz., colour rot, wilt and dry root rot, can be done effectively with seed treatment with carboxin at 3 g/kg seed or thiram 2 g + 1 g carbendazim and alternatively follow crop rotation and use resistant varieties. Foliar diseases like *Ascochyta* blight and botrytis grey mould can be controlled with the application of Dithan M 45 or Bavistin. Among various insect pests, Gram pod borer is the most devastating one. The incidence of gram pod borer can be minimised with the application of indoxacarb at 1 ml/l or emamectin benzoate 5 SG at 2 g/l water at 10–15 day intervals at the podding stage. These insecticides are of a broad spectrum in nature and take care of miner insect pests as well.

5.1.5 Harvesting

When plants and pods turn yellow and start shedding the leaves, the chickpea crop can be harvested manually or mechanically. Recently, efforts have been made to develop varieties having tall and erect plant types which facilitate machine/combined harvesting [37]. A delay in harvesting led to the pod shattering. The crop is generally harvested at 15% seed moisture, and manually harvested crops are to be allowed 2–4 days to dry up in an open field before thrashing. Freshly thrashed seeds need to be dried to bring down moisture to the level of 9–10% for safe storage. The metallic seed bins are best for safe storage. The warehouses need to be fumigated to protect grains from bruchid infestation.

5.1.6 Usage and composition

The chickpea green leaves/ immature seeds are consumed for vegetable/salad purposes [38, 39], and matured seeds are boiled, roasted or split or whole seed cooked for dhal or soup purposes. In the Mediterranean region, Hummus is prepared by grinding the chickpea grains and eaten as a paste with bread. The chickpea grains are rich in protein (18–26%), vitamins (C, B1, B2, B3, B6 and K), minerals (Fe: 6–24 mg/100 g, Zn: 3–43 mg/100 g, Ca: 160 mg/100 g, Mg: 138 mg/100 g, K: 875 mg/100 g, Na: 24 mg/100 g, P: 366 mg/100 g), lipids (6.6–6.73 g/100 g), folates (351.5–588.8 μg/100 g), ash (3.15–3.48 g/100 g) and dietary fibres [40].

5.2 Lentil

Lentil (*Lens culinaris*) is an annual herbaceous plant belonging to the family Fabaceae. Lentil seeds are lens-shaped and varied in colour. Its seeds are rich sources of protein, minerals, dietary fibres, etc. Two types of lentils, namely, small-seeded called microsperma and large-seeded called macrosperma, are grown globally. Lentil is known by as many as 30 vernacular names in different parts of the world [41].

5.2.1 Distribution and climate

Lentil is the crop of the ancient world and is well adapted to temperate climates. Lentils have been grown in southwestern Asia since 7000 BC. It is widely grown in entire subtropics of North Africa, Central and South America and temperate countries like Canada, Europe and Mediterranean regions like Syria, Lebanon, Jordan, Ethiopia and the Indian sub-continent, India, Nepal, Bangladesh and Myanmar. The primary centre of diversity of cultivated species is believed to be the Middle East, and lentils spread from here to other countries. The area, production and yield of major lentil-producing countries is given in **Table 7**.

5.2.2 Agronomic practices

Lentil grows well in different types of soils but strives well in heavy clay soils, which are neutral in pH and well-drained. Plant population maintenance is a key factor in getting higher yields. It is recommended to follow 30x5 cm matrices for rows and plant spacing, respectively. Lentil seeding is also done with zero/minimum tillage in lines 25 cm apart after rice harvest. Sowing is usually done in the months of October and November, which is an ideal time for the Indian sub-continent, whereas in countries like Australia, lentils are usually sown in mid-May. The seed rate for small seeded lentils is 40–45 kg/ha and for large seeded varieties, 50–60 kg/ha seed is required. For the broadcasted method of sowing or for *Utera/Paira* cropping, 1.5 times more seed is required in comparison to normal sowing. Seeds are often treated with Rhizobium culture, which helps in early nodulation initiation. An initial dose of 20 kg N, 40 kg P2O5 and 20 kg sulphur per hectare fertilisers promotes crop establishment and its growth. Rice fallow crops do not need any irrigation while in other situations one irrigation at 40–45 days after planting gives a better yield. Lentil crops

Country	Area (mha)	Production (mt)	Yield kg/ha
Canada	1.72	2.30	1341
India	1.41	1.27	899
Australia	0.58	1.00	1738
Türkiye	0.34	0.45	1299
Russian Federation	0.22	0.26	1189
United States of America	0.24	0.25	1022
Nepal	0.20	0.25	1271
Bangladesh	0.14	0.19	1320
China	0.07	0.17	2570
Kazakhstan	0.14	0.15	1019
Total of Top 10 countries	5.06	6.29	
Other Countries	0.51	0.53	1039
Total	5.57	6.82	1224

Source: https://www.fao.org/faostat/en/#data/QCL (Accessed on 3 January 2024).

Table 7.
Area, production and yield of lentil of top 10 countries of world during 2022.

grow well at $18C^0$ to $30C^0$ temperature. Lentil crops can be intercropped with linseed, rapeseed-mustard and autumn planted sugarcane.

5.2.3 Improved varieties

Lentil is one of the important commercial crops as it covers a significant area and production. Improved varieties insulated against biotic and abiotic stresses are the prerequisite to high yield and quality produce. A number of varieties have been developed for different ecologies, keeping in view the crop requirements of the region. Some of the Indian varieties are Pusa Viabhav, IPL 534, Pant L406, PL639, Mallika (K-75), NDL2, WBL 58, HUL 57, WBL77, Arun (PL777–12), IPL-81 (Noori), JL 3, L 4076, IPL 316, DPL62 (Sheri), Pant L 639, Pant L 4, DPL 15 (Priya), Sapna, L-4147, LL 147, LL 931, L 4076, HUL 57, DPL 15, Pant L 5, Pant L 6, VL 103, VL 129, VL 133, VL 507, VL 514, [39], BARI 1 to BARI 8 and BINA 1 to BINA 7 [42]. Red lentil varieties grown in Canada include Robin, Blaze, Crimson, Imax, Redcap, Redberry, Rouleau and Rosetown, whereas large green-seeded varieties include Laird, Grandora, Glamis, Impower, Impress, Imvincible, Plato, Sedley and Sovereign. Medium green varieties include Vantage, Richlea and Meteor and the small green type includes Milestone, Eston and Viceroy. Other lentil varieties grown in Canada include French green, Spanish brown, LeMay and Peridot. Joyce L Boye. Teshale (FLIP 96-46 L), Alemtena (FLIP 96-49 L), Gudo (Flip 84-78 L) and one local variety (EL-142)] gown in Ethiopia [43, 44].

5.2.4 Plant protection

Before sowing, treat the seed with thiram (2 gm) + carbendazim (1gm) or thiram at 3 gm or carbendazim at 2.5 g per kg of seed; insecticide: chlorpyriphos 20 EC at 8 ml/kg of seed; culture: rhizobium + PSB, one packet each for 10 kg seed. Seed treatment gives protection against wilt and rust diseases. Pod borer causes heavy losses while eating immature grains in pods, and a timely spray of neem seed extract (5%) at 50 ml/litre of water, or spray of Profenphos 50 at 2 ml/litre or emamectin benzoate 5 SG at 0.2 g/litre of water provide protection against pod borer damage. Aphids is a minor insect and can be managed with the application of dimethoate 30 EC at 1.7 ml/litre or imidacloprid 17.8 SL at 0.2 ml/litre of water.

5.2.5 Harvesting

The lentil crop can be harvested when plants start shedding leaves and pods turn brown. Harvesting is usually done manually or using combines. Before thrashing, the crop needs to be allowed for 2–3 days of sun drying, if harvested manually. Various types of threshers are available to separate seeds from the pods or plants. Seeds are to be dried up to bring down seed moisture level to 9–10%. Store the seed in clean bins or Jute bags properly treated with insecticide against stored grain pests.

5.2.6 Usage and composition

The lentil grains are consumed as whole grains, decorticated and split cotyledons after cooking in the form of *dhal* (split cotyledons are cooked in water) or soup.

The remaining parts of the plant like discarded seeds and seed coat are used as animal feed. Lentils are good sources of protein, carbohydrates and micronutrients. Lentil proteins' aminogram is near ideal as its grains contain a good amount of essential amino acids like leucine, lysine, threonine and phenylalanine. However, as usual, pulse protein is deficient in the sulphur-containing amino acids, namely methionine and cysteine. Lentil seeds contain 24–26% protein, 50–60% carbohydrates, 1.3% fat and 3.2% fibre and per 100 g seed consists of 300 mg phosphorous, 7 mg iron, 10–15 mg vitamin C, 69 mg calcium, 343 mg calorific value and 450 IU vitamin A.

5.3 Peas

Peas (*Pisum sativum*) are a viny annual herb with either determinate growth habit (bush type) or indeterminate growth habit (climbing type). Peas is an important pulse crop grown for edible immature green grains, green pods and dry seeds. The pea seeds are a rich source of protein, minerals and dietary fibres. Generally, matured dry grains are decorticated and cooked after splitting to make *dhal* (cooked in water) or soup, salty snacks and other preparations. Pea is identified as field pea, green peas, yellow pea, blue pea, garden pea and wrinkled pea.

5.3.1 Distribution and climate

The primary centre of diversity of peas is considered near the Eastern centre, and from here, the genetic diversity of peas shifted to the Mediterranean and African regions, which is today known as the secondary centre of diversity [45]. Peas are the most important cool season crop, which is widely grown in Russia and China,

Country	Area (mha)	Production (mt)	Yield (kg/ha)
Canada	1.35	3.42	2540
China	0.93	1.48	1580
India	0.76	1.00	1318
United States of America	0.35	0.68	1962
France	0.13	0.40	2996
Ethiopia	0.23	0.40	1781
Germany	0.11	0.32	3018
Argentina	0.11	0.30	2641
Ukraine	0.13	0.26	2066
Australia	0.19	0.26	1360
Total of Top 10 countries	5.90	12.14	-
Other Countries	2.19	3.50	1598
Total	8.09	15.64	1933

Source: https://www.fao.org/faostat/en/#data/QCL (Accessed on 3 January 2024).

Table 8.
Area, production and yield of peas of top 10 countries of world during 2022.

followed by Canada, Europe, Australia and the United States. In India, pea is grown as a winter-season crop. Area, production and yield of peas are given in **Table 8**.

5.3.2 Agronomy practices

Peas are grown both under irrigated and rainfed conditions. The sandy loams and loam soils are good for pea cultivation. Pea crop is sown in lines maintaining row-to-row spacing of 30 cm and plant-to-plant spacing of 10 cm. Non-tendril (dwarf) type peas are more productive because they are short in stature and accommodate more plant population per unit area than tall-type peas. The tall types of varieties need a seed rate of 70–80 kg, while 100 kg/ha of seed is required for dwarf varieties to get better productivity. The best time for planting starts from mid-October to mid-November. Peas required a low dose of 20–30 N kg fertiliser for dwarf varieties and 40 kg/ha for tall varieties for the initial establishment of the crop. Occasionally, additional application of 40 kg N and 40–60 kg P2O5 per ha is recommended in case soil is low in nitrogen and phosphorus. The application of potash and sulphur on a soil test basis is pertinent to yield increment. Need basis irrigation application helps to save the crop in the dry season, and generally, one or two irrigations are sufficient. Weed management with the application of pre-emergence weedicide, Pendimethalin (STOMP) 30 EC at 0.75–1 kg a.i./ha followed by post-emergence weedicide metribuzin 70% WP at 0.25 kg a.i./ha in 400–600 litre of water at 15–20 days after sowing [46] ensure better yield from per unit area of cultivation.

5.3.3 Improved varieties

A number of high-yielding varieties having tall and dwarf growth habits have been developed for cultivation in different agro-climatic conditions. The dwarf-type varieties are generally leafless or have very few leaves and are generally converted into tendrils. These varieties do better under agri-inputs management conditions, have resistance against diseases and perform better than tall varieties. The maturity duration of most of the varieties from 110 to 140 days but few varieties mature in less than 100 days also. The prominent varieties developed are JP 885, Adarsh (IPF 99–25), IPF 4–9, Prakash (IPFD 1–10), Vikas (IPFD 99–13), Aman (IPF5–19), Ambika, Indra (KPMR 400), IPFD 10–12 (green seeded), Jaya (KPMR 522), Pant Pea 14, Pant Pea 25, Pant Pea 42, V L Matar 47, V L Matar 42, KFP 103 (Shikha), Uttra, DDR-27 (Pusa Panna), Hariyal (HFP-9907 B), HFP 9426, Alankar, Jayanti (HFP 8712), Swati (KFPD 24), Malviya Matar 15, HUP 2, Shubhra and Paras [47].

5.3.4 Plant protection

Diseases like wilt, rust and powdery mildew attack the pea crop. In India, most of the pea varieties are resistant to powdery mildew (PM) disease. The seed treatment starts with thirum (2 gm) + carbendazim (1 gm)/kg of seed, and the adoption of crop rotation and delayed sowing help to manage the wilt disease. Spray the crop with Mancozeb 75 WP at 2 g/litre of water to save it from rust and powdery mildew. Application of Karathane at 1 ml/litre or wettable sulphur at 3 gm/litre or dinocap at 1 ml/litre of water minimises disease incidence, particularly powdery mildew. Pea stem fly can be managed with the application of a mix of 30 kg/ha carbofuran (Furadon) 3% granules or 10 kg /ha phorate (Thimet) 10% granules in the soils before sowing the crop. Spray 1 litre of oxydemeton methyl (Metasystox) 25 EC in

1000 litre of water per hectare and repeat at 15-day intervals, if required to control the leaf miner. Sucking pests like the aphids of a pea can be managed with the application of 1.25 litre of dimethoate 30 EC or oxydemeton methyl (Metasystox) 25 EC in 1000 litre of water per hectare. Repeat spray after 10–12 days, if required. The spiny pod borer can be managed safely and effectively by spraying 1.25 litre of cypermethrin in 1000 litre of water per ha.

5.3.5 Harvest

When leaves and stems are dried up and pod turns brown, a crop can be harvested manually or mechanically using combines. The threshing can be done with the different kinds of threshers that are available in most of the countries where pea is grown. Harvested seeds are to be dried up to bring down moisture level to 9.10% for safe storage. Pea seeds can be stored in store bins or gunny bags to save grains from store grain pests. If required, fumigation with recommended insecticides should be adopted.

5.3.6 Usage and composition

Pea is used as green pods, immature green seeds as salad and vegetables. Dried seeds are decorticated to make dhal or other preparations. The pea straw is used to feed cattle. The pea seeds contain protein (22.5%), vitamins A, C and K, fat (1.8%), minerals (calcium 64 mg/100 g, iron 4.8 mg/100 g), carbohydrate (62.1%) and dietary fibre.

5.4 Minor legumes crops

There are a number of locally important legume crops grown in different parts of the world. Some of these legumes are cluster bean (*Cyamopsis tetragonoloba* L.), Cowpea (*Vigna unguiculata* L.), pigeon pea or *Tur* (*Cajanus cajan*), fababean or *bakla* (*Vicia faba*), lupin (*Lupinus* sp.), Lathyrus (*Lathyrus sativus*), green gram (*Vigna radiata*) and black gram (*Vigna mungo*). The cluster bean is an annual herbaceous plant and it grows well in warm environments of arid and semi-arid regions of the world. The major cluster bean-growing countries are India, Pakistan, Indonesia, Myanmar, parts of Central Africa and the arid region of the southwestern United States. It is used for gum extraction, green pods as vegetables, green manuring, animal fodder and feed. Cluster bean seeds contain protein (18%), fibre (32%) and gum (30–33%). Cowpea is widely grown for animal feed and food crops in dry areas like Africa, Asia, Europe, the United States and South America. Cowpea seeds are an important source of protein (22–24%), carbohydrates (55–66%), iron (0.005%) and calcium (0.08–0.11%). Green pods of cowpea are also consumed as vegetables, and grains are cooked to prepare dhal/soup. It is an important green manuring crop as it attains very fast growth and produces high biomass in a short period. Pigeon pea is a very hardy crop grown for food as a pulse crop in different parts of the world. Pigeon pea seeds are dehulled and used for dhal purposes in India, and also immature green grains are consumed as vegetables. Pigeon pea sticks are used to cook food and make huts. Its seeds contain protein (22%), carbohydrates (57%), minerals and vitamins. Faba bean (broad bean) is mainly grown in drier regions for animal feed and human consumption. Faba bean seeds contain protein (26%) and carbohydrates (56%). Its green and tender pods are consumed as vegetables, and green plants form nutrient-rich fodder for animals. Lupin is cultivated in water-stressed climate of Australia, and its seeds are a good source of iron and zinc. It is generally used for animal feed. Raw lupin grains contain protein (36%) and carbohydrates (40%). Lathyrus is also known as grass pea and is a very hardy crop. It is grown for dual purposes under harsh climatic conditions during the winter season. Its seeds are used as food and green plants are used as fodder for livestock. Green gram and Black gram are important pulse crops and are grown mainly in India, Pakistan, Nepal and Myanmar. Their seeds are rich in protein, dietary fibres, iron, potassium, magnesium and zinc. Besides, legume crops like broad bean or kidney bean or French bean (*Phaseolus vulgaris*), moth bean (*Vigna acointifolia*) and laplap bean (*Laplap purpureus*) are also grown in some pockets of the world. They are mainly used for vegetable and dhal purposes.

6. Soil health

Legume crops have unique characteristics that fix atmospheric nitrogen through symbiosis, and it is called biological nitrogen fixation (BNF). Most legume crops root nodules harbour different species of bacteria, and they are crop-specific. The most common bacteria are Rhizobia and Bradyrhizobium and have crop-specific species. The legume plants' roots developed specialised structures to harbour the bacteria called nodules. Thus, legume plants provide shelter to bacteria and in return, bacteria fix the nitrogen for plants to use as nutrients. Both host and bacteria are important to each other, and this relationship is called symbiosis. It has been proven that almost 50% of the total N used in the agriculture production system has proven to be fixed through BNF by the legume plants [48]. So, the significance of BNF is that it is key to the maintenance of soil fertility and the sustainability of the cropping system. In this process, greenhouse gas (GHG) emissions such as CO_2 and N_2O emissions are reduced by five to seven per cent in legume crops compared to other crops. Legumes play an important role in carbon sequestration in soil and reduce the overall use of fossil energy in the system. Thus, indirectly, legume cultivation slows down climate change through mitigation. Legume crops endowed by nature have tap root systems which penetrate deep in soil profiles as it accentuates to withstand moisture stress conditions as root systems draw the water available in lower soil profiles. Secondly, the tap system also makes soil porous to improve the physical properties of the soil as well. Thirdly, legumes are grown mostly in sandy-loam soil under rainfed conditions, and their well-developed root system holds soil firmly and does not allow soil erosion. So, soil erosion is checked, which is a serious problem in arid zones like the *Thar Desert* of Rajasthan in India. Some legume crops like pigeon pea root exudates have pisidic acid, which makes non-available phosphorous available phosphorus from plants [49]. Similarly, other legume crops root exudates made them stand in adverse conditions. Legumes crops are often used as green manuring as they produce enough biomass in a short period. Green manure adds N and organic C to the soil and improves quality by increasing the soil organic carbon (SOC), macro- and micronutrients and humus content [48, 50, 51]. The legume crops are short-duration and require less irrigation water in comparison to other crops. Thus, there is net saving water as a natural resource. Secondly, the short duration and tap root system of legume crops give an advantage in growing as an intercrop to provide system stability. Lastly, crop residues of these crops are important sources of high-value fodder in drier regions of the world and help to reduce animal migration in the off-season. The details of the amount of nitrogen fixed by different legume crops are summarised in **Table 9**.

Sr. No.	Legume name	Botanical/Scientific name	Associated organism	N fixing (Kg/ha/Year)	References
1.	Soybean	*Glycine max L.*	Bacteria (Bradyrhizobium)	100–150	[52–54]
2.	Chickpea	*Cicer arietinum L.*	Bacteria Rhizobium	40–50	[52, 55]
3.	Lentil	*Lens culinaris L.*	Bacteria (Rhizobium)	40–68	[53, 56]
4.	Groundnut	*Arachis hypogea L.*	Bacteria (Bradyrhizobium)	150	[53, 55]
5	Peas	*Pisum sativum L.*	Bacteria (Rhizobium)	65–100	[57]
6	Pigeon pea	*Cajanus cajan L.*	Bacteria (Bradyrhizobium)	100. 200	[53]
7.	Mungbean	*Vigna radiate L.*	Bacteria (Rhizobium)	60, 112	[55, 56]
8.	Urdbean	*Vigna mungo L.*	Bacteria (Rhizobium)	100	[52]
9.	Cow pea	*Vigna unguiculata L.*	Bacteria (Bradyrhizobium)	90	[52]
10	Alfalfa	*Medicago sativa L*	Bacteria (Rhizobium)	103, 209	[53, 58, 59]
11.	Cluster bean	*Cyamopsis tetragonoloba L.*	Bacteria (Rhizobium)	60–150	[52, 54]
12.	Faba bean	*Vicia faba L.*	Bacteria (Rhizobium)	130	[55, 57]
13	Lupins	*Lupinus sp. L.*	Bacteria (Rhizobium)	60–100	[60]

Source: https://doi.org/10.1007/978-981-13-0253-4_9.

Table 9.
Range in quantity of N_2 fixed by selected legumes.

7. Summary

Legumes crops are grown in drier regions of semi-arid, arid and temperate regions under conserved moisture conditions of the world. These crops are valuable sources of food, feed, fibre and essential minerals for human beings and livestock. A package of practices for legume crop production and improved technology are applied for sustainable crop production. Legume crops play an important role in improving soil fertility and climate mitigation and enhancing soil organic matter for system sustainability. It is felt that safe storage of legume seeds for the short and long term is lacking—improper way of storage of legume seeds spoiled by Bruchids—a store grain pest. Secondly, legume crops like groundnut seeds get infested with alfalfa toxin, which is a health hazard. Future studies must focus on these aspects.

Author details

Jeet Singh Sandhu[1]* and Sushil Kumar Chaturvedi[2]

1 SKN Agriculture University, Jobner, India

2 Rani Laxmi Bai Central Agricultural University, Jhansi, Uttar Pradesh, India

*Address all correspondence to: js_sandhuin@yahoo.com

References

[1] Paul G, Beavis WD, Brummer EC, Shoemaker RC, Stalker HT, Weeden NF, et al. Legumes as a model plant family. Genomics for food and feed report of the cross-legume advances through genomics conference. Plant Physiology. 2005;**137**(4):1228-1235. DOI: 10.1104/pp.105.0608712

[2] Rubel F, Kottek M. Comments on: "The thermal zones of the Earth" by Wladimir Köppen (1884). Meteorologische Zeitschrift. 2001;**20**:361-365. DOI: 10.1127/0941-2948/2011/0285

[3] Sanwan S, Rao DV, Sharma RA. A review on *Pongamia Pinnata* (L) pierre: A great versatile leguminous. Plant Nature and Science. 2010;**8**(11):130-139

[4] Lewis G, Schrire B, Mackinder B, Lock M. Legumes of the World. Kew: Royal Botanic Gardens; 2005

[5] Agarwal DK, Billmore SD, Sharma AN, Dupare UB, Srivastiva SK. Soybean: Introduction, improvement and utilization in India—Problems and prospects. Agricultural Research. 2013;**2**(4):293-300

[6] Krishnamurthy K, Shivashankar K. Soybean production in Karnataka. UAS Tech. Series, No. 12. University of Agricultural Sciences, Bangalore; 1975

[7] Hymowitz T. On domestication of soybean. Economic Botany. 1970;**24**:408-421

[8] Singh BB. Success of Soybean in India: The Early Challenges and Pioneer Promoters. International Institute of Tropical Agriculture (IITA), Kano Station, Sabo BakinZuwo Road, PMB 3112, Kano, Nigeria (email: bbsingh@cgiar.org)

[9] Hymowitz T, Harlan JR. Introduction of soybeans to North America by Samuel Bowen in 1765. Economic Botany. 1983;**37**:371-379

[10] Anonymous. Improved technologies and technical recommendations maximizing soybean productivity. Extension Bulletin (Revised) no. 18. ICAR-Indian Institute of Soybean Research, Indore, India; 2023

[11] Konno UK, Shoshin. Soybean cultivation in Japan. JARQ. 1970;**5**(2):21-27

[12] Pattee HE, Young CY. Peanut Science and Technology. Yoakum, Texas, USA: American Peanut Research and Education Society, Inc.; 1982

[13] Anonymous. Oilseed Division, Department of Agriculture, Cooperation and Farmers Welfare, Ministry of Agriculture, GOI; 2017

[14] Tiwari S, Kumar N, Pramanik A, Joshi A, Sasode DS, Sikarwar RS, et al. Breeding for foliar disease resistance in groundnut using conventional and molecular approaches. In: Proceedings of PSMB, 15-16 February 2018. New Delhi: National Environmental Science Academy (NESA); 2018. pp. 56-62

[15] Singh L, Vidya C. Macronutrient requirement of groundnut: Effects on growth and yield components. Indian Js. Plant Physiology. 2005;**11**(4):401-409

[16] Nautiyal PC. GROUNDNUT: Post-harvest Operations (nautiyal@nrcg.guj.nic.in or pnaut@ad1.vsnl.nic.in) Edited by AGSI/FAO: Danilo Mejia, PhD (Technical), Beverly Lewis (Language & Style). National Research Centre for Groundnut (ICAR) (www.icar.org.in); 2002

[17] Jambunathan R. Groundnut quality characteristics. In: Uses of Tropical Grain Legumes: Proceedings of a Consultants Meeting, March 27-30, 1989 (Patancheru: ICRISAT), Andhra Pradesh, India. 1991. pp. 267-275

[18] Lacefield G, Henning J, Rasnake M, Collins M. University of Kentucky; Cooperative Extension Service: Alfalfa, The Queen of Forage Crops, AGR-76; 2014

[19] Ecoport. Feedipedia - Animal Feed Resources Information System - INRAE CIRAD AFZ and FAO © 2012-22; 2009

[20] Cook BG, Pengelly BC, Brown SD, Donnelly JL, Eagles DA, Franco MA, et al. Tropical Forages. DPI and F(Qld), CIAT and ILRI, Brisbane, Australia: CSIRO; 2005

[21] Radović J, Sokolović D, Marković J. Alfalfa-most important perennial forage legume in animal husbandry. Biotechnology in Animal Husbandry. 2009;**25**(5-6-1):465-475

[22] Anonymous. Lucerne-Packages and Practices. All India Coordinated Research Project on Forages, ICAR-Indian Grassland and Fodder Research Institute, Jhansi, India; 2022

[23] Sullivan J. Medicago sativa. In: Fire Effects Information System. U.S. Department of Agriculture, Forest Service, Rocky Mountain Research Station, Fire Sciences Laboratory (Producer). Available from: https://www.fs.usda.gov/database/feis/plants/forb/medsat/all.html; 1992 [Accessed: 10 December 2023]

[24] Piotr S, Wasielewska-Michniewska K, Maria G, Marcin P. Topical Classification of Food Safety Publications with a Knowledge Base. Poland: Systems Research Institute, Polish Academy of Sciences Warsaw; 2022

[25] Sreenath HK, Koegel RG, Belén MA, Thomas J. Ethanol production from alfalfa fiber fractions by saccharification and fermentation June. Process Biochemistry. 2001;**36**(12):1199-1204. DOI: 10.1016/S0032-9592(01)00162-5

[26] Gebhardt SE, Thomas RG. Nutritive Value of Foods. U.S. Department of Agriculture, Agricultural Research Service, Home and Garden Bulletin 72 Nutritive Value of Foods—USDA ARS. 2002. Available from: https://www.ars.usda.gov/ARSUserFiles/80400525/Data/hg72/hg72_2002.pdf

[27] Van der Maesen LJG. Cicer L. Origin, History and Taxonomy of Chickpea. In: Saxena MC, Singh KB, editors. The Chickpea. Wallingford, Oxon, UK: CABI; 1987. pp. 11-34

[28] Berger JD, Ali M, Basu PS, Choudhary BD, Chaturvedi SK, Deshmukh PS, et al. Genotype by environment studies demonstrate the critical role of phenology in adaptation of chickpea (*Cicer arietinum* L.) to high and low yielding environments of India. Field Crops Research. 2006;**98**:230-244

[29] Ladizinsky G. A new Cicer from Turkey. Notes of the Royal Botanic Garden Edinburgh. 1975;**34**:201-202

[30] Muehlbauer FJ, Singh KB. Genetics of chickpea. In: Saxena MC, Singh KB, editors. The Chickpea. Wallingford, Oxon, UK: CABI; 1987. pp. 99-125

[31] Malhotra RS, Pundir RPS, Slinkard AE. Genetic resources of chickpea. In: Saxena MC, Singh KB, editors. The Chickpea. Aberystwyth, UK: International Cambrian News Ltd; 1987. pp. 67-81

[32] Gaur PM, Kumar J, Gowda CLL, Pande S, Siddique KHM, Khan TN, et al. Breeding chickpea for early phenology: Perspectives, progress and prospects. In: Kharkwal MC, editor. Food Legumes for Nutritional Security and Sustainable Agriculture, Proceedings of Fourth International Food Legumes Research Conference, 18-22 October 2005. New Delhi, India: Indian Agricultural Research Institute (IARI); 2005

[33] Singh KB, Malhotra RS, Saxena MC, Bejiga G. Superiority of winter sowing over traditional spring sowing of chickpea in the Mediterranean region. Agronomy Journal. 1997;**89**(1):112-118

[34] Venkatesh MS, Basu PS. Effect of foliar application of urea on growth, yield and quality of chickpea under rainfed conditions. Journal of Food Legumes. 2011;**24**(2):110-112

[35] Anonymous. AICRP- Chickpea, Project Coordinator annual Report of 2021-22; 2022

[36] Dadi L, Regassa S, Mitiku AFD. Adoption studies on improved chickpea varieties in Ethiopia; 2005

[37] Munirathnam P, Jayalakshmi V, Ashok Kumar K, Padmalatha Y. Suitability of Chickpea NBeG47 for mechanical harvesting under rainfed conditions. Journal of Food Legumes. 2015;**28**(2):64-66

[38] Lonagre AG, Thakare AY, Salunke PB. Efficacy of different pyrethroid alone and in combination with dimethoate against girdle beetle of soybean. Journal of Food Legumes. 2015;**28**(2):64-66

[39] Sandhu JS, Gupta SK, Gaur PM, Saxena AK, Sharma S, Prabhjit K. Studies on early podding varieties and post-harvest management of immature green grains of chickpea to be used as vegetable. Acta Horticulturae. 2007;**752**:353-358

[40] Sandhu JS, Shailesh T, Chaturvedi SK. Chickpea nutritional status and value chain for sustainable development. In: Narula S, editor. Sustainable Food Value Development. London Borough of Camden: Palgrave Macmillan; 2023

[41] Kay D. Food Pulses. Tropical Development and Research Institute (TPI). TPI Crop and Product Digest No. 31979. pp. 48-71

[42] Anonymous. Seednet, Ministry of Agriculture and Farmers Welfare Government of India and ICAR- Indian Institute of Pulses Research, Kanpur; 2022

[43] Yigezu YA, Rahman MW, El-Shater T, Alene AD, Sarker A, Shiv K, et al. Plot-level impacts of improved lentil varieties in Bangladesh. 2022. DOI: 10-1371/journal

[44] Takele Z, AberaTolera HI. Response of improved lentil (*Lens culinaris* Medik) varieties to phosphorus nutrition on Vertisols of west Showa, central highlands of Ethiopia. Advances in Crop Science and Technology. 2017;**5**(6):315. DOI: 10.4172/2329-8863-100315

[45] Zeven A, Zhukovsky P. Dictionary of Cultivated Plants and their Centres of Diversity: Excluding Most Ornamentals, Forest Trees and Lower Plants. Wageningen: CAPD; 1975

[46] Mikić A, Mihailović V, Duc G, Ćupina B, Étévé G, Lejeune-Hénaut I, et al. Evaluation of winter protein pea cultivars in the conditions of Serbia. ZbornikradovaInstitu-tazaratarstvoipovrtarstvo / A Periodical of Scientific Research on Field and Vegetable Crops. 2007;**44**:107-112

[47] Anonymous. Pea- Packages and Practices, ICAR-Indian Institute of Pulses Research, Kanpur, India; 2023

[48] Graham PH, Vance CP. Legumes: Importance and constraints to greater use. Plant Physiology. 2003;**131**:872-877

[49] Ae N, Arihara J, Okada K, Yoshihara T, Otani T, Johansen C. The role of piscidic acid secreted by pigeonpea roots grown in an Alfisol with low-P fertility. In: Randall PJ, Delhaize E, Richards RA, Munns R, editors. Genetic Aspects of Plant Mineral Nutrition. Developments in Plant and Soil Sciences. Vol. 50. Dordrecht: Springer; 2010. DOI: 10.1007/978-94-011-1650-3_34

[50] Jensen ES, Peoples MB, Boddey RM, Gresshoff PM, Hauggaard-Nielsen H, Alves BJ, et al. Legumes for mitigation of climate change and the provision of feedstock for biofuels and bio refineries. A Review. Agronomy for Sustainable Development. 2012;**32**:329-364

[51] Hajduk E, Właśniewski S, Szpunar KE. Influence of legume crops on content of organic carbon in sandy soil. Soil Science Annual. 2015;**66**:52-56

[52] Ahlawat IPS, Gangaiah B. Grain legumes farmers own nitrogen fertilizer units' role revisited. In: Souvenir: National Symposium on Resource Conservation and Agricultural Productivity. New Delhi: Indian Society of Agronomy; 2004. pp. 22-25

[53] Mugwe J, Mugendi DN, Mucheru-Muna M, Kungu J. Soil inorganic N and N uptake by maize following application of legume biomass, tithonia, manure and mineral fertilizer in Central Kenya. In: Bationo A, editor. Innovations as Key to the Green Revolution in Africa. London Borough of Camden: Palgrave Macmillan; 2011. pp. 605-612

[54] Meena RS, Kumar V, Yadav GS, Mitran T. Response and interaction of *Bradyrhizobium japonicum* and Arbuscular mycorrhizal fungi in the soybean rhizosphere: A review. In: Plant Growth Regulation. 2018;**84**:207-223

[55] Seymour N, McKenzie K, Krosch S. Fixing more nitrogen in pulse crops. In: Rachaputi RCN, editor. Nindigully GRDC Grains Research Update. Queensland, Australia: Department of Agriculture and Fisheries; 2015

[56] Shaha Z, Shahb SH, Peoplesc MB, Schwenked GD, Herridge DF. Crop residue and fertilizer N effects on nitrogen fixation and yields of legume-cereal rotations and soil organic fertility. Field Crops Research. 2003;**83**:1-11

[57] Peoples MB, Brockwell J, Herridge DF, Rochester IJ, Alves BJR, Urquiaga S, et al. The contributions of nitrogen-fixing crop legumes to the productivity of agricultural systems. Symbiosis. 2009;**48**(1-3):1-17

[58] Carlsson G, Huss-Danell K. Nitrogen fixation in perennial forage legumes in the field. Plant and Soil. 2003;**253**:353-372. DOI: 10.1023/A:1024847017371

[59] Aranjuelo I, Irigoyen JJ, Nogués S, Sánchez DM. Elevated CO_2 and water availability effect on gas exchange and nodule development in N_2-fixing alfalfa plants. Environmental and Experimental Botany. 2009;**65**:18-26

[60] Havlin JL, Tisdale SL, Nelson WL, Beaton JD. Soil Fertility and Fertilizers; an Introduction to Nutrient Management. 8th edition. India: Pearson Education India; 31 March 2016. ISBN: 978-81-203-4868-4

Chapter 5

Food Legumes for Sustainable Crop Intensification of Rice-Based Cropping Systems: Bolstering Food and Nutritional Security in South Asia

Smita Kaul, Akanksha Singh, Chandrashekhar Biradar and Shiv Kumar

Abstract

Food legume crops are gaining significant recognition as a crucial option for sustainable intensification of rice fallows in rainfed lowland agroecology, offering substantial potential to enhance food and nutritional security while enhancing soil health and ecosystem services. In South Asia, 22.3 million hectares of fallow land offer opportunities to boost food legume production, nutrition, soil health, and smallholder farmer income by leveraging legumes' nitrogen-fixing ability and integrating them into cropping systems to improve soil organic carbon and reduce greenhouse gas emissions. Legume-based intensification and diversification have been accompanied by innovations that include short-duration, nutrient-dense, high-yielding varieties that fit well in the rice fallow system. Innovative seed systems, coupled with effective value addition support, market linkages, and intensive extension efforts, are positioned to enable farmers to choose food legumes as a sustainable option for crop intensification and diversification. Despite advancements, challenges such as location-specific innovations and scaling up adoption of modern technologies must be addressed to fully realize the impact of food legumes in cereal-based cropping systems. This chapter highlights the importance of introducing food legumes to sustainably intensify rice-based cropping systems in South Asia, covering existing approaches, potential scope, challenges, and associated socioeconomic and environmental impacts.

Keywords: cropping systems, diversification, legumes, rice fallow, sustainable intensification

1. Introduction

Food legumes serve as the most crucial source of cheap dietary protein and other essential micronutrients in South Asia, where a significant portion of the population is impoverished, and continuous population growth in the region demands increased food production. This challenge of meeting food and nutritional demands in an environmentally and socially sustainable manner, with a priority on natural ecosystems, can be addressed by enhancing the productivity of existing available agricultural land through sustainable intensification [1]. In the world's most densely populated region where the availability of arable land is limited, it is projected that future enhancements in crop production must originate from intensification: higher yields, increased multiple cropping, and replacement of fallow periods [2]. Therefore, sustainable agricultural intensification is recognized as a major research focus aiming to feed humanity toward 2030 [2] and 2050 [3].

In South Asia, 22.3 million hectares of land remain fallow after rice cultivation, out of which 88.3% (19.6 M ha) of land is in India, 0.5% in Pakistan, 1.1% in Sri Lanka, 8.7% in Bangladesh, 1.4% in Nepal, and 0.02% in Bhutan [4]. In a mono-crop-based rice agroecosystem, rice fallow represents unutilized cropland that remains uncropped after the rice is harvested. These unutilized areas have recently garnered significant attention for their potential in sustainable intensification with legumes to boost food production, nutrition, and soil health and generate extra income for smallholder farmers. Food legumes such as mung bean (*Vigna radiata*), urd bean (*Vigna mungo*), cowpea (*Vigna unguiculata*), pea (*Pisum sativum*), chickpea (*Cicer arietinum*), grass pea (*Lathyrus sativus* L.), and lentil (*Lens culinaris*) being hardy, low maintenance, and short duration are considered suitable crops for the intensification of rice fallows. They offer a tremendous opportunity to use residual soil moisture in rice fallow [5]. Moreover, owing to their distinctive attributes such as biological N-fixation (BNF), deep root morphology, potential to establish surface broadcast in standing rice fields, and soil-fertility restoration property, they can be best fitted in rice fallows [6]. Their role in the improvement of overall soil health (physical, chemical, and biological) and curtailing GHG emissions are reported by many researchers [7]. Further, food legumes being rich in protein and micronutrients could make a major contribution to elevating the hidden hunger widespread among the rural poor. In addition to providing ecosystem services, the incorporation of legumes in rice fallow intensification also has the potential to boost farmers' income and generate additional employment opportunities for the farming communities in South Asia.

The current technological innovations deployed for sustainable intensification of rice fallow lands encompass a multifaceted approach, including remote sensing and geospatial technologies, genetic advancements, agronomic innovations, seed systems, and intensive extension efforts. These technological innovations are effectively utilized to accurately target rice fallow lands for intensification and suggest legume crops and suitable cultivars based on soil properties, soil residual moisture, and other parameters to policymakers and scientists to enhance land productivity [8]. Researchers and social scientists have utilized these mapped datasets to develop and deploy suitable legume varieties, fostering the sustainable intensification of rice fallows. This all has been achieved by realizing faster genetic gains using modern breeding tools such as genomic tools, high throughput precision phenotyping, and molecular breeding to discover target traits and develop and deploy legume varieties that are high-yielding, short-duration, resistant to major biotic and abiotic stresses, and loaded with high nutritional quality [9], which has resulted in growth in terms of legume production and productivity in South Asia over the last few years. Furthermore, agronomic innovations such as resource-conserving technologies (RCTs), primarily zero-tillage, have provided options for more timely planting, potentially leading to higher productivity, input-use efficiency, and profitability; all these improved agricultural technologies have been disseminated using both vibrant formal and informal seed systems, and there is an effort to create integrated seed systems, specifically those based on communities like farmer-producers' groups, self-help groups, village-based seed hubs, and so on, which are designed to be more viable and sustainable. These community-based seed systems contribute to overall quality seed availability of improved varieties and thus support the livelihoods and resilience of smallholder farmers.

Despite various technological innovations, several socioeconomic and production constraints currently impede the realization of its full potential within the rice fallow production system. These include the lack of life-saving irrigation, poor soil moisture after the harvest of rice, soil hardiness, soil acidity, unavailability of quality seeds of appropriate varieties, storage infrastructure limitations, lack of legume varieties resistant to biotic and abiotic stresses, poor technical knowledge and awareness among farmers, and inefficient support for value addition and marketing of their produce. Addressing these challenges necessitates comprehensive interventions, such as site-specific technological interventions, research on fallow-land management, consolidation of R and D activities, policy support, and introduction of good agronomic practices and high-yielding stress-tolerant crop seeds to catalyze the intensification of rice fallows across South Asia for strengthening nutritional and food security.

2. Role of food legumes in rice fallow intensification

Metrics of the system outputs best quantify the success of implementing a Sustainable Intensification (SI) approach. SI must seek to improve production, measured in terms of yield, while optimizing economic returns per unit of time and land. Crucially, this objective should be attained without causing adverse effects on soil, water, and other natural resources. Legume-based intensification serves these objectives by providing multifaceted benefits to humans and the ecosystem (**Figure 1**).

2.1 Soil health improvement

Sustainable intensification of rice fallow is crucial to increase productivity and strengthen the resilience of small farms. However, agricultural transformation through intensification must prioritize crops that have the capacity to enhance soil organic carbon (SOC). The SOC pool is the key indicator of soil health and quality, which in turn plays a critical role in ensuring long-term soil sustainability. The unique capacity of legumes for biological nitrogen fixation (BNF), coupled with their superior aboveground and belowground biomass quality, fosters microbial growth efficiency and soil aggregation, directly contributing to the soil organic carbon (SOC) pool [10–13]. Previous studies have suggested that food legumes exhibit an exceptional ability to store 30% more soil organic carbon (SOC) when compared to other crops [13].

Kuyah et al. [14] found that adopting cropping systems with food legumes resulted in the highest increase in soil organic carbon (SOC) concentrations and estimated carbon stock of 2.29 ± 0.10 mg C ha^{-1} in post-harvest residues of legumes in South Asia. Similarly, Ghosh et al. [15], in their studies conducted in the Indo-Gangetic plains of India, demonstrated the positive impact of including food legumes in rice-based cropping systems on soil organic carbon. Additionally, the World

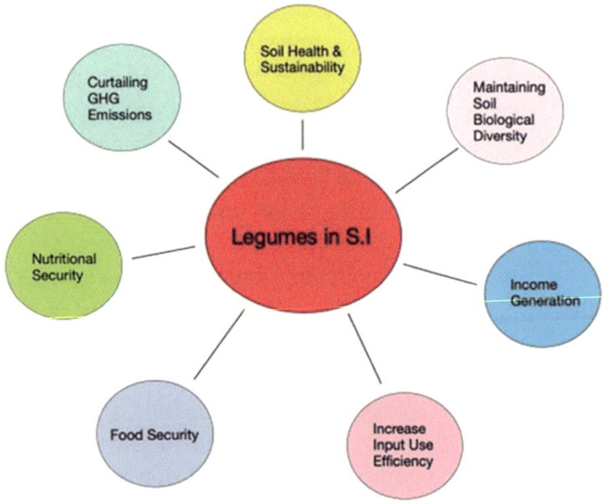

Figure 1.
Role of legumes in sustainable intensification of rice fallows.

Bank's model-based prediction anticipates substantial cumulative soil carbon (C) sequestration from food legumes in Asia and Africa. By 2030, it is projected to reach 33.0 mg ha^{-1} in Asia and 35.12 mg ha^{-1} in Africa. However, the carbon sequestration potential and the amount of organic C returned by leguminous species to soil depend largely on specific legume species, growth behavior, root morphology and physiology, leaf morphology, climatic conditions, structure and aggregation, prevailing cropping system, and agronomic interventions during the crop growth period. Collectively, the unique characteristics of legumes, including BNF, and legume-promoted microbial biomass C contribute directly to the enrichment of the SOC pool. This, in turn, enhances overall soil health and quality, making sustainable intensification with food legumes a promising strategy for smallholder farming.

2.2 Reducing carbon/GHG emissions

Food legumes through their biological nitrogen fixation (BNF) capacity have the potential to lower GHG emissions such as CO_2 and nitrous oxide (N_2O) by reducing the use of industrial nitrogenous fertilizers and CO_2 emitted during the manufacturing of these fertilizers. Globally, land-based biological nitrogen fixation (BNF) exceeds the amount of industrial nitrogen (N) production by 1.1 times [16]. This fixed nitrogen is partly utilized by the subsequent nonlegumes, allowing for the reduction in nitrogen fertilizer application for crops cultivated after food legumes. Therefore, the main advantage of using food legumes in the systems is to ensure the benefits of BNF to the succeeding cereal crops, which in turn reduces the amount of nitrogen (N) fertilizer required for raising a successful crop. Stagnari et al. [17] reported that savings of N from BNF could reduce carbon dioxide (CO_2) emission by approximately 200 kg ha^{-1}. Furthermore, Jeuffroy et al. [18] demonstrated that legume crops emitted around 5–7 times less GHG per unit area as compared with other crops. Similarly, Schwenke et al. [19] demonstrated that the cumulative N_2O emissions from N-fertilized canola greatly exceeded those from chickpeas, faba beans, and field peas. They also highlighted that food legumes significantly reduced their emission factors, suggesting that legume-fixed N is a less-emissive form of N input to the soil than fertilizer N.

2.3 Elevating food security and livelihoods

Global climate change is expected to significantly influence the agricultural production of rainfed regions in South Asia, owing to the severe water crisis and uneven rainfall distribution in the region. This impact appears to be more pronounced in food legumes due to a shrinking land base for agriculture and a preference for cereals under irrigation. In this scenario, fallow land represents a promising opportunity to boost food legume production, land productivity, and income for smallholder communities. It has been identified as particularly suitable for food legumes [6], such as lentil, urd beans, chickpea, pea, mung bean, and Lathyrus as they can survive in post-rice harvest conditions, owing to their ability to withstand and perform well under harsh environments, including water, temperature, fertility, and nutritional stress as well as their less demand on resources. An economic analysis has shown that cultivating food legumes in rice fallows is profitable for the farmers with a benefit-cost ratio (B/C) exceeding 3.0 for many food legumes [20]. Also, employing rice fallows for growing food legumes has the potential to generate 584 million man-days of employment in South Asia [20]. Many studies have documented primary candidate legumes in SA to improve the fallow land productivity and income for farmers (**Table 1**).

2.3.1 Legume-based SI in India

About 19.6 million hectares of land in India remain fallow after the rice harvest, with about 80% of this area located in eastern states [4, 11]. It is assumed that the present and future demand for food legumes can be met through intensive production with an increase in productivity per unit area and time, especially under the rainfed agroecosystems of the Northeast region. Usually, spring (before kharif rice sowing) and autumn (after kharif rice harvesting) seasons in this region remain fallow, and short-duration legumes offer a huge opportunity to use this land and enhance food legume production. In recent years, the area under food legume crops has increased sharply in the Northeast regions of India from 112,000 hectares in 2000–2001 to about 217,000 hectares in 2013–2014, and the productivity during this period also enhanced from 647 to 848 kg/ha [23].

The selection of legume crops for intensification and diversification always depends upon the soil type, texture, temperature, and residual moisture in the fallow land. Ghosh et al. [30] observed that appropriate agronomic interventions

Rice fallow region SA	Primary legume crops	References
India—North and North-eastern states (Uttar Pradesh, Bihar, Odisha, West Bengal, Tripura, and Assam)	Lentil, Lathyrus, Pea, Chickpea, Black gram	[5, 21–24]
India—Central states (Chhattisgarh, Madhya Pradesh, and Maharashtra)	Lathyrus, Lentil, Chickpea, Black gram	[25]
India—Southern states (Andhra Pradesh, Tamil Nadu, and Karnataka)	Black gram, Mung bean	
Eastern Gangetic Plains of Bangladesh and Nepal	Lentil, Lathyrus, Chickpea, Pea	[26–29]

Table 1.
Primary legume crops for rice fallow regions in South Asia to enhance land productivity.

and varietal screening under the rice fallow conditions could popularize pea and lentil crops at mid-altitude, contributing to the food and nutritional security of resource-poor farmers in the Northeast region of India. Similarly, various studies in the Northeast Hill Region (NEH) have consistently shown that incorporating food legumes like pea and lentil significantly enhances farmers' income and employment compared to rice fallow or other systems. Das et al. [31] observed that farmers' net income increased by INR 29,000/ha and INR 21,500/ha over the rice-fallow system owing to the inclusion of pea and lentil, respectively, in zero-tillage cultivation. Similarly, Ngachan et al. [32] indicated net returns of INR 25,000/ha and INR 20,000/ha due to the cultivation of pea and lentil in rice fallow. Another experiment conducted by researchers at the ICAR Research Complex for NEH region [33] on *utera* cultivation (a method of cultivation in which the sowing of a second crop is done just after the harvesting of paddy crop) showed that relay cropping of lentil with rice resulted in high land productivity. In addition to that, studies by researchers revealed that the inclusion of lentil and pea in the rice fallow system not only enhanced overall system productivity but also improved soil health and farm profitability in the Northeast regions of India [21].

Chickpea is one of the major food legumes widely grown in India and is suitable for rainfed rice fallow lands (RRFL) of Uttar Pradesh, Uttarakhand, Rajasthan, Madhya Pradesh, Chhattisgarh, Bihar, Jharkhand, West Bengal, Assam, Orissa, Maharashtra, Gujarat, Andhra Pradesh, Karnataka, and Tamil Nadu. Sequential and *paira* cropping of RRFL has witnessed drastic changes with advances in chickpea cultivation. Researchers in their studies reported that the additional production of chickpea from the *utera* cropping system, based on an average productivity of 300 kg ha^{-1}, could be expected to be 165,000 tons [25]. In recent years, the development of early-maturing heat-tolerant chickpea cultivars, along with appropriate agronomic practices, has increased late-planted chickpea productivity, enabling the adoption of rice-chickpea systems in fallow areas of eastern Uttar Pradesh and Bihar; this trend is also gaining popularity in Chhattisgarh, and the successful introduction and adoption of high-yielding chickpea cultivars (ICCV 2 and ICCV 10) in rice fallows has been observed in Andhra Pradesh.

Khesari or Lathyrus is a robust legume crop commonly grown after rice in eastern India; it is also suitable for RRFL in Uttar Pradesh, Madhya Pradesh, Bihar, Chhattisgarh West Bengal, and Assam. It has wide adaptability to grow in paddy soils as it suffers much less in adverse soil moisture as compared to other food legumes.

Black gram or urd bean is extensively grown after rice in peninsular India (Orissa, Madhya Pradesh, Chhattisgarh, Jharkhand, Maharashtra, Andhra Pradesh, Tamil Nadu, and Karnataka) during the Rabi season. It has good adaptation to paddy soils, and it is also tolerant to waterlogging during the early phases of crop growth. Commercialization of urd bean production in RRFL of Andhra Pradesh contributed to area expansion and higher yields from 410 kg ha^{-1} on 0.22 m ha in 1981–1982 to 576 kg ha^{-1} on 0.42 m ha in 1999–2000 [25]. Similarly, other legume crops such as green gram, field pea, and so on can also be effectively utilized for intensifying fallow lands in India to increase legume production and generate additional income for farmers.

2.3.2 Legume-based SI in Bangladesh

In Bangladesh, approximately 20% of cultivable land lies fallow and can be utilized for growing winter legumes like grass pea, lentil, chickpea, and field pea.

Ahmed et al. [29] reported that out of 1.8 million hectares in the dominant rotation (monsoon rice–fallow–irrigated spring rice), around 5–7% or approximately 100,000 hectares have suitable upland soil for pea production. Malik et al. [27] demonstrated that incorporating peas as a green pod vegetable could boost farm productivity by 1.4-fold and increase farm net income by fourfold compared to the dominant cropping sequence (rice–fallow–rice). Similarly, another researcher [6] emphasized that integrating lentil cultivation in monsoon rice fields could enhance lentil production, contribute to human nutrition, and promote soil health, supporting a sustainable production system. Mazid et al. [34] showed that farmers could double farm income by growing chickpea instead of leaving land fallow in high *Barind* tracts of Bangladesh and suggested that the cropping system of monsoon rice followed by a second crop of legumes (pea, lentil, Lathyrus, or chickpea) and a third crop of spring rice is the most suitable for food, forage, vegetable production, and net returns for resource-poor farmers in Bangladesh.

2.3.3 Legume-based SI in Nepal

In Nepal, the eastern and central terai regions with an extent of 0.3 M ha are the potential rice fallow areas for short-season food legume intensification [4]. Legumes play a pivotal role in the Nepalese farming system and diets. The highly diverse environment in Nepal allows the cultivation of different species of legumes. Winter food legumes, including lentil, chickpea, grasspea, and field pea, collectively account for 70 and 72% in terms of area and production, respectively, while summer legumes such as black gram, soybean, pigeon pea, cowpea, mung bean, etc., contribute approximately 27.56% in area and about 26.19% in production [35]. Lentil alone dominates all other pulses grown in Nepal, sharing 63.5% in area and 67.2% of output; moreover, Nepalese lentils account for 90% of the total export of pulses [36]. Smallholder farmers, whose livelihoods are primarily dependent on agriculture, are anticipated to benefit from the intensification of legumes in rice-based cropping systems [37]. Studies suggest that sustainable intensification can reduce crop yield gaps in Nepal, helping livelihood security and socio-economic conditions [38]. In recent years, owing to various national and international legume improvement programs, Nepal has witnessed an impressive growth in pulse production (6.54%). The overall productivity increased from 762 kg/ha in 2000 to 1052 kg/ha in 2018 [35].

2.4 Elevating nutritional security

Widespread malnutrition, notably protein-calorie malnutrition (PEM), is a significant issue, especially in South Asia, where the lower-income population is highly vulnerable due to limited purchasing power, as conventional protein sources like meat and milk are often expensive. Food legumes also called pulses, are nutrient-dense food and generally contain twice the amount of protein (18–30%); they are also rich sources of 15 essential macro and micronutrient elements, low glycemic index carbohydrates, dietary fiber, vitamins, folic acid, oligosaccharides (primarily raffinose), necessary fatty acids, carotenoids, and polyphenols [39, 40]. Consumption of legumes can fulfill the daily requirements for various minerals. For instance, meeting the daily zinc requirement of 3.0 mg/day for women and 4.2 mg/day for men can be achieved by consuming 100–200 g of lentil, cowpea, and chickpea. Similarly, the daily iron requirement (1.46 mg/day for women and 1.05 mg/day for men) can be satisfied by consuming 100 g of most types of food legumes [35, 41].

Legumes are regarded as "health food" because they have a significant role in disease protection. Regular intake of legumes improves cardio-metabolic health and reduces the risk of type 2 diabetes, obesity, and cancers [42, 43]. Moreover, sustainable intensification and genetic innovations in food legumes empower farmers to reach desired yield targets and ensure nutritional security through biofortified varieties of food legumes especially in lentils (IPL 220, Pusa Vaibhav, Pusa Ageti Masoor in India, and Barimasur 4, 5, 6, 7, 8, and 9 in Bangladesh) which are rich in iron and zinc aiding in addressing hidden hunger among humans especially women and children [44]. Thus, food legumes can play a crucial role in food and nutritional security by alleviating protein hunger and malnutrition among the poorest of the poor.

3. Current approaches for legume-based SI of rice fallow lands

3.1 Genetic innovations

The necessity of deploying food legumes for the sustainable intensification (SI) of rice fallow lands is critical and underlines a need for focused and innovative breeding programs. Realizing this, considerable research efforts have been directed toward developing varieties with improved traits such as grain yield, short crop duration, stress tolerance, and nutritional quality. Genetic gains have been recorded for traits imparting resistance to major biotic and abiotic stresses, seed size, yield, and improved nutritional quality [9]. This has resulted in growth, in terms of production and productivity, in South Asia. For instance, in the last 15 years, India has witnessed the highest growth in the production of mung bean (178%), followed by chickpea (125%), urd bean (90%), pigeon pea (51%), and lentil (34%) [45]. This progress has been realized through farmers' development, demonstration, and adoption of improved varieties, fostering a sustainable rice-based production system in South Asia (**Table 2**).

While breeding efforts over the past decades have yielded various practical benefits; current modernized breeding projects, facilitated by genomic breeding tools, have led to faster genetic gains in legumes. Deciphering the genome sequence of lentil (4000 Mbp), pea (5000 Mbp), chickpea (738 Mbp), mung bean (548 Mbp), faba bean (13,000 Mbp), cowpea (587 Mb), and other legumes has generated sufficient genomic resources with associated phenotypic data to discover target traits and breed superior varieties [51, 52]. Pre-breeding is considered one of the potential genetic diversity-enhancing approaches to broadening the genetic base of legumes, with the introgression of desirable genes from wild or exotic species into current cultivars to develop improved climate-resilient genotypes. Genome-wide association studies (GWAS) are also being efficiently employed to dissect and integrate the complex climate-smart traits in legumes. Additionally, modern genetic enhancement tools, including high throughput precision phenotyping; molecular breeding such as marker-assisted selection (MAS), marker-assisted backcrossing (MABC), and marker-assisted recurrent selection (MARS); molecular dissection of qualitative traits; omics; and genome editing tools have also been deployed to expedite wider adaptability, resilience, and increased productivity in legumes. The successful development and scale-up of extra early biofortified lentil varieties (BARIMasur-3, BARIMasur-4, BARIMasur-5, BARIMasur-6, BARIMasur-7, BARIMasur-8, and

Crop	Varieties	Country	Reference
Lentil	HUL 57, KLS 218, Subrata, PL 639, JL 3, IPL 81, Moitree, DPL 81, IPL 220, NDL 1, WBL 58, LG 225, LG 178, SKL 251, L 4147, PL 4, VL 1, IPL 406, Sapana, DPL 62, DPL 15, L 303, L 304, L 4727, IPL 315, IPL 321, KLB 345, KLS 1322	India	[22, 23, 39, 46–49]
	BARIMasur3, BARIMasur4, BARIMasur5, BARIMasur6, BARIMasur7, BARIMasur8, BARIMasur9	Bangladesh	[50]
	Sikhar, Simal, Sindur, Khajura Masuro2, Khajura Masuro3, Khajura Masuro4	Nepal	[47]
Grasspea	Ratan, Prateek, Nirmal, Mahateora	India	[47, 49]
	BARIKhesari2, BARIKhesari3, BARIKhesari4	Bangladesh	[47]
Chickpea	Anuradha, Udai, CSJ 515, GNG 1958, LBG17, GCP 105, Pusa 372, JG16, Rajas, Pant G 186, JG 14, JG 11, JG 16, Pusa 547, Vaibhav, JG 315, ICCV 10, ICCV 2, Samrat	India	[39, 47–49]
Green gram	Virat, PDM 14, Pusa 109, MH 309, PDM 62, K851, PDM 11, PDM 54, Pant M 4, HUM 12, Pusa Vishal, Tripura Mung, Samrat, ML 5	India	[23, 39, 49]
Black gram	DU 4, UG 203, UG 302, Azad Urd 1, Uttara, PDU 1 (Basant Bahar), KU 300, Pant U 35, Tripura Maskoli 1, Kalindi, LBG 752, Pant U 31, LBG 402, LBG 709	India	[23, 49]
Pea	DMR 3, DMR 4, DMR 4, DMR 5, P 81-10, KMPR 84-2, DMR 7, HUP 1, HUP 2, DDR 50, TRCP 8, TRCP 9, Prakash, Vikash (Field pea); Arkel, Rachna, Azad, Ambika (Vegetable pea)	India	[23, 49]

Table 2.
Improved legume varieties demonstrated and recommended for rice fallow intensification.

BARIMasur-9) to enhance the nutritional and productivity gains in rice-based cropping systems of Bangladesh [50] is one such example where 99% of the lentil area under improved cultivars is now covered by ICARDA varieties, delivering an additional 55,000 t of lentil, valued at US$ 38 million annually [53]. Similarly, the Grain Legumes Research Program (GLRP) in Nepal, in collaboration with CGIAR centers, has enhanced genetic improvement for lentil, chickpea, pigeon pea, soybean, mung bean, black gram, and cowpea, producing improved varieties and production technologies. This effort has resulted in a significant 142% increase in grain legume production, driven by a 47% expansion in cultivation area and a 65% yield boost over the past 2.5 decades [54].

3.2 Geospatial technologies for mapping rice fallows

Geospatial technologies play a pivotal role in providing policymakers and scientists with accurate information about the spatial distribution and extent of rice fallows for the effective targeting of agro technologies for sustainable crop intensification and enhanced productivity and profitability [8]. They have been used in mapping cropping intensity, agricultural practices identifying cropping cycles in a year, and assessing the land that remains fallow thereafter for intensification [55, 56]. MODIS normalized difference vegetation index (NDVI) and season-wise intensive ground survey data were used to map rice systems and the subsequent fallow periods (rabi-fallow) in South Asia [57]. Similarly, in India, the dynamics of rice fallow in Orissa were analyzed using Landsat-8 OLI (optical) and Sentinel-1 (Synthetic Aperture Radar—SAR) data. They illustrated the time-series profiles of SMAP (soil moisture) datasets, which can be utilized to select crops and cultivars to target the rice fallows [58]. In other studies, satellite data, acquired through diverse sensors such as optical, thermal infrared, and microwave sensors, are used to assess residual soil moisture and plan for the subsequent rabi crop in rainfed fallow regions [59]. Geospatial science has also been employed to suggest crop suitability based on soil properties, soil moisture, and other parameters of fallow lands as well as to enhance resource conservation and agricultural productivity in the underutilized lands of South Asia [60].

3.3 Use of resource conservation technology

Conservation agriculture (CA) is an agroecological approach that aims to achieve sustainable and profitable intensification of agricultural systems [61]. Resource-conserving technologies (RCTs) such as zero-tillage are economical, time-saving, eco-friendly, and less capital-intensive options for smallholder farmers. It offers an opportunity to utilize soil moisture in succeeding rabi crops after paddy harvest, presenting options for more timely planting potentially leading to higher productivity, input-use efficiency, and profitability. Conservation agriculture-based resource conservation technologies have also proven to promote conjunctive use of organics (avoids residue burning), improve soil health, and help adapt to climatic risks. Reduced tillage has increased the yield of food legumes (Lathyrus, green gram, black gram, field pea) by 33–44% over conventional tillage, and retention of rice stubble/mulching and ZT sowing of food legume crops significantly enhanced productivity of food legumes in rice fallows [5, 49]. Several similar studies have demonstrated that the inclusion of food legumes in a conservation agriculture (CA)-based production system can be a sustainable approach to enhance crop intensification in rice-fallow agroecosystems, thereby contributing to the sustainability of food, nutritional, economic, and environmental security [62, 63]. Realizing this, significant efforts have been made over the years to advance these technologies through various eco-regional programs.

3.4 Innovations in seed systems

The successful adoption of agri-innovations relies on appropriate technologies and a conducive environment. In agrifood systems, establishing a viable and sustainable innovative seed system is crucial for facilitating technology adoption, as the absence of access to improved quality seeds diminishes agricultural productivity and competitiveness and leaves rural communities vulnerable, marginalized, and food insecure. Understanding the value of both established formal and informal seed systems in South Asia and taking comparative advantage of both systems, the approach of an integrated seed system has been recognized [64]. Community-based seed systems (CBSS) are emerging as a complementary approach to formal and informal seed systems for making seed systems inclusive and dynamic to meet the diverse needs of smallholder farmers in marginal farming systems. This system offers the potential to integrate informal and formal seed systems as it deals with all the crops and commodities of the locality, including local and improved varieties, neglected and underutilized species, and future smart food crops [65].

4. Challenges and opportunities

Sustainable Intensification (SI) in rice fallows offers ample potential to improve land productivity, increase employment opportunities, and boost income for farming communities in South Asia. However, several socioeconomic and production constraints currently impede the realization of its full potential within the rice-fallow production system. Some of the major limiting factors for the slow intensification of rice fallows comprise the soil moisture stress at the time of planting due to the early withdrawal of the monsoon, lack of life-saving irrigation, late harvesting of long-duration rice varieties, or waterlogging in coastal tracts [66]. Several studies have reported that among the abiotic factors, the rapid drop in soil moisture leading to terminal drought overlapping with flowering and pod-filling stages is a drawback in cultivating food legumes in rice fallows [25, 67]. Insufficient irrigation infrastructure and a lack of technical knowledge within the farming community, including limited information on crop management, soil moisture conservation, and sowing technologies for germinating seeds in low moisture conditions, also contribute to farmers' reluctance to plant subsequent crops. Furthermore, the public extension system requires strengthening to effectively deliver agro-technologies, inputs, and information to the farmers for enhanced adoption.

Moreover, farmers in South Asia are resource-poor and lack sufficient capital to purchase critical inputs, so the non-availability of critical inputs, primarily high-quality seeds of improved varieties that can ensure an adequate yield for economic gains, is also one of the constraints compelling farmers to leave their lands fallow. Additionally, the unavailability of suitable legume varieties for fallow lands, which are short duration, high yielding, and tolerant to biotic and abiotic stresses, further exacerbates these challenges. Another constraint is soil hardness in puddled fields, where the physical disruption caused by rice puddling decreases soil bulk density and leads to frequent drying, crack formation, and the unavailability of moisture for winter crops; a decrease in the subsequent crops has been reported due to hardness of these soils, perhaps due to restricted root growth [5]. Other constraints include grazing of crops by stray animals, scarcity, and diversion of labor during the sowing and intercultural operations of food legumes following rice harvest, lack of improved post-harvest technologies (and value addition) to reduce post-harvest losses, severe weed menaces, poor crop management, low volume of produce, and lack of market linkage to the farmers to get a fair support price for their produce.

These constraints can be addressed through comprehensive interventions such as collaborative research, site-specific technological and policy interventions, incentives, and the implementation of viable seed supply chain systems at the village level. The development of suitable storage structures and techniques, such as community seed banks, for the continual storage and supply of quality seed, value addition support, timely access to information, and swift institutional credit systems can significantly contribute to the increased adoption of food legumes as the second crop in rice fallows [13, 40]. In addition, situation-specific improved production technology dealing with soil moisture conservation, adequacy in timely plant population management, mitigation or termination of drought and heat stresses, optimum seed rate, improved seeding techniques (e.g., with zero till drill), and improvement in soil physical condition could effectively address the production constraints. Further in-depth research needs to be done on fallow challenges (biophysical and socioeconomic), including farmers' behavior along with potential and feasible solutions and stakeholders' responsibility.

5. Conclusion

Food legumes in the context of sustainable crop intensification in rice-based cropping systems, particularly in the rice-fallow regions of South Asia, as detailed in this chapter, underline their pivotal role in enhancing food and nutritional security, restoring soil health, reducing carbon footprints, and providing long-term environmental and socioeconomic benefits. In recent years, SI has offered significant benefits to farmers by integrating modern breeding methods to develop improved varieties, geospatial technologies for effective targeting of innovations, resource conservation technologies, and innovative seed systems. However, inclusive site-specific research along with reinforcement of extension systems is required to realize its full potential. Thus, the present study serves as a focal resource for scientists to strategically design and implement collaborative research programs to address the challenges faced by farmers. It also guides policymakers in providing timely policy support to promote SI for the overall benefit of marginalized communities in South Asia.

Conflict of interest

All authors wish to confirm that there is no potential conflict of interest.

Author details

Smita Kaul[1*], Akanksha Singh[1], Chandrashekhar Biradar[2] and Shiv Kumar[1]

1 South Asia and China Regional Program, International Centre for Agricultural Research in the Dry Area (ICARDA), New Delhi, India

2 Asia Continental Program, Center for International Forestry Research and World Agroforestry (CIFOR-ICRAF), New Delhi, India

*Address all correspondence to: s.kaul@cgiar.org

References

[1] Dias LC, Pimenta FM, Santos AB, Costa MH, Ladle RJ. Patterns of land use, extensification, and intensification of Brazilian agriculture. Global Change Biology. 2016;**22**(8):2887-2903

[2] FAO. World agriculture: Towards 2015/30. In: Bruinsma J, editor. An FAO Study, Rome, Italy. 1st ed. FAO; 2002

[3] Chartres CJ, Noble A. Sustainable intensification: Overcoming land and water constraints on food production. Food Security. 2015;**7**:235-245

[4] Gumma MK, Thenkabail PS, Teluguntla P, Rao MN, Mohammed IA, Whitbread AM. Mapping rice-fallow cropland areas for short-season grain legumes intensification in South Asia using MODIS 250 m time-series data. International Journal of Digital Earth. 2016;**9**(10):981-1003

[5] Kar G, Kumar A. Evaluation of post-rainy season crops with residual soil moisture and different tillage methods in rice–fallow of eastern India. Agricultural Water Management. 2009;**96**:931-938

[6] Ali M, Ghosh PK, Hazra KK. Resource conservation technologies in rice fallow. In: Ghosh PK, Kumar N, Venkatesh MS, Hazra KK, Nadarajan N, editors. Resource Conservation Technology in Pulses. Jodhpur, Rajasthan, India: Scientific Publishers; 2014. pp. 83-88

[7] Kumar N, Nath CP, Hazra KK, Das K, Venkatesh MS, Singh MK, et al. Impact of zero–till residue management and crop diversification with legumes on soil aggregation and carbon sequestration. Soil and Tillage Research. 2019;**189**(6):158-167

[8] Srivastava A, Borah S, Dastidar P, Sharma A, Gogoi D, Goswami P, et al. Rice-fallow targeting for cropping intensification through geospatial technologies in the rice belt of northeast India. Agriculture. 2023;**13**:1509

[9] Kumar S, Gupta P, Choukri H, Siddique KHM. Efficient breeding of pulse crops. In: Gosal SS, Wani SH, editors. Accelerated Plant Breeding. Vol. 3. Cham: Springer; 2020. pp. 1-30

[10] McDaniel MD, Tiemann LK, Grandy AS. Does agricultural crop diversity enhance soil microbial biomass and organic matter dynamics? A meta-analysis. Ecological Applications. 2014;**24**:560-570

[11] Kumar R, Mishra JS, Hans H. Enhancing productivity of rice–fallows of eastern India through inclusion of pulses and oilseeds. Indian Farming. 2018;**68**(8):7-10

[12] Meena RS, Kumar S, Yadav GS. Soil carbon sequestration in crop production. In: Nutrient Dynamics for Sustainable Crop Production. Singapore: Springer; 2020. pp. 1-39

[13] Kumar S, Meena RS, Lal R, Singh Yadav G, Mitran T, Meena BL, et al. Role of legumes in soil carbon sequestration. In: Legumes for Soil Health and Sustainable Management. Singapore: Springer; 2018. pp. 109-138

[14] Kuyah S, Muoni T, Bayala J, Chopin P, Dahlin S, Hughes K, et al. Grain legumes and dryland cereals contribute to carbon sequestration in the drylands of Africa and South Asia. Agriculture Ecosystems and Environment. 2023;**355**:1-11

[15] Ghosh P, Venkatesh M, Hazra K, Kumar N. Long-term effect of pulses and

nutrient management on soil organic carbon dynamics and sustainability on an inceptisol of Indo-Gangetic plains of India. Experimental Agriculture. 2012;**48**(4):473-487

[16] Rahman MM, Alam MS, Islam MM, Kamal MZ, Rahman GM, Haque MM, et al. Potential of legume-based cropping systems for climate change adaptation and mitigation. In: Advances in Legumes for Sustainable Intensification. Cambridge, MA, USA: Academic Press (Elsevier); 2022. pp. 381-402

[17] Stagnari F, Maggio A, Galieni A, et al. Multiple benefits of legumes for agriculture sustainability: An overview. Chemical and Biological Technologies in Agriculture. 2017;**4**:2

[18] Jeuffroy MH, Baranger E, Carrouée B, Chezelles ED, Gosme M, Hénault C. Nitrous oxide emissions from crop rotations including wheat, oilseed rape and dry peas. Biogeosciences. 2013;**10**:1787-1797

[19] Schwenke GD, Herridge DF, Scheer C, Rowlings DW, Haigh BM, McMullen KG. Soil N2O emissions under N2-fixing legumes and N-fertilised canola: A reappraisal of emissions factor calculations. Agriculture Ecosystem and Environment. 2015;**202**:232-242

[20] Wani SP, Chander G, Patil MD, Sawargavkar G, Kumar S. Scaling up food legume production through genetic gain and improved management. In: Genetic Enhancement in Major Food Legumes: Advances in Major Food Legumes. Switzerland: Springer; 2021. pp. 303-338

[21] Baishya LK, Rathore SS, Sarkar D, Jamir T, Rajkhowa DJ. Crop and varietal diversification for enhancing productivity and profitability of rice fallow system in eastern Himalayan region. Indian Journal of Agricultural Sciences. 2019;**89**(5):800-805

[22] Yadav GS, Devi AG, Kandpal B, Das A, Barman KK, Babu S. Minimum till lentil (*Lens culinaris*): An efficient way for rice fallow utilization and income enhancement in subtropical Tripura. Indian Journal of Agricultural Sciences. 2020;**90**(1):133-137

[23] Das A, Subhash B, Yadav GS, Ansari M, Singh R, Baishya L, et al. Status, and strategies for pulses production for food and nutritional security in the northeastern region of India. Indian Journal of Agronomy. 2016;**61**:43-57

[24] Das A, Chowdhury S, Jayanta L, Ramkrushna GI, Panwar AS, Ngachan SV. Agro-physical assessment of different pea (*Pisum sativum*) cultivars in lowland rice (*Oryza sativa*) fallow under no-till system for enhancing cropping intensity and productivity in mid hills of northeast India. Indian Journal of Agricultural Sciences. 2017;**87**(12):1612-1618

[25] Pande S, Sharma M, Ghosh R. Role of pulses in sustaining agricultural productivity in the rainfed rice-fallow lands of India in changing climatic scenario. In: Climate Change and Food Security in India: Proceeding of National Symposium on Food Security in Context of Changing Climate. Kanpur, India: CSAUAT; 2012. pp. 53-70

[26] Malik AI, Ali MO, Zaman MS, Flower K, Rahman MM, Erskine W. Relay sowing of lentil (*Lens culinaris* subsp. *culinaris*) to intensify rice-based cropping. Journal of Agricultural Science. 2016;**154**:850-857

[27] Malik AI, Nasim M, Flower K, Hossain M, Rahman M, Anwar B, et al. Cropping system intensification: Vegetable pea can replace fallow between

rainfed monsoon rice and irrigated spring rice. The Journal of Agricultural Science. 2017;**155**:1-12

[28] Ali MO, Ashutosh S. More benefit from less land: A new rice-pea-rice cropping pattern for resource-poor farmers of Bangladesh. Journal of Agricultural Science and Technology. 2013;**3**:204-210

[29] Ahmed A. Bangladesh Integrated Household Survey, 2011-12. Washington, DC: IFPRI; 2016. Available from: https://dataverse.harvard.edu/dataset.xhtml?persistentId=hdl:1902.1/21266

[30] Ghosh PK, Das A, Munda GC, Patel DP, Saha R. Prospects, and opportunities for conservation agriculture in rice-based cropping systems of northeast India. Conservation Agriculture Newsletters (PACA). 2011;**18**:4-5

[31] Das A, Ramkrushna GI, Patel DP, Choudhury BU, Munda GC, Rajkhowa DJ, et al. Zero Tillage Pea, Lentil and Toria Cultivation in Rice Fallow for Diversification and Resource Conservation in Hills. Umiam Meghalaya: ICAR Research Complex for NEH Region; 2012

[32] Ngachan SV, Das A, Ramkrushna GI, Layek J. Potentialities of rabi pulses in rice fallow under conservation agriculture in northeast India. Agricomplex Newsletter. 2013;**28**(1):1

[33] Das A, Patel DP, Ramkrushna GI, Munda GC, Ngachan SV, Kumar M, et al. Crop diversification, crop and energy productivity under raised and sunken beds: Results from a seven-year study in a high rainfall organic production system. Biological Agriculture and Horticulture. 2014;**30**(2):73-87

[34] Mazid MA, Wade LJ, Saleque MA, Sarkar ABS, Mollah MIU, Olea AB, et al. Nutrient management in rainfed lowland rice for the high barind tract of Bangladesh. In: Ladha JK, Wade L, Dobermann A, Reichardt W, Kirk GJD, Piggin C, editors. Rainfed Lowland Rice: Advances in Nutrient Management Research. Manila: International Rice Research Institute; 1998. pp. 217-227

[35] Darai R, Sarker A, Aryal L, Gaur P, Neupane RK. Adoption and impact of pulses research and development strategies for Nepal. Journal of Horticulture Science and Forestry. 2021;**104**(3):1-18

[36] Acharya P, Basnet DB, Sanjel PK. Pulses value chain development activities for achieving food and nutrition security and contributing to SDGs: Present status, challenges and way forward in Nepal. In: Pulses Value Chain Development for Achieving Food and Nutrition Security in South Asia: Current Status and Future Prospects. Bangladesh: SAARC Agriculture Centre; 2019. p. 94

[37] Khanal NP, Khanal NN, Gurung GB, Thapa S, Gupta KP, Sherpa LT, et al. Mungbean (*Vigna radiata* (L.) Wilczek) in cereal fallows: Experience of farmers' participatory research and development activities in foothills and tarai of Nepal. In: Kharkawal MC, editor. Proceedings of the 4th International Food Legumes Research Conference. Indian Society of Genetics and Plant Breeding. New Delhi: IARI; 2006. pp. 357-366

[38] Shrestha J, Subedi S. Improving crop productivity through sustainable intensification. South Asian Research Journal of Agriculture and Fisheries. 2019;**1**(1):8-11

[39] Praharaj CS, Ali M, Kumar N, Dutta A, Singh U, Singh R. Pulses for crop intensification and sustainable livelihood. Indian Journal of Agronomy. 2021;**66**:60-72

[40] Kumar S, Bamboriya S, Rani K, Meena RS, Sheoran S, Loyal A, et al.

Grain legumes: A diversified diet for sustainable livelihood, food, and nutritional security. In: Advances in Legumes for Sustainable Intensification. Cambridge, MA, USA: Academic Press (Elsevier); 2022. pp. 157-178

[41] FAO. Human Vitamin and Mineral Requirement. Report of a Joint FAO/WHO Expert Consultation, Bangkok, Thailand. [Internet]. 2002. Available from: http://www.fao.org/DOCREP/004/Y2809E/y2809e00.html

[42] Bazzano LA, He J, Ogden LG, Loria C, Vupputuri S, Myers L, et al. Legume consumption and risk of coronary heart disease in US men and women: NHANES 1 epidemiologic follow-up study. Archives of Internal Medicine. 2001;**161**:2573-2578

[43] Kouris-Blazos A, Belski R. Health benefits of legumes and pulses with a focus on Australian sweet lupins. Asian Pacific Journal of Clinical Nutrition. 2016;**21**(1):1-17

[44] Kumar S, Dikshit HK, Mishra GP, Singh A, Aski M, Virk PS. Biofortification of staple crops: Present status and future strategies. In: Biofortification of Staple Crops. Singapore: Springer; 2022. pp. 1-30

[45] Gaur P. Can India sustain high growth of pulses production. Journal of Food Legumes. 2021;**33**(1):1-3

[46] Yadav GS, Datta M, Saha P. Crop diversification and food security through lentil cultivation in lowland rice fallow. In: Das A, Mohapatra KP, Ngachan SV, Panwar AS, Rajkhowa DJ, Ramkrushna GI, Layek L, editors. Conservation Agriculture for Advancing Food Security in Changing Climate. 2017. pp. 103-121

[47] Sarker A, Swain N, Nath R, Darai R, Ali MO. Sustainable intensification and diversification of cropping and food systems through lentil and grass peas in South Asia. In: Wani SP, Raju K, Bhattacharyya T, editors. Scaling-Up Solutions for Farmers. Cham: Springer; 2021. pp. 265-277

[48] Jayanta L, Samik C, Ramkrushna GI, Das A. Evaluation of different lentil cultivars in lowland rice fallow under no-till system for enhancing cropping intensity and productivity. Indian Journal of Hill Farming. 2014;**27**(2):4-9

[49] Ghosh PK, Hazra KK, Nath CP, Das A, Acharya CL. Scope, constraints, and challenges of intensifying rice (*Oryza sativa*) fallows through pulses. Indian Journal of Agronomy. 2016;**61**(4):122-128

[50] Kumar J, Sen Gupta D, Baum M, Varshney RK, Kumar S. Genomics-assisted lentil breeding: Status and future strategies. Legume Science. 2021;**3**(3):1-20

[51] Varshney RK, Thudi M, Pandey MK, Tardieu F, Ojiewo C, Vadez V, et al. Accelerating genetic gains in legumes for the development of prosperous smallholder agriculture: Integrating genomics, phenotyping, systems modelling and agronomy. Journal of Experimental Botany. 2018;**69**(13):3293-3312

[52] Khan M, Pandey A, Athar T, Hamurcu M, Gezgin S, Sassi G, et al. Current trends in genetic enhancement of legumes in the genomics era for a sustainable future. In: Advances in Legumes for Sustainable Intensification. Cambridge, MA, USA: Academic Press (Elsevier); 2022. pp. 533-552

[53] Yigezu YA, Alwang J, Rahman MW, Mollah MBR, El-Shater T, Aw-Hassan A, et al. Is DNA fingerprinting the gold standard for estimation of adoption and

impacts of improved lentil varieties? Food Policy. 2019;**83**:48-59

[54] Gharti DB, Darai R, Subedi S, Sarker A, Kumar S. Grain legumes in Nepal: Present scenario and future prospects. World Journal of Agricultural Research. 2014;**2**(5):216-222

[55] Bégué A, Arvor D, Bellon B, Betbeder J, de Abelleyra DP, Ferraz R, et al. Remote sensing and cropping practices: A review. Remote Sensing. 2018;**10**:99

[56] Yan H, Xiao X, Huang H, Liu J, Chen J, Bai X. Multiple cropping intensity in China derived from agro-meteorological observations and MODIS data. Chinese Geographical Science. 2014;**24**:205-219

[57] Gumma MK, Thenkabail PS, Teluguntla P, Whitbread AM. Monitoring of spatiotemporal dynamics of rabi rice fallows in South Asia using remote sensing. In: Reddy G, Singh S, editors. Geospatial Technologies in Land Resources Mapping, Monitoring and Management. Geotechnologies and the Environment. Cham, Switzerland: Springer; 2018. pp. 425-449

[58] Chandna P, Mondal S. Analyzing multi-year rice-fallow dynamics in Odisha using multi-temporal landsat-8 OLI and sentinel-1 data. GIScience and Remote Sensing. 2020;**57**:1-19

[59] Kolassa J, Reichle RH, Draper CS. Merging active and passive microwave observations in soil moisture data assimilation. Remote Sensing of Environment. 2017;**191**:117-130

[60] Bandyopadhyay KK, Sahoo RN, Singh R, Pradhan S, Singh S, Krishna G, et al. Characterization and crop planning of rabi fallows using remote sensing and GIS. Current Science. 2015;**108**:2051-2062

[61] Dang Y. Special issue "conservation agriculture". Agriculture. 2019. Available from: https://www.mdpi.com/journal/agriculture/special_issues/conservation_agriculture

[62] Mishra JS, Kumar R. Conservation agriculture, and its role in rice-fallow production system of eastern India. In: Rice Fallow Management in India. Delhi: International Books and Periodical Supply Service; 2022. pp. 27-34

[63] Ariful Islam M, Sarkar D, Robiul Alam M, Jahangir MMR, Ali MO, et al. Legumes in conservation agriculture: A sustainable approach in rice-based ecology of the eastern Indo-Gangetic plain of South Asia—An overview. Technology in Agronomy. 2023;**3**:1-17

[64] Louwaars NP, de Boef WS. Integrated seed sector development in Africa: A conceptual framework for creating coherence between practices, programs, and policies. Journal of Crop Improvement. 2012;**26**:39-59

[65] Gauchan D, Shrestha R. Community based seed systems for agrobiodiversity and resilient farming of smallholder agriculture in South Asia. In: Strengthening Seed Systems and Farmers' Rights for Improving Food and Nutrition Security in South Asia. Dhaka, Bangladesh: SAARC Agriculture Centre (SAC); 2020. pp. 29-53

[66] Mohapatra BK, Veettil PC, Kumar A, Kumar V. Rice-fallow management in eastern India: Challenges and opportunities for enhancing system productivity and profitability. Economic Affairs. 2022;**67**(05):859-867

[67] Singh S, Kumar N, Praharaj C, Singh N. Agrotechnology for Enhancing Pulses Production in Rice-Fallows. ICAR-Indian Institute of Pulses Research, Kanpur: Technical Bulletin; 2017

Chapter 6

Nutritional Value of Some Algerian Legumes from the *Medicago* Genus

Alane Farida and Chabaca Rabha

Abstract

In this work, we studied about some populations of spontaneous annual alfalfa from Algeria: *Medicago ciliaris* (Population.: S and C); *Medicago intertexta* (Population.: I); *Medicago truncatula* (Population.: Tr); *Medicago polymorpha* (Population.: Poly) and incidentally *Medicago granadensis*, *Medicago muricoleptis*, and *Medicago sativa*. All crops are grown under rainfed conditions. The vegetative parameters were harvested at the early flowering stage and chemical analyses: nitrogenous matters (CP), walls, and mineral matter were carried out at the early flowering stage and *in vitro* digestibility for some of these populations is also carried out. The number of stems per m^2 measures the vigor of plant development, approaching on average 1000 units with a reasonable standard deviation of 0.33 on 21 populations. Tr 27 stands out with 1340 stems/m^2 close to the control *M. muricoleptis* (1300 stems), and the minimum recorded is for another *M. truncu-lata*, Tr238: 410 stems. The 2015 crop, although their growth and development were weaker, their stems had more leaves than stems compared to that of 2013. The average *in vitro* digestibility corrected by the control of the *in vivo* reference digestibility varies from 73.37 to 77.14%.

Keywords: annual alfalfa, morphometry, chemical composition, digestibility, plant genetic resources

1. Introduction

Following independence, through successive agricultural development plans, up to the national agricultural development plan in 2000 and its extensions, national rural agricultural development plan, regional rural agricultural plan, and the national regional agricultural development fund for financing, the search for food security has always been the objective.

Among the dormant sources of productivity, the most important is the absorption of fallow land [1–3]. Indeed, Algerian agriculture is characterized by a traditional agricultural system based on the triptych: wheat-fallow-livestock (brought by the Romans almost 2000 years ago). In 1981, fallow represented 47% of the land and still 43% (more than 3 million ha) today [3, 4]. For grazing animals, it would only provide between 200 and 360 fodder unit/ha [1, 2], while the deficit in fodder units amounts to 3–4 billion UF out of a need of 10–12 billion.

For years, voices have been raised to encourage the profession to adopt a cereal alfalfa rotation, which has the advantage of increasing wheat yield at lower cost (alfalfa as an antecedent enriching the soil with nitrogen), but also that of fodder unit production that could reach up to 1500 units/ha [5]. An animal weight gain of 400 kg without counting the weight of the wool is indicated by Gachet [6].

This production system called Ley-Farming has been, and continues to be, very successful in Australia. Experimented in Algeria in the 1970s, with Australian alfalfa varieties, unsuitable for Algerian conditions, the results were not convincing. Hope then focused on the country's very numerous species and populations of spontaneous alfalfa. For 40 years, they have been inventoried and harvested and their qualities have been studied, in order to offer the best ecotypes to a selection program with a view to obtaining local varieties intended to be used in the alfalfa-cereal rotation to absorb the fallow all by intensifying agricultural production: killing two birds with one stone.

It is within this ambitious framework that our work begins with the heavy work of prospecting and harvesting carried out by our predecessors already in the nineteenth century (1873) by Urban Trabut (in Trabut) [7]. But also and above all for almost 50 years: the 70s and 80s as mentioned in Refs. [8–11]. This fruitful prospecting still continues, since, recently, Boulaacheb *et al.* [12] identified *M. sativa* L in DjebelMegriss, north of Sétif in fields and bushes as well as two annual species (*M. hispida* and *M. orbicularis L*). In another plant survey in DjebelYoucef and DjebelZdim south of Sétif, Chermat *et al.* [13] recorded seven species of annual alfalfa (*M. hispida, M. minima, M. tuberculata, M. laciniata, M. secondiflora, M. intertexta,* and *M. tribuloides*).

At the same time, numerous works were devoted to the adaptation of foreign varieties in our country before independence (works of Laumont [14] and Couranjou [15]) and after independence by research establishments such as: national higher agronomic school, technical institute for field crops (ITGC), national agronomic research institute, which for the latter cases, probably, did not allow constant support to advance research toward obtaining varieties from local populations adapting to various pedoclimatic conditions of the territory: at least in the North of the country, because in the South local varieties are cultivated on a large scale with success (Aoulef, In Salah, El-Menia, Tamentit and Tamacine, Magali, Lodi, Chott, as witnesses).

This work, modestly, aims to continue research on the potential of our local populations of annual alfalfa with a view to submitting the best for selection, the final stage of this journey that began 50 years ago. It comes largely from the research work that we have been carrying out for 15 years at national agronomic research institute (INRA), but started at national higher agronomic school (ENSA) as part of Magister. Our objective at this stage of the work is not to propose populations of alfalfa to be popularized but to provide phenological, biometric, and nutritional parameters capable of being integrated into a selection program in order to configure, according to needs, the variety which would come out.

After a brief overview of plant classification and a more in-depth look at legumes, we will discuss the importance of legumes in the human economy, focusing more particularly on forage legumes from the *Medicago* genus, which are more particularly annual.

We present the collection of species and populations of spontaneous alfalfa from Algeria, collected and identified by national institutes (INRAA, ITGC, etc.) and school (ENSA) from which we will extract those whose characteristics and qualities will be studied under the 'corner [16]:

- Biomass production parameters,
- Linked to different phenological and biometric characteristics;
- Their chemical composition, and their nutritional value for the animal.

Our study methods having been specified, we will develop our results and discuss them.

2. Materials and methods

2.1 Plant material, sowing device, cultivation management

The plant material (species and populations of alfalfa) used in this work is taken from the collection of the Ecole National Supérieure Agricole which has been built up over the years (**Table 1**).

The annual alfalfa species studied are categorized according to their color and size into three groups:

- Species with large seeds, *M. intertexta, M. ciliaris,* black in color,
- Species with medium seed weight, *M. muricoleptis*, black in color,
- Species with small seeds, *M. polymorpha, M. truncatula*, yellow-brown in color.

The species	Population code	1000 grain weight (Species)	Seed color	Native country
M. ciliaris	C_2	13.21 g	Black	Algérie (2004)
	C204			
	C52			
	C59			
	S3			
	S5			
M. truncatula	Tr27	3.58 g	Yellow brown	Algérie (2004)
	Tr55			
	Tr221			
	Tr238			
	Tr 334			
	Tr407			
M. intertexta	I11	15.77 g	Black	Algérie (2004)
	I31			
	I4O			
	I107			
	I756			
Witness	—	3.90 g	Black	Australia (−)

Table 1.
Codes and origins of the populations studied as well as the average weight of 1000 grains.

In legumes, there is often a relationship between the coloring of the seed coats and the percentage of hard seeds [17]. The seed shape of all species is kidney-shaped. The first test was installed on 13/12/2012 (for data collection in 2013) and the second on 24/12/2014 (for data collection in 2015) at the ENSA experimental farm.

The test is of the complete randomized block type with three repetitions. Each block is spaced 1.50 m apart from the other for the first test and 60 cm for the second test. The surface area of the tests in 2013 was 3585.6 m^2 (228 m x 15.77 m) and 31,488 m^2 in 2012 (5248 m x 6 m). The tests were carried out dry and permanently maintained, against weed infestations.

2.2 Soil analysis

Soil analyses of samples taken from the soil of the two plots at a depth of 20 cm and 40 cm (P20 and P40) before sowing method [18] and AISS (International Association of Soil Science) carried out in 2012 and repeated in 2013 show a loamy-sandy texture (clay: 23%, sand: 41%, and silt 36%), for the two horizons 20 and 40 cm. This soil has a neutral pH (potentiometric method on an earth/liquid suspension equal to 1/2.5) to slightly alkaline (>7.23), a very low electrical conductivity (C.E. < 0.2dS/m: method: soil - 1504: total conductivity on extract 1/25 (m/v) at 25°c), indicating the absence of salinity, a limestone rate of 7% (moderately calcareous soils: Bernard calcimeter volumetric method).

For fertility parameters, the soil is well provided with organic matter (method: Walkley-Black): respectively 5.53% and 5.15% for the surface horizon and the subsurface horizon. This richness is the result of burying organic amendments.

The total nitrogen rate (method: Kjeldahl) of the surface and subsurface horizon is respectively 0.11 and 0.09, and this result is that of a soil very poor in nitrogen (**Table 2**).

The soil on both plots is poor, and we have not made any amendments. The analyses were carried out at the soil science department of ENSA in 2013 and 2015. The comparison with an experimental soil for alfalfa trials reported by Larbre *et al.* [19], whose previous crop of sorghum, is comparatively as follows (in ppm):

- Organic matter (in %): 3.21
- Phosphorus (ppm): 110
- Potassium (ppm): 288
- Magnesium (in ppm): 166

2.3 Climatic conditions

Table 3 shows the climatic conditions for the two years 2013 and 2015. The average temperature in the month of May for 2013 and 2015 was comparable (around 18°C) but colder in January and February in 2015 (9.4°C compared to 11.7 for 2013). On the other hand, while on average, for the first 4 months of 2013, 83 mm of rain fell, the quantity is only 69 mm in 2015 with a strong deficit in the month of May for 2013 and a near drought in April for the year 2015. More details are given in **Table 3**.

Settings	Year 2014		Year 2012	
	P 20	P40	P20	P40
—	—	—	10.05	6.85
Average	**0.7220**		8.67	
Potassium mg/100 g	0.1750	0.1621	5.62	3.57
Average	**0.1685**		**4.59**	
Ph	7.22	7.75	7.23	7.0
Average	**7485**		**7115**	
CEE	0.1526	0.0654	—	—
Average	**0.109**			
Azote %	0.1121	0.0934	0.075	0.058
Average	**0.1027**		**0.0665**	
Organic material (%)	—	—	1.15	0.68
CaCo3 (%)	73.330	70.000	2.6	—
Average	**71.665**			
Argile (%)	21.2500	21.5000	25.0	25.5
Average	**21.375**		**25.25**	
Limon (%)	40.4900	39.7200	41.89	55.66
Average	**39.605**		**48.77**	
Sable (%)	38.2500	38.7800	27.75	8.06
Average	**38.51**		17.90	
Texture	L	L	LA	LA

Bold in tables indicates means and standard deviation.

Table 2.
Sown soil analysis results.

Month	2013		2015	
	Average temperature in °c	Rainfall in mm	Average temperature in °c	Rainfall in mm
January	11.7	120.4	9.4	98.9
February	11.5	89.8	9.6	80.0
March	16.3	45.9	14.8	96.3
April	15.6	76.0	17.7	1.6
May	17.8	5.3	18.8	28.4
June	13.0	71.5	22.4	14.4

Table 3.
Climate data for 2013 and 2015.

2.4 Data records for vegetative parameters

Vegetative parameters are important to know in order to discriminate between alfalfa populations with a view to obtaining high-performance local varieties in the medium term, adapted to the needs of breeders. Thus, the leaf-to-stem ratio (F/T) is known to be related to the nutritional value of the plant as is the number of stems for biomass production. The production of green and dry matter, the width of the branches, the number of stems per square meter, the length of the cycle are correlated. The % viability is given by the following formula:

$$\%\text{viability} = \text{number of plants mowed} / \text{number of seeds sown} \times 100$$

These data are taken at the start of flowering stage for 23 days: March 30 to April 21, 2015, explained by the completely logical variations in precocity observed, between blocks, between species, and between populations of the same species.

The instruments and tools used for the different operations are: pruners, measuring tapes, and scales. About 5-kg paper bags receive the collected samples.

- The mowing height chosen is 8–10 cm from the ground to avoid exhaustion of still young alfalfa plants and contamination by soil. This height is also that of animal grazing.

- Mowing begins after sunrise between 10 a.m. and 10:30 a.m. to minimize humidity on the plant.

- Weighing (whole plant, stems, and leaves separately) takes place on the same day as harvesting as well as putting it in the oven to measure the dry matter.

- The precocity of alfalfa species and populations is calculated in the number of days at the stage: beginning of flowering.

2.4.1 Evaluation of the nutritional value of alfalfa

Good quality alfalfa, in terms of content of chemical components and nutritional value, is strictly essential since the aim is to make it into nutritious fodder, knowing also that a good diet can partly compensate for a genetic deficiency.

In ruminants, the digestibility of forage is the most significant marker of its nutritional value [20].

For more than a century, indirect methods, called laboratory methods, which are faster and much less expensive, have been proposed. Among these, the most used microbiological test worldwide is inspired by the method recommended by Tilley and Terry [21]. It gives reliable digestibility values that correlate very well with the reference *in vivo* method [22, 23]. The samples that will be used for this test come from the 2015 crops developed above. Its total accomplishment requires:

- A Bain Marie system with 100-ml test tubes,

- An artificial saliva buffer solution

- A solution of trace elements is added at the end to the artificial saliva

- A control forage whose *in vivo* digestibility is known with precision and accuracy, obtained with the international reference method [24], and which gave an average of 61.36% ± 2.61. It is used to assess the activity of the rumen juice

- *Rumen inoculum* (called "rumen juice").

Products	Quantities (g)
$Na_2HPO_4, 12H_2O$	23.25
$NaHCO_3$	24.5
NaCl	1.18
KCl	1.43
$CaCl_2$	0.1
$MgCl_2$	0.15

Table 4.
Composition of Mc Dougall artificial saliva (for 0.5 liter of solution) [25].

Elements	Quantity (mg)	Salts used	Weight of salt (g)
Zn	300	$ZnSO_4 7H_2O$	13.196
Co	3	$CoCl_2, 6H_2O$	0.012
Mn	615	$MnSO_4, H_2O$	1.889
Fe	575	$FeSO_4,7H_2O$	2.862
Se	0.6	Se	0.0006
Cu	55	$CuSO_4,5H_2O$	0.216
Mo	0.7	$MO_2 (NH_4)_5, 5 H_2O$	0.009
I	1.4	I	0.0014

Table 5.
Composition of the trace element solution (l).

Details of their composition are given in **Tables 4** and **5**, respectively.

To this 0.5 liter of mother solution are added 2 liters of distilled water and 5 mL of trace elements. The pH is adjusted with CO_2 gas.

All results were subjected to simple descriptive statistical analysis: mean, standard deviation, coefficient of variation, and analysis of variance (ANOVA).

3. Results and discussion

3.1 Precocity of populations

The notion of precocity is understood by specialists as the arrival at the heading stage for grasses and at the beginning of flowering stage for legumes. In the latter, the early flowering stage corresponds to 5–10% of the stems examined on a 1 m line with at least one flower in bloom [26].

The values that appear in **Table 6** result from this definition. They do not call for any particular comments. Variations between populations of the same species are not very discriminating, as one might expect, nor between species. Thus, the coefficient of variation varies from 3 to 6%, although 71% of the variations observed are explained by precocity ($p < 0.05$). However, the species *M. ciliaris* would be 5 to 6 days earlier compared to *M. truncatula* and *M. intertexta.*

Species	Populations	Precocity(day)
M. ciliaris	C204	116 a
	C2	113 ac
	C11	116 a
	C58	121 b
	S7	114 ac
	S15	113 ac
	S5	110 ac
	S3	116 a
Average		**115.0 a**
Standard deviation		**3.35**
CV		**0.03**
M. truncatula	Tr238	117 a
	Tr27	130 b
	Tr334	127 b
Average		**124.83 b**
Standard deviation		**6.90**
CV		**0.06**
M. intertexta	I52	130 ac
	I756	125 b
	I107	124 b
	I58	127 b
	I31	127 b
	I11	126 b
Average		**126.39 b**
Standard deviation		**3.19**
CV		**0.03**
Overall average		**120.43**
Standard deviation		**6.47**
CV		**0.05**
RSE		**3.82**
R2		**0.71**
P		**0.0001**

a, b, c homogeneous groups.

Table 6.
Precocity of the different populations studied.

3.2 Biometric parameters determining dry matter production

3.2.1 Culture 2013

The results appear in **Table** 7. Of the 80 seeds sown per line and per population, on average, 19 plants remained for data collection. The average viability percentage is 26% for our endemic populations. It is comparable for *M. ciliaris, M truncatula, and M. polymorpha* (24 to 25%) and a little higher of 15% for *M. intertextra* (29%). The

Species	Populations	Number of plants/m^2	% de viabilities	Number of stems per m^2	Stem height (cm)	Green material g/m^2	Dry matter g/m^2	Leaves on stems in dry
M. ciliaris	S5	17.67	22.08	209.33	18.00	160.00	53.79	4.33
	S15	16.50	20.62	154.00	13.50	91.20	12.56	2.43
	C2	25.00	31.25	244.00	19.50	129.61	27.25	4.02
	S7	16.00	20.00	260.00	22.25	205.46	37.06	3.16
	S3	11.00	13.75	156.00	20.00	80.00	32.00	5.00
	C11	16.00	20.00	276.00	18.00	244.40	47.68	3.21
	C204	14.00	17.50	136.00	23.00	158.04	33.00	3.64
	C58	13.67	18.12	189.33	20.33	252.87	91.57	2.33
	average ± sd	**16.23 4.11**	**20.42 5.05**	**203.08 52.89**	**19.32 2.95**	**165.20 65.09**	**43.27 25.19**	**3.51 0.93**
	CV	**0.25**	**0.25**	**0.26**	**0.15**	**0.39**	**1.29**	**0.26**
M. truncatula	Tr238	19.00	23.75	312.00	18.50	256.54	74.06	2.81
	Tr334	14.50	18.13	209.33	13.00	80.00	51.72	1.03
	Tr27	7.00	18.50	172.00	19.00	140.00	36.00	1.28
	average ± sd	**13.50**	**20.13**	**231.11**	**16.83**	**158.85**	**53.93**	**1.86**
		6.06	**3.14**	**72.50**	**3.33**	**89.77**	**19.13**	**0.96**
	CV	**0.45**	**0.16**	**0.31**	**0.19**	**0.56**	**0.35**	**0.52**

Species	Populations	Number of plants/m²	% de viabilities	Number of stems per m²	Stem height (cm)	Green material g/m²	Dry matter g/m²	Leaves on stems in dry
M. intertexta	I253	28.00	35.00	420.00	38.33	1072.00	196.00	2.07
	I756	23.67	23.88	353.33	36.00	749.33	141.33	1.94
	I107	15.00	25.79	314.67	40.67	942.67	174.67	1.66
	I11	21.67	27.08	332.00	27.00	41.50	166.00	2.53
	I58	15.33	19.17	248.00	29.67	530.67	117.33	1.55
	I31	18.50	25.00	290.00	28.33	113.33	453.33	1.32
	I52	21.67	27.08	420.00	44.00	1277.06	260.00	1.11
	average ± sd	**20.55**	**26.14**	**339.71**	**34.86**	**675.22**	**215.52**	**1.74**
		4.65	4.75	**64.09**	**6.61**	**471.68**	**114.15**	**0.48**
	CV	**0.23**	**0.18**	**0.19**	**0.19**	**0.70**	**0.50**	**0.28**
Overall average	—	**16.76**	**22.23**	**257.97**	**20.34**	**362.31**	**117.57**	**2.37**
JemalongTémoin	**average ± sd**	**10.00**	**12.50**	**152.00**	**17.67**	**103.65**	**25.60**	**1.72**
		2.65	**0.88**	**31.24**	**1.15**	**32.15**	**11.78**	**0.94**
	CV	**0.26**	**0.26**	**0.21**	**0.06**	**0.31**	**0.46**	**0.54**

Bold in tables indicates means and standard deviation.

Table 7.
Biometric parameters determining dry matter production (2015).

behavior of the two witnesses for this viability criterion is mixed. Based on 26% average viability for our populations, *M. muricoleptis* (Aus 106 cultivar of Turkish origin) is better at 33% (+ 21%). As for the Australian cultivar (Jemalong), it almost did not germinate (**Table 8**).

The height of the stems, significant parameters of biomass production on average, is distributed as follows: 57, 47, 42, and 22 cm respectively for *M. intertextra, M. ciliaris, M. polymorpha*, and *M. trancatula*, which gives us almost the same hierarchy for the production of dry matter with quantities of: 751, 468, 502, and 234 g/m^2. The witness *M. muricoleptis* with a stem height and dry matter production of 41 cm and 534 g/m^2 respectively is very honorably placed. The height of the stems is in fact a well-known indicator of dry matter production, just like the number of stems.

As for the leaf-stem ratio (F/T), accepted as a reliable criterion of digestibility, *M. trunatula* (with an average value of 2.33) would carry more leaves than stems, compared to *M. ciliaris* (1.55), *M. intertexta* (1.30), and *M. polymorpha* (1.22) and *M. muricoleptis* (1.44) (**Table 8**).

3.2.2 Culture 2015

The same data collection protocol was applied to the 2015 crop which, however, only included the Jemalong witness. The results appear in **Table 7**.

The first observation is that the values are lower than for the 2013 crop. The calculations made from **Tables 9** and **10** reveal: 11% for the number of plants; 14% for the viability rate; 73% for the number of rods/m^2; 52% for stem height; and 76% for MS production. On the other hand, an increase of 55% is noted for the F/T, which admits that the alfalfas of the 2015 crop, although their growth and development were lower, their stems carried more leaves than stems compared to that from 2013. Perhaps not surprisingly, these stems were two times shorter. The number of stems being an indicator of the vigor of the crop.

3.2.3 Summary of the years 2013 + 2015 for dry matter production

We have synthesized the two years of cultivation in the sense that each year represents a repetition, and the results are presented. These average data do not call for any particular comments, other than that the 2013 culture, with its higher performances compared to that of 2015, prints the results. All the parameters studied clearly discriminate between alfalfa populations with R2 values varying between 25 and 65%. This is for F/T where variations are less well explained by alfalfa populations (weighing and leaf loss are in fact sources of error). On the basis of this synthesis, we have drawn up the correlation matrix between the parameters that determine biomass production (**Table 9**).

It appears that the correlations cover all possibilities existing in statistical terms: positive connections, significant or not—negative connections, significant or not.

The connections observed do not pose any particular constraints on understanding. So:

- It is logical that the number of stems/m^2 (NT) varies in the same direction as the number of plants/m^2 (NP) also, when the latter increases, NT does the same,

- NP is also positively linked to stem height (HT) and the F/T ratio, although this link is not significant. However, it can be assumed that under high seed density

Species	Populations	Umber of plants/m²	% of viability	Number of stems per m²	Stem height cm	Green material /m²	Dry matter g/m²	Leaf on stems Sec
M. ciliaris	C52	17.50	21.88	1165.0	40.34	3350.0	446.7	1.32
	C204	17.00	21.25	746.7	49.89	3670.0	567.9	1.73
	S5	21.50	26.88	950.0	43.93	2926.7	401.1	1.51
	C2	23.17	28.96	1166.7	59.76	5056.7	574.2	1.38
	S3	19.17	23.96	996.7	53.08	3233.3	451.9	1.37
	C58	18.67	23.33	1125.0	35.94	3050.0	367.3	1.68
	Average. ±DS	**19.50**	**24.38**	**1025.0**	**47.16**	**3547.78**	**468.21**	**1.55**
		2.38	**2.99**	**163.4**	**8.76**	**782.72**	**85.54**	**0.17**
	CV	**0.12**	**0.12**	**0.16**	**0.19**	**0.22**	**0.18**	**0.11**
M. truncatula	Tr334	29.67	37.08	1610.0	22.83	2123.3	469.4	2.25
	Tr55	15.33	19.17	1128.3	29.23	1500.0	204.8	2.31
	Tr238	5.50	20.60	410.0	10.53	1392.1	41.3	2.70
	Tr407	15.75	19.69	870.0	17.50	2015.0	227.2	2.15
	Tr27	25.83	32.29	1340.0	32.11	1674.2	247.2	2.27
	Tr221	12.33	23.13	580.0	21.07	1852.5	216.8	2.30
	Average. ±DS	**17.40**	**25.33**	**989.7**	**22.21**	**1759.52**	**234. 37**	**2.33**
		8.90	**7.53**	**457.1**	**8.76**	**288.41**	**137.06**	**0.19**
	CV	**0.51**	**0.29**	**0.46**	**0.35**	**0.16**	**0.58**	**0.08**

Species	Populations	Umber of plants/m^2	% of viability	Number of stems per m^2	Stem height cm	Green material /m^2	Dry matter g/m^2	Leaf on stems Sec
M. intertexta	I11	10.67	20.00	1036.7	50.51	6593.3	622.3	1.31
	I756	24.50	30.63	958.3	60.69	4920.0	706.7	1.39
	I107	24.83	31.04	1116.7	54.15	6630.0	610.6	1.09
	I253	22.50	28.13	1030.0	63.71	6545.0	849.8	1.42
	I31	26.33	32.92	1268.3	50.23	5826.7	798.4	—
	I52	27.33	34.17	975.0	62.02	6730.0	920.4	1.17
	Average ± DS	**22.69**	**29.48**	**1064.17**	**56.89**	**6207.50**	**751.35**	**1.30**
		6.12	**5.08**	**114.47**	**6.00**	**709.50**	**125.67**	**0.14**
	CV	**0.27**	**0.17**	**0.11**	**0.10**	**0.11**	**0.17**	**0.51**
M. polymorpha	Poly205	18.00	22.50	1091.1	58.65	636.7	856.2	1.31
	Poly27	13.67	25.62	433.3	26.25	203.6	148.9	1.13
	Average ± DS	**15.83**	**24.06**	**762.22**	**42.45**	**419.91**	**502.5**	**1.22**
		3.06	**2.21**	**465.12**	**22.91**	**435.8**	**500.1**	**0.13**
	CV	**0.19**	**0.09**	**0.61**	**0.54**	**1.04**	**0.99**	**0.11**
Overall average	—	**18.85**	**25.81**	**960.28**	**42.17**	**2983.68**	**489.12**	**1.53**
M. muricoleptis	**Average ± DS**	**26.33**	**32.92**	**1300.0**	**41.20**	**3626.67**	**534.6**	**1.44**
Aus 106		**3.79**	**4.73**	**173.49**	**15.59**	**611.58**	**90.18**	**0.02**
Witness 1	**CV**	**0.14**	**0.14**	**0.13**	**0.38**	**0.17**	**0.17**	**0.01**
Jemalong witness 2	**Average**	**6.5**	**8.0**	**0**	**0**	**0**	**0**	**0**

SD: standard deviation; CV: coefficient of variation.
Bold in tables indicates means and standard deviation.

Table 8.
Biometric parameters (2013) determining dry matter production.

	Number of plants/ m²	Number of stems per m²	Stem height (cm)	Leaves on stems in dry	Green material g/ m²	Dry matter g /m²
Number of plants/ m²	1					
Number of stems per m²	0.685***	1				
Stem height (cm)	0.188	0.481***	1			
Leaves on stems in dry	0.127	−0.167	−0.291*	1		

**correlated*
***highly correlated*
****very highly correlated*

Table 9.
Correlation matrix between biometric parameters.

conditions, the binding could be negative. On the other hand, it is logical that the production of dry matter (DM) under normal growing conditions is a function of the NP (therefore indirectly of the sowing density).

- The NT is positively correlated with the HT: We can think that the NT increasing with age leads in the same direction to the growth in height of the stems. Likewise, NT is in negative connection with F/T (but $p > 0$); indeed if the NT increases with age, the number of leaves could decrease. The NT is logically in a positive and significant connection with the production of biomass, represented here by MS;

- HT is negatively and significantly related to F/T; in fact, HT, being a function of the age of the alfalfa, impacts the stem-leaf distribution of the plant. The sign of the connection would indicate that when the stems progress in height, they take precedence over the weight of the leaves. We find the same trend (but $p > 0$) between F/T and MS: When the production of MS increases, the share of stems in this biomass increases, causing F/T to fall.

- Finally, the very strong correlation (0.934) recorded between green and dry biomass does not call for comment, the latter being the result of the former.

3.2.3.1 Chemical composition and digestibility of the alfalfa populations studied

3.2.3.1.1 Chemical composition

The chemical composition of the populations of the two 2013/2015 trials is close to the 2007 trial.

The total nitrogenous matter of the populations varies between 12.47 (I125) and 37.43 (C35) with an average of 26.09%. The CP contents of the introduced species that vary from 22.83% (AUS100) to 28.79% (AUS103) are lower than those of the perennial population El Méniaa (31.5%).

Thus, the *M. ciliaris* (C) is with a higher nitrogenous content of 32.49% and a moderate fiber content (NDF: 30.08% and CB1: 18.18%), while the species poorest in walls are

Populations	MS2%	MM%	MAT%	CB1 (%MS)	NDF%	ADF%	Hc%	CB2%	Ln%
S15	98.79	15.26	28.83	12.70	27.56	16.80	10.76	11.55	5.06
S9	98.75	15.33	25.27	20.51	35.63	22.77	12.86	7.18	10.67
S6	98.81	14.82	30.51	16.72	20.39	14.11	6.28	7.62	7.37
S8	98.36	17.35	23.94	16.15	41.53	18.90	22.63	10.00	8.97
S5	98.80	15.68	33.77	17.62	31.68	14.23	17.46	8.34	7.62
S4	99.16	14.30	24.94	20.76	37.80	18.56	19.25	13.16	5.90
S11	98.69	16.98	28.34	20.50	36.36	15.27	21.09	7.38	8.88
S7	98.59	15.87	26.49	16.75	36.21	18.05	17.25	9.96	9.52
S3	98.75	17.22	23.49	15.14	33.36	17.46	15.84	8.21	7.89
POL205	98.66	19.71	32.94	15.71	33.69	16.82	17.79	10.41	7.46
POLY58	99.40	18.80	25.68	23.84	36.77	19.43	17.34	15.04	8.18
POL218	99.33	20.40	24.56	17.20	42.63	24.15	18.48	6.86	6.11
POLY236	99.10	16.82	28.05	22.96	54.31	25.96	28.35	11.23	16.66
POLY60	99.18	17.59	35.44	18.64	38.17	17.89	20.27	18.90	6.99
POLY27	99.23	14.97	22.79	20.59	40.12	16.63	23.50	5.54	4.65
TR55	98.92	15.86	29.15	20.20	45.62	24.48	21.14	5.22	15.54
TR201	98.96	21.14	28.28	15.48	43.29	25.04	18.26	1.29	16.67
TR238	99.21	16.84	26.59	20.88	48.97	23.55	25.43	8.54	21.52
TR27	99.02	14.06	21.34	28.17	43.69	28.98	14.71	7.84	18.39
TR221	99.33	21.55	21.68	22.15	38.34	20.03	18.31	11.97	12.09
C242	98.95	15.84	31.24	16.85	45.11	20.96	24.15	13.58	9.82
C2	98.97	15.55	30.01	15.78	38.07	17.77	20.31	8.82	3.45
C11	98.88	13.60	28.83	18.39	26.04	12.17	13.87	6.80	3.94
C58	99.04	15.73		16.59	21.37	15.74	5.63	12.26	5.63
C204	99.06	20.12	27.48	21.06	54.80	18.64	36.16	12.05	15.13
C35	98.84	20.30	37.43	19.96	49.00	23.04	25.97	6.67	8.78
C52	98.95	14.59	28.85	18.67	37.75	19.30	18.45	9.98	9.90
I11	99.14	15.96	26.26	21.06	35.06	21.78	13.28	12.67	15.22
I31	99.07	12.68	24.11	28.46	74.91	27.58	47.32	12.22	15.22
I756	99.22	14.28	18.89	21.96	39.15	22.09	17.06	10.50	13.71
I58	99.05	16.27	22.91	24.50	36.58	21.68	14.90	17.67	6.24
I253	99.14	12.47	19.57	27.17	44.02	29.32	14.70	19.88	6.13
I52	99.42	13.31	28.50	27.01	40.49	21.42	19.08	14.30	7.96
I107	99.44	15.13	17.18	24.92	38.48	20.80	17.68	14.96	4.82
I755	99.16	23.81	22.70	15.57	42.42	16.81	25.61	5.90	10.10
AUS100	99.57	17.79	22.83	18.92	35.30	18.47	16.83	11.20	6.10
AUS106	99.05	16.49	26.06	20.40	39.18	21.87	17.31	12.33	6.35

Populations	MS2%	MM%	MAT%	CB1 (%MS)	NDF%	ADF%	Hc%	CB2%	Ln%
AUS103	99.39	18.05	28.79	22.70	41.69	21.33	20.36	4.29	7.08
AUS107	99.36	14.79	28.32	20.95	46.54	15.34	31.20	4.57	11.63
ElMéniaa	99.30	14.29	31.51	16.62	33.46	18.40	15.06	11.06	7.71
Tamentit	99.34	14.81	23.47	24.34	31.24	18.33	12.91	8.36	8.00
Temacine	99.05	16.40	25.56	21.35	30.87	19.09	11.19	11.0	8.83
Moyenne	96.72	16.05	26.09	19.65	39.25	19.78	18.20	10.10	9.28

Table 10.
Chemical composition of populations of 42 annual alfalfas studied in 2007.

the *M. ciliaris* (S) (ADF: 17.46%, CB1: 17.43%). The highest rate of indigestible chemical elements (lignin) is found in *M. intertexta* (6.10%). The correlation of fibers and nitrogenous matter is negative (CB1: r = −0.45** and ADF: r = −0.36*) (**Table 10**).

In alfalfa (as in all forages), the chemical composition follows the stage of development of the plant. Thus, Guy [24] notes the increase in the MAT content from 24.6% to 13.90% from April 20 to June 1. Similarly, Mauriès [27] indicates that, when moving from the vegetative stage to the flowering stage, the MAT content dropped from 27 to 15%. At the same time, the leaf-to-stem ratio decreases, while the DM content increases. This ratio can go from 1.4 at the flower bud stage to 0.7 in full flowering.

This author also specifies that the weight of the leaves increases linearly with the height of the stems, while the weight of the stems increases as the square of the height. The chemical composition of the leaves being little variable, it is that of the stem which is responsible for the reduction in the leaf-stem ratio and therefore in the MAT content, since the content of the leaves varies between 21.9 and 24.4% compared to 11.9 and 15.0% [24]. Other factors of genetic variability between species or populations are added, such as leaf surface area. Thus, *M. falcata* cultivars have a leaf surface area of 98 m^2 compared to 138 m^2 for *M. sativa.* [24].

Little attention has also been paid to stem thickness, which is associated with the physical distribution of structural components including lignin. Thus, the less rigid stems are more digestible just as the base of the stem is more lignified than the top.

In vitro digestibility of dry matter from the populations studied:

The results appear in **Table 11** and the digestibility values on average 75%. The average digestibility per species is 77, 73, and 75% for *M. ciliaris, M. truncatula*, and *M. intertexta*.

The aim of analyzing a table of chemical composition of fodder as we did above is to be able to assess the nutritional value of the fodder through its content of the main components, which determines this nutritional value. Since the eighteenth century, crude cellulose (replaced more or less from the 1970s by NDF or ADF in light of the work of Van Soest [28]) has been recognized as a factor negatively impacting food value as well as lignin at a higher level. Its content in the plant increases with the age of the plant, while, on the other hand, the nitrogenous materials, a positive factor in the nutritional value of the fodder, quantitatively decrease with the age of the plant [27, 29].

On this basis, thousands of works around the world have established general models making it possible to achieve the nutritional value of fodder by introducing into these mathematical models, the chemical components of the fodder as explanatory

Species	Populations	Average trials	Average species
M. ciliaris	C2	78.75	77.14
	C58	77.20	
	S5	75.46	
M. truncatula	Tr238	69.94	73.34
	Tr334	76.75	
M. intertexta	I11	74.04	74.60
	I31	76.33	
	I52	76.34	
	I253	75.02	
	I58	73.98	
	I107	70.63	
	I756	75.88	
	Avg ± ET	75.3 ± 2.60	75.17 ± 2.34
	CV	0.034	0.031
	RSE	6.01	—
	R^2	0.51	—
	P	0.44	—
	—	**	—

R^2: coefficient of determination; RSE: residual standard deviation:
**P<0.05; significant*
***P<0.01; highly significant*
****P<0.001; very highly significant*

Table 11.
Digestibility value of the alfalfa populations studied.

variables of their nutritional value. For example internationally: INRA France (1978, 1988); and ARC, Great Britain (1980, 1993) NRC: USA (2001; 2007) and in Algeria, very recently [30].

3.2.3.1.2 Performance in DM and MAT (CP)

The quantity of DM produced per unit of soil used is an important nutritional and economic criterion. It contains all the nutrients that justified cultivation. Annual and perennial alfalfas are rich in nutritional compounds: minerals, nitrogenous materials rich in amino acids, carotenes, vitamins [31, 32] and minerals including calcium, phosphorus, manganese, magnesium and many important trace elements such as zinc, boron, and selenium [26].

We have previously commented on the correlations between the different parameters of DM production from biomass: The number of stems/m^2 and the height of the stems are significant explanatory variables (**Table 9**). The highest DM productions were found within *M. intertexta*: more than half of the populations (4 out of 7) exceed 4 tons of DM/ha.

The general average for all species combined is 2.8 tons (**Table 12**). The situation is different for MS production in Syria, Cocks *et al.* [33] measured the production of 6

Species	Populations	DM		Total nitrogen material (TNM)
		g/m^2	Tons/ha	Tons/ha
M. ciliaris	C2	300.75	3.01	*0.747*
	C58	229.45	2.29	*0.636*
	S5	227.46	2.27	*0.666*
	C52	446.70	4.47	—
	C204	333.62	3.34	0.827
	S3	385.97	3.86	0.852
	S15	12.56	0.13	0.032
	S7	53.79	0.54	0.156
	C11	12.56	0.13	0.029
Average species		**222.57**	**2.23**	**0.523**
M. truncatula	Tr238	57.66	0.58	*0.153*
	Tr334	260.56	2.61	*0.580*
	Tr55	204.84	2.85	—
	Tr407	227.23	2.27	—
	Tr27	141.58	1.42	0.326
	Tr221	216.78	2.17	—
Average species		**175.81**	**1.76**	**0.346**
M. intertexta	I11	331.89	3.32	*0.728*
	I31	625.85	6.25	*1428*
	I52	590.21	5.90	*1.386*
	I253	522.88	5.23	*1.132*
	I58	117.33	1.17	*0.280*
	I107	392.62	3.93	*0.800*
	I756	424.00	4.24	*1.086*
Average species		**429.25**	**4.29**	**0.977**
Overall average		**275.25**	**2.76**	**0.615**
Jemalong: T1	**CV**	**25.6**	**0.26**	**0.077**
Muricoleptis: T2	**Aus**	**534.57**	**5.35**	**1.246**

Bold in tables indicates means and standard deviation.

Table 12.
Theoretical production of DM and nitrogenous materials per hectare for the populations studied.

to 8 tons under normal rainy conditions (200–300 mm) and up to 9–10 tons in wetter conditions. These same authors, in Australia, with a view to a Ley farming project, compared the production of dry matter in three species of candidate fodder legumes: *Medicago littoralis, Trifolium subterranewn*, and *Trifolium repens,* the respective annual DM production was 9.9; 6.0 and 3.4 tons per hectare.

Papastilianou [34] for his part showed with annual Alfalfa in Cyprus (rainfall of 200–260 mm) that DM production could vary between 4.6 and 7.5 tons depending on the variety and bacterial strain. For M. *rigidula* for example, this quantity is between 5.0 and 6.0 kg ha/year.

In Tunisia, Seklani and Hassen [35] evaluated the DM production of 5 annual *Medicago: M. rugosa; M. polymorpha; M. orbicularis; M. truncatula; M. turbinata* at different sites throughout the country: field yield in DM varied between 1 and 5 tons.

In Morocco, Birouk *et al.* [36] on around twenty local populations and in 5 different regions and on local populations of alfalfa obtained average yields per cut in DM of 2.8 to 3.12 tons per hectare, that is, on the 3 cuts practiced 8.4 to 9.4 tons of MS. These values are close to those of Cocks *et al.* [33], obtained in Syria.

What emerges from these series of experiments in the Mediterranean basin is the very wide dispersion of MS yields. It is due to the variability of pedoclimatic conditions but also to the specific characteristics of the alfalfa varieties [37].

According to Sultan-Tubeileh [38], one of the fairly common causes of the variability of productive yields of alfalfa in the Mediterranean regions is the unadaptability and inefficiency of the strains of *Sinorhizobium meliloti* whose annual alfalfas are more demanding in terms of strain-specific than perennial alfalfa. According to this author, in addition to classic selection methods, a look at the quality of the suitability of the alfalfa-*Sinorhizobium meliloti* association is an important reserve for productive improvement of annual alfalfa.

Regarding the production of total nitrogenous materials of the alfalfa populations studied, the results appear in **Table 13**. The MAT content being little dispersed, respectively on average: 27, 25, and 23% for *M. ciliaris, M. truncatula,* and *M. intertexta* (**Table 12**) and being expressed in relation to the DM, it is therefore the quantity of DM produced, which discriminates the populations in terms of their production of nitrogenous materials. The species *M. intertexta* with almost 1 ton of MAT produces two times and three times more than *M. ciliaris* and *M. truncatula*, respectively. The population I52 exceeds 1.4 tons, follows: I31 (1400), I253, and I756 with each 1.1 tons.

What can we ultimately say about the quality of the populations studied on the two criteria: production of MS and MAT?

	Div	Lig	TNM	GT	NT	F/T
Div	1					
Lig	**-0.52***	1				
MAT	**0.63****	**-0.53***	1			
GT	0.05	0.33	−0.29	1		
NT	0.31	0.15	−0.01	**0.78*****	1	
F/T	0.21	−0.29	**0.63****	−0.34	0.22	1
LT	**- 0.51***	**0.55***	**-0.66*****	0.33	0.04	- 0.43

Div: in vitro digestibility; Lig: lignin; NT: number of stems/m^2; GT: stem size; TNM: total nitrogen material:
**p > 0.05; significant*
***p < 0.02; highly significant*
****0.01; very highly significant*
Bold in tables indicates means and standard deviation.

Table 13.
Correlation matrix of parameters determining the Div.

- For dry matter, on the basis of one harvest compared to 3 or 4 in the literature, we would be at a good level: with two harvests we would approach on average 6 tons annually,

- For MAT, production is very high with *M. intertexta* (up to 1.4 tons given that it is a single harvest). On average for the three species: 0.6 tons for a harvest is more than respectable.

3.2.3.2 Parameters that can impact digestibility

It is presented in the variable correlation matrix (R) (**Table 13**). Here, we find results already presented previously: Div correlated negatively with lignin and positively with nitrogen. The age of the plant being related to the width of the branches and its lignin content, we see it appear in the form of a negative R between Div and LT and positive between lignin and LT.

4. Conclusion

This work aimed to contribute to the deepening of knowledge in the species and populations of spontaneous annual alfalfa in Algeria.

Our work focused on six species of medics: (*M. ciliaris; M. truncatula; M. intertexta; M. polymorpha; M. sativa;* and *M. muricoleptis*), bringing together up to 22 populations. But, all of our investigations particularly concerned the three species: *M. ciliaris; M. truncatula* and *M. intertexta* whose availability of plant material was greater. Most of the phenological or biometric samples were taken at the beginning of flowering. Several points emerge from the results of this work:

- In addition to the very broad spectrum of genetic variability contained in our populations, the quality criteria for alfalfa are already at high levels in our alfalfa. Thus, the MAT content averages 27% for *M. ciliaris* (peak at 30% for the S5 population).

- For these three quality criteria, the Australian control *M. truncatula* (CV Jemalong), collected in Algeria and acclimatized in Australia, showed itself to be totally unsuitable with 0.3 tons of DM per hectare and infertile (0 g of seed)

- The M. *ciliaris* species is also well placed for the F/T ratio, another indicator of digestibility. There appears to be a presumption of good digestibility of the *M. ciliaris* species.

Finally, we provide the breeder with a whole series of phenological and biometric criteria and statistical combinations of this quality (for the zootechnician) and quantity (for the farmer) criteria:

- Number of stems per square meter, stem height (cm), MS g/m^2 F/T Dry; NDF, ADF, HC, ADL, MSg/m^2, MAT%MS; Div, lignin, MAT, stem size, number of $stems/m^2$, stem length; length of stems (cm), number of $stems/m^2$.

Author details

Alane Farida[1*] and Chabaca Rabha[2]

1 National Agricultural Research Institute, Algerian

2 National Higher Agronomic School, Algerian

*Address all correspondence to: alanefarida@hotmail.fr

References

[1] Nedjraoui D. Document FAO Les ressources pastorales en Algérie; 2003. [en ligne] www.fao.org/ag/agp/agpc/doc/counprof/Algeria/Algerie.htm

[2] Abbas K, Abdelguerfi A. Perspectives d'avenir de la jachère pâturée dans les zones céréalières semi-arides. Fourrages. 2005;**184**:533-546

[3] MADR. Le Renouveau Agricole et Rural en marche Revue et Perspectives; 2012. disponible on-line: http://www.minagri.dz/pdf/Divers/Juillet/LE_RAR-FR.pdf

[4] Yahioui A. Capital nutritionnel pour *Ovisaries*de différents supports alimentaires en zones agro-pastorales de l'ouest Algérien. [thesis] doctorat Ecole Nationale Supérieure Agronomique; 2011, p. 130

[5] Van Swinderen H. Développement et aménagement forestiers. Algérie-FAO-FO; 1973: SF/ALG/68/515

[6] Gachet JP. Résultats d'engraissement d'agneaux sur prairies de luzernes annuelles. Lab. Production fourragère et Zootechnie. Institut National de la Recherche Agronomique de Tunis; 1979, INRAT ISSN: 0365-4761

[7] Urban Trabut VL. Le Medicagosativa L. et le Phalaris stenoptera Hackel dans le Nord de l'Afrique. Leur origine hybride. Bulletin de la Société Botanique de France. 1914;**61**(5):XXX-XXXIV. DOI: 10.1080/00378941.1914.10832609. Available from: http://www.tandfonline.com/doi/pdf/10.1080/00378941.1914.10832609

[8] Adem L. Etude du comportement des annuelles (écotypes locaux et populations étrangères) dans les régions de Sétif, Médéa, Tiaret et alger. INA El Harrach: Thèse d'ingénieur Agro; 1974. p. 95

[9] Conesa AP, Hadj-Miloud D, Djeddou R. Etude écologique, Essais de comportement, Sélection (de luzernes annuelles). Alger: Lab. Agric, I. N. A. El-Harrach; 1974. p. 97

[10] Abdelguerfi A. Contribution à l'étude de la répartition des espèces locales des luzernes annuelles en fonction des facteurs du milieu (200 stations) liaison entre les caractères de ces 600 populations étudiées à Beni Slimane et leur milieu d'origine. Mémoire Ing. INA; 1976, p. 74

[11] Zeghida A. La rotation céréale-Medicago dans les zones à vocation céréale-élevage Revue céréaliculture, n°16; 1987, p. 52-56

[12] Boulaacheb N, Clement B, Djellouli Y, Gharzouli R. Les paysages végétaux du djebel Mégriss (Tell septentrional, Algérie): diversité des écosystèmes, richesse floristique, ampleur de l'anthropisation _ Laouer H4 Revue des Régions Arides-Numéro spécial; 2006. Actes du séminaire international « les Plantes à Parfum, Aromatiques et Médicinales » SIPAM. Les plantes médicinales du Djebel Megriss (Algérie, Nord Afrique)-Famille des lamiacées

[13] Chermat S, Djellouli Y, Gharzouli R, Djebel Zdimm en Algérie nord-orientale. Revue Internationale d'Ecologie Méditerranéenne. International Journal of Mediterranean Ecology. 2016;**42**(1):85-100

[14] Laumont PN. Orientation à donner à l'agriculture algérienne, pendant la

compagne 1941-1942. Doc.rec, Agri Calger. Bull. 1941;**48**:1-14

[15] Couranjou J. Essai de classification physiologique des luzernes en fonction de leur comportement et de leur adaptation en différents milieux algériens. Comportement phénologique comparé des divers types de luzernes suivant le milieu. Fourrages. 1965;**23**:127-142

[16] Lemaire G, Allirand JM. Relation entre croissance et qualité de la luzerne: interaction génotype-mode d'exploitation. Fourrages. 1993;**134**:183-198

[17] Côme D, Semadeni IA. Dégazage des enveloppes séminales lors de leur imbibition. III. Application à l'étude de la dureté des graine s *d'Hedysarumcoronarium L.* The Physiology of Vegetable Crops. 1973;**11**:171-179

[18] Atterberg A. Die Plastizitat der Tone. Int. Mitt. Bodenk. Berlin: Verlagfür Fachliteratur; 1911. pp. 10-43

[19] Larbre D, Coulmier D, Pacquetet H, Joya R. Etude de la courbe de réponse de la luzerne à l'apport de potasse et de soufre COOP de France; 2015; @ coopdefrance.coop. Visité le 15/10/2018

[20] Peyraud JL, Baumont R. Qualité des fourrages: de la plante à la ration alimentaire. Fourrages. 2002;**171**:241-251

[21] Tilley JM, Terry RA. A two –stage technique for the in vitro digestion of forage crops. Journal of the British Grassland Society. 1963;**18**:104-111

[22] Varel VH, Kreikemeier K. Comparison of *in vitro* and *in situ* digestibility methods. Journal of Animal Science. 1995;**73**:578.17

[23] Stern MD, Bath A, Calsanigha S. Alternative techniques for measuring nutrient digestion in ruminants. The Journal of Animal Science. 1997;**75**:2256-2276

[24] Guy P. Principes généraux de production de la semence de luzerne. Revue Fourrages. 1961;**5**:136-145

[25] McDougall E. Studies on ruminant saliva. The composition and output of Sheep's saliva. Biochemical Journal. 1948;**43**:99-109

[26] Andrieu J, Demarquilly C. Composition et valeur alimentaire des foins et des pailles. Les fourrages secs: récolte traitement. Paris: INRA; 1969. pp. 142-161

[27] Mauriès M. Luzerne: Culture, Récolte, Conservation, Utilisation; 2003

[28] Van Soest PJ, Wine RH. Use of Detergent in Analysis Fibruns Feed. An of Agrch; 1963, p. 466-829

[29] Alimentation des ruminants, editor. Institut National de Recherche Agronomique de France. France, Versailles: INRA; 1978. p. 597

[30] Chibani C. Eléments princeps pour une table de valeur alimentaire, pour le ruminant, des fourrages algériens: modèles de prédiction. Thèse de Doctorat. El Harrach (Alger): Ecole Nationale supérieure agronomique; 2013. p. 177

[31] Porqueddu C. Screening germplasm and varieties for forage quality: Cons traints and potentials in annual medics. In: Delgado I, Lloveras J, editors. Quality in Lucerne and Medics for Animal Production. Vol. 45. Zaragoza: CIHEAM Options Méditerranéennes: Série A. Séminaires Méditerranéens; 2000. pp. 89-98

[32] Alane F. Valeur nutritive des légumineuses fourragères cas des

luzernes (Genre Medicago). [thesis] Thèse de Magister. INA; 2007, p. 96

[33] Cocks PS, Mawlawi B, Sawmy-Edo H. Introduction of ley farming in Syrian villages. In: Christiansen S, Materon L, Falcinelli M, Cocks P, editors. Introducing Ley-Farming to the Mediterranean Basin. Proceedings of an International Workshop, Perugia, Italy. Montpellier, Cedex, France: CIHEAM Montpellier; 1993. pp. 52-64

[34] Papastilianou I. Nitrogen fixation by medics and their role in rotations and in marginal areas in Cyprus. In: Christiansen S, Materon L, Falcinelli M, Cocks P, editors. Introducing Ley Farming to the Mediterranean Basin. Proceedings, International Workshop, 26-30 June 1989, Perugia, Italy. Aleppo, Syria: ICARDA; 1993. pp. 99-105

[35] Seklani H, Hassen H. Contribution à l'étude des espèces spontanées du genre *Medicago* en Tunisie. Annales de I'INRAT. 1990;**63**(Fasc. 20):1-15

[36] Birouk A, Dattee Y, Sadiki M, Roumet P. Evaluation agronomique et adaptation de populations marocaines de luzerne (*Medicagosativa L.*). Agronomie. 1989;**9**(4):363-376

[37] Demarly Y. Les variétés de luzerne catalogue actuel et avenir de la sélection. Exposé présenté le 13 janvier 1966, lors de la réunion d'hiver de l'A.F.P.F (Association française de producteurs de fourrages), p. 11

[38] Sultan-Tubeileh K. Etude du comportement hivernal d'associations: Luzerne annuelles-rhizobium en régions méditerranées. Elaboration de la matière sèche et fixation d'azote. [thesis] Thèse de Doctorat. Université de Nancy ENSAIA-INPL. Global ETD Search; 2001. p. 167

Section 3

Legume Crops Breeding

Chapter 7

Breeding Legumes Crop for Low-Input Organic Farming under Rice Fallow

Anirban Roy, Suman Dutta, Tapash Dasgupta, Payabrata Roy, Satyaki Nandan Maity, Samita Saha, Sukamal Sarkar and Sourav Garai

Abstract

Enhancing the productivity of low-input farming needs managing nutrient availability through crop management and enhancing genetic potentiality of crops. Cereal-based mono-cropping systems need legume interventions to improve overall farm productivity. Developing climate-smart nutrient-responsive pulses for organic farming conditions for enhanced market linkage with value-added products enhances the sustainability of pulse addition. Breeding for intense root system, disease resistance, higher water and nutrient use efficiency, and quick growing habits with early flowering nature may produce a sustainable pulse under such slow nutrient-releasing soils. The addition of mungbean, lentil, chickpea, lathyrus, blackgram, and pigeonpea to the rice fallow coastal system may improve the total productivity and sustainability of the system. Pre-breeding with potential donors may generate a wide adaptable breeding population for easy selection of desirable recombinant using phenotyic, biochemical, and molecular marker. Wide genomic information and next-generation sequencing advancement for quick genetic gain is utmost required for low-input farming responsive variety.

Keywords: low-input responsive breeding, allele mining, legumes, climate resilience, organic breeding

1. Introduction

Challenges for growing crops for organic farming condition is characterized by low-input, unstable yield, heterogeneous environmental variation, and multiple complex interactions of organisms. Quality products remain one of the major drivers for organic product development. Organic production, considering organic farm management, organic inputs, and chemical-free cropping, causes a severe challenge for crop protection due to prevalent disease and pests. Major objectives for breeding crops for organic

IntechOpen

conditions include multidimensional. Organic farming is becoming more and more popular as a substitute for conventional agriculture because of its emphasis on ecosystem-friendly and sustainable practices. Organic farming practices have gained popularity as a sustainable and environmentally friendly approach to agriculture, with potential implications for the nutritional status of crops, including pulses. Pulses, which encompass a variety of leguminous crops like lentils, chickpeas, and beans, are known for their high protein content and nutritional value. Under organic farming, the nutritional status of pulses is influenced by several key factors, including soil fertility management, pest and disease control, crop rotation, and environmental sustainability [1]. One of the fundamental principles of organic farming is maintaining soil health through natural means. This involves strategies such as crop rotation, cover cropping, and the use of organic amendments like compost [2]. These practices aim to enhance soil structure, increase water retention, and promote a diverse and thriving microbial community [3].

2. Importance of legumes in rice fallow

Healthy soils contribute to the overall well-being of plants, influencing their ability to absorb nutrients from the soil. In the case of pulses, which are often consumed for their protein content, the organic approach to soil fertility can positively impact the nutritional composition of these crops [4]. An essential aspect of organic farming is the avoidance of synthetic pesticides and fertilizers. Instead, organic systems rely on natural methods for pest and disease control, such as the use of beneficial insects, companion planting, and cultural practices. By steering clear of chemical inputs, organic farming reduces the risk of pesticide residues in the final product. This can be particularly significant for pulses, as the avoidance of synthetic pesticides may contribute to a cleaner and more natural end-product [5]. Consumers increasingly seek products with lower pesticide residues, making organic pulses an appealing choice for health-conscious individuals. The ability of legumes to fix atmospheric nitrogen is another crucial factor in the nutritional dynamics of pulses under organic farming [6]. Pulses form symbiotic relationships with nitrogen-fixing bacteria, housed in root nodules, to convert atmospheric nitrogen into a form that plants can use. This natural process reduces the reliance on synthetic nitrogen fertilizers, which are commonly used in conventional agriculture [7]. The reduced use of synthetic fertilizers not only aligns with organic farming principles but also has implications for the nutritional content of pulses. Organic pulses may benefit from the nitrogen-fixing ability of legumes, leading to improved protein content and overall nutritional quality [8]. Diverse crop rotations are a hallmark of organic farming practices. This approach involves alternating different crops in a sequence over time, which can break pest and disease cycles, improve soil health, and enhance overall sustainability [9]. The rotation of crops can also contribute to the nutritional status of pulses by preventing nutrient depletion in the soil. Each crop has specific nutrient requirements, and by diversifying the types of crops grown in a field, organic farmers can maintain a more balanced nutrient profile in the soil [10]. This, in turn, can positively influence the nutritional composition of pulses grown in rotation with other crops. Weed management is another critical aspect of organic farming. Instead of relying on synthetic herbicides, organic farmers use mechanical and cultural methods to control weeds [11]. While this approach may be more labor-intensive, it has implications for the overall cleanliness and purity of organic pulses. The absence of herbicide residues in organic systems may contribute to a healthier and more natural end-product, aligning with consumer preferences for minimally processed and chemical-free foods [12]. The environmental impact of farming practices is a significant consideration in organic agriculture. By avoiding synthetic inputs and promoting biodiversity through techniques like agroforestry and polyculture, organic farming contributes to conserving natural resources and ecosystems. This commitment to environmental stewardship indirectly affects the nutritional status of pulses. Sustainable farming practices can lead to healthier ecosystems, which, in turn, may positively influence the nutritional composition of crops like pulses [13]. In conclusion, the nutritional status of pulses under organic farming is influenced by a combination of factors that prioritize soil health, avoid synthetic inputs, and promote environmental sustainability. The organic approach to soil fertility, pest and disease control, crop rotation, and weed management can collectively contribute to the overall quality and nutritional content of pulses. While individual organic farming practices may vary, the overarching principles of organic agriculture align to produce nutritious and environmentally responsible food. As consumer awareness of sustainable and healthy food choices continues to grow, organic pulses represent a compelling option that combines nutritional value with eco-friendly farming practices.

3. Organic pulse variety development: Improving soil and plant quality traits

Rice fallow remains one of the major potential areas for pulse area expansion in South Asia. Developing short-duration variety in legumes is suitable in South Asia rice fallow for sustainability (**Figure 1**). Organic soil is characterized by low nitrogen availability. Rice fallow situation is characterized by depletion of moisture and temperature rise toward reproduction transition in crops in rice fallow succeeding rabi crops. Legumes are the major candidate crops for rice fallow situation. Early flowering is observed in organic or low-input farming characterized by difficulties in nutrient acquisition from soil [14]. Areas converted to monocropping rice were one of the potential conventional legume growing areas usually cultivated in low-input conditions that need further intervention for regain of rice fallow areas for cropping intensification. Exploiting genomic regions controlling disease resistance, important

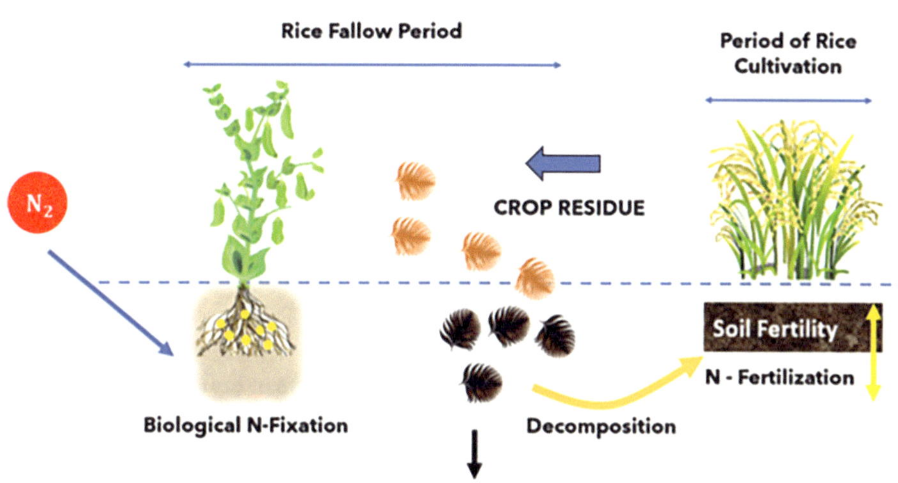

Figure 1.
Efficient use of rice fallow through pulse cultivation.

physiological process using their evolutionary relationship can lead to develop new variety [15]. Breeding lentils using functional markers may enhance developing selection of adaptive suitable lentils in rice fallow [16].

4. Need for developing suitable for organic pulse production

The promotion and adoption of organic farming are experiencing rapid growth, particularly among individuals with limited access to agricultural inputs who have demonstrated increased interest in cultivating their fields organically. Given that a substantial portion of pulse-growing areas is characterized by marginal and rainfed conditions, there exists a noteworthy opportunity to cultivate pulses under organic management. Pulse crops, by nature, rely on residual soil fertility and necessitate fewer inputs, both in terms of chemical plant nutrients and water. Consequently, among the various field crops, pulses emerge as an ideal choice for cultivation under organic nutrient management, facilitating successful and sustainable crop production. In the context of organic pulse cultivation, paramount importance is accorded to soil health. This involves the systematic restoration of nutrient levels and the application of organic amendments such as compost and organic matter to augment fertility. Additionally, the implementation of crop rotation is deemed essential to mitigate the accumulation of pests and diseases, foster a balanced ecosystem, and mitigate risks associated with specific crops. Efficient water management constitutes another pivotal aspect, with practices such as mulching, rainwater harvesting, and drip irrigation being instrumental in water conservation, and the maintenance of optimal soil moisture and organic pulse cultivation also reduces the depilation of groundwater resources [17]. The utilization of compost and organic fertilizers assumes significance in providing essential nutrients, thereby contributing to soil fertility and overall crop health. Further enhancement of soil nutrient content and the promotion of a healthy soil microbiome are achieved through the incorporation of natural amendments such as biofertilizers, green manure, and crop residues. The employment of *in situ* organic matter production, requiring nearly half the energy compared to chemical plant nutrients [17], not only represents a sustainable option but also generates employment opportunities for farming families. Research on organic pulse cultivation underscores the potential of organic foliar sprays in augmenting seed yield and quality [18]. This assumes particular significance in the Indian pulse production system, where diminishing productivity is attributed to nutrient stress. The application of organic manures and biofertilizers, coupled with nutrient supplementation, has been observed to increase flowering branches and enhance soil health [19]. However, addressing the need for increased productivity and pulse production in India necessitates the development of short-duration varieties and the engagement of rural youths in seed production. The careful selection of organic seeds adapted to local conditions is imperative for ensuring resilience and disease resistance. Furthermore, the implementation of natural farming practices, which involve the strategic cultivation of different plants to enhance growth and deter pests, contributes to a healthier crop environment. Preferential adoption of integrated pest management (IPM) techniques, utilizing natural predators and traps, over synthetic pesticides further align with sustainable practices. In essence, there exists ample opportunity to cultivate pulses organically, yielding dual benefits in terms of economic viability and sustainability.

5. Breeding quality traits for organic pulse

Enhancing quality food intake considering nutritional value reducing non-communicable lifestyle disease may be integrated in breeding pipeline for low-caloric food source [20]. Multiple quality traits have been considered for organic pulse development (**Table 1**). Breeding for slow release and low digestible carbohydrates in dry pea with beneficial minerals for the low organic system has been developed, viz. CDC Striker, Fiddle, Hamton etc. [24]. Selection for prebiotic carbohydrates may enhance stress tolerance and human health. Developing an organic variety of legumes that considers these nutritional factors may act as a strategic development for quality breeding [23]. Breeding for raffinose like oligosaccharide, carbohydrates need more diversity of germplasm to be explored as these traits are influenced by precipitation and temperature of the growing cycle [31].

6. Genes for enhancing nutrient acquisition in pulse

Low-input farming methods, like organic agriculture, are receiving more and more attention when it comes to environmentally friendly and sustainable agricultural techniques. Pulses are a broad category of leguminous crops that include common beans, chickpeas, lentils, and peas. Because of their nutritional value and capacity to fix nitrogen, pulses are important components of these systems. Comprehending the genetic foundation of pulse crops grown in low-input agricultural environments is essential to creating robust and high-yielding cultivars that require little outside assistance to flourish [32]. Introduction of legume species including *Vicia ervilia* in low organic matter soil improves soil nitrogen fertilization status after decomposition [33]. In this context, genes and quantitative trait locus (QTLs) linked to characteristics like stress resilience, nutrient use efficiency, and water use efficiency are particularly interesting. These genetic components provide information on the molecular processes that control pulse performance and adaptability in agroecosystems with no chemical input. Increasing demand for phosphatic

Crop	Traits	References
Dry pea (*Pisum sativum* L.)	Quality traits (protein)	[21]
Chickpea (*Cicer arietinum* L.)	Quality traits (fatty acids)	[22]
Lentil (*Lens culinaris* Medikus)	Quality traits (protein and micronutrients)	[23]
Dry pea (*Pisum sativum* L.)	Quality traits (nutrients)	[24]
Lentil (*Lens culinaris* Medik.)	Quality traits (nutrients)	[25]
Kale (*Brassica oleracea* var. *acephala*)	Quality traits (nutrients)	[26]
Lentil (*Lens culinaris* Medikus)	Quality traits (nutrients)	[27]
Kale (*Brassica oleracea* var. *acephala*)	Quality traits (nutrients)	[28]
Other legumes	Nitrogen use efficiency, phosphorus use efficiency	[29]
Chickpea (*Cicer arietinum* L.)	Quality traits (grain yield and protein)	[30]
Chickpea (*Cicer arietinum* L.)	Root traits	[19]

Table 1.
Pulse breeding objective in low-input organic condition for quality traits.

fertilizer and depleting natural resources for phosphatic fertilizer invite strategies to deploy crop breeding with higher phosphate use efficiency, PUE [34]. Improving PUE and NUE contributing toward grain quality must be addressed for organic crop production. Enhancing root surface area and root dry weight for enhancing phosphate uptake under phosphate-limited conditions may enhance breeding for higher phosphate uptake variety in legumes [35]. Selection of genotypes with higher phosphate remobilization efficiency is suitable for genetic improvement of phosphate uptake, which is directly correlated with seed phosphate content in field peas [36]. A hypothetical ideotype in phosphate uptake, use, and translocation efficiency considering catch and release mechanism may enhance breeding low phosphate soil-adopted plants [37]. Utilizing new breeding techniques, genomic selection, genomic-assisted breeding, and CRISPR to improve PUE have been emphasized in multiple legumes, which can restrict phosphate use (**Table 2**) [38].

7. Breeding for disease resistance in legumes

Fungus being major growth and devastating pathogen for legumes cultivation, reducing fungal pathogen load in legume root and enhancing population of beneficial microbes need to be addressed for disease resistance in legumes [46]. Plant genotype and microbial community have enhanced overall legumes are mostly affected by soil-borne disease at early growth stages viz. seedling rot and foliar blight like *Alternaria*, *Stemphyllium* during pre-flowering to post-flowering situations. As organic farming is low-input condition which lacks chemical plant protection measure, improving plant pathogen interactions for immune reaction is necessary. Identification of resistant lines for root disease and foliar blight and deciphering their resistance mechanism need intervention for enhancing breeding initiatives in legumes for disease resistance through introgression breeding [47]. Genotypes like EC866132, IC267667, and IC201778 in lentils have been identified for lentil rice fallow areas with root rot and Alternaria blight.

8. Breeding for sustainable root system for organic pulse

A plant's root system is its main point of contact with the soil, where it influences the intake of nutrients and water as well as the general health of the plant. For a

Gene/QTL	Function/role	Associated pulse	References
NRT1	Nitrogen use efficiency	Common bean	[39]
PHO1	Phosphorus use efficiency	Chickpea	[40]
PHR1	Phosphorus uptake efficiency	Chickpea	[41]
NOD2	Nitrogen fixation	Lentil	[42]
PSTOL1	Phosphorus uptake	Common bean	[43]
PsENOD12B	Phosphate deficiency	Lucerne	[44]
PsNRT2;3	Nitrogen uptake efficiency	Pea	[32]
PvNADP-ME2	Nitrogen use efficiency	Common bean	[45]

Table 2.
Major genes enhancing nutrient acquisition in legumes.

number of reasons, knowing the genetic foundation of pulse root properties is practically significant when it comes to organic farming [48]. First and foremost, because organic systems frequently rely on organic amendments and cover crops as natural sources of nitrogen, the availability of nitrogen may be erratic and restricted. To ensure optimal growth and yield, it is necessary to produce pulse types that are more adaptable to low-nitrogen environments. Under low nitrogen input circumstances in organic farming, the genetics of root features in pulses (leguminous crops like lentils, chickpeas, and peas) involves intricate interactions between genes, the environment, and management techniques. For plants to absorb nutrients and water, as well as to be healthy overall, root characteristics are essential. Here are some crucial things to remember. Lateral root branching is controlled by genetic regulation to promote root biomass and enhanced nutrient uptake in organic soil [49, 50]. Certain plants have the potential to search a greater area of soil for materials because of their more branching and fibrous root systems. Root hairs are extensions of the root epidermal cells that are important for multiple physiological responses as well as nutrient acquisition,

Legume name (scientific name)	Root type	Fruit development under low nitrogen	Fruit development under low phosphorus	Fruit development under low potash	References
Lentils (*Lens culinaris*)	Taproot	Reduced, smaller fruits	Reduced, altered development	Altered development	[48]
Chickpeas (*Cicer arietinum*)	Taproot with lateral roots	Delayed, smaller fruits	Reduced, modified development	Altered development	[55]
Peas (*Pisum sativum*)	Fibrous root system	Altered, smaller fruits	Reduced, modified development	Delayed development	[56]
Soybeans (*Glycine max*)	Taproot with nodules	Resilient fruiting, reduced size	Adapted development	Resilient development	[57]
Faba Beans (*Vicia faba*)	Taproot with lateral roots	Good fruiting under nitrogen stress	Moderate impact on development	Moderate impact on development	[48]
Pigeonpea (*Cajanus cajan*)	Taproot with lateral roots	Adapted fruiting under low nitrogen	Resilient development	Moderate impact on development	[58]
Garden Pea (*Pisum sativum* var. *sativum*)	Fibrous root system	Good fruiting under nitrogen stress	Moderate impact on development	Resilient development	[59]
Blackgram (*Vigna mungo*)	Taproot with lateral roots	Resilient fruiting, reduced size	Resilient development	Altered development	[60]
Greengram (*Vigna radiata*)	Taproot with lateral roots	Adapted fruiting under low nitrogen	Moderate impact on development	Resilient development	[61]

Table 3.
The root types of different legumes and their fruiting behavior under low inputs of nitrogen, phosphorus, and potassium.

determined by genetic factors. Intensive root hair causes enhancement of nutrient acquisition surface area for more nutrient absorption. For example, as a hereditary characteristic, some genotypes possess a higher density of root hairs. Legume genotypes having denser root hairs can be suitable for experimental materials for extracting genomic studies. Genes involved in gravitropism control the plant's ability to grow roots downward in response to gravity. Mutations in these genes can lead to altered root growth orientations. The LAZY gene family is a key regulator of gravitropism in plants. LAZY genes are also known as negative gravitropic response of roots (NGR) and deeper rooting (DRO) [50]. Additionally, tropic responses, such as phototropism (response to light), can influence root architecture in specific directions involved in multidimensional nutrient acquisition. Genetic factors determine the thickness and overall morphology of roots. Some plants may have fleshy tap roots like lentils and chickpeas adapted to nutrient storage, while others like peas may have fine, fibrous roots optimized for nutrient absorption. Cytokinin is an important hormone characterized for root development in multiple crops. A common gene, ARR is involved in cytokinin signaling pathways, affecting root growth and branching [51]. AUX1 is involved in the gravity-sensing mechanism in roots. It plays a role in auxin transport, contributing to the downward growth of roots in response to gravity [52]. AGR1 is another gene associated with gravitropism in roots, influencing the orientation of root growth in relation to gravity [53]. Nodulation Genes (e.g., NodA, NodB, and NodC): In leguminous plants, genes for nodulation development play a crucial role in establishing a symbiotic relationship with nitrogen-fixing bacteria for symbiotic nitrogen fixation. Another important gene, EXPANSIN is involved in cell wall loosening and extension, contributing to root elongation (**Table 3**) [54].

9. Microbial diversity

A significant obstacle encountered by organic farmers is the restricted supply of nitrogen, an essential ingredient for the growth and development of plants. Pulses, which include legumes like lentils, chickpeas, and peas, are essential to organic farming practices. These crops are special because they can coexist peacefully with bacteria that fix nitrogen, which makes it easier for them to take nitrogen from the atmosphere. The genetics of root characteristics in pulses grown in organic farming systems with little nitrogen input, however, continues to be a major and relevant problem. Nitrogen fixation and nutritional availability improvement can be done by ensuring that legumes and rhizobia possess a robust symbiotic interaction. Legumes and root microorganisms' association can be enhanced by breeding for particular root features. Characteristics like fine root hairs, increased root surface area, and root exudation might promote the growth of good microorganisms.

10. Molecular crop improvement strategies

Precise selection of desirable traits in crop improvement has been one of the most important achievements of molecular marker-assisted breeding. By integrating allele mining and marker-assisted selection (MAS), breeders can enhance the resilience, productivity, and nutritional quality of legume crops [62] while adhering to the principles of low-input organic farming [63].

11. Allele mining

The process of identifying and describing various alternative forms of genes that are connected to particular desirable features is known as allele mining [64]. This strategy is essential for making use of the genetic variety seen in populations of legumes, which may be used to target features like enhanced nutritional profiles [65], abiotic stress tolerance like drought, heat, cold, flash floods, resistance to disease caused by bacteria, fungus or losses due to insect pests [63, 66]. Next-generation sequencing (NGS) and other advanced sequencing technologies can be utilized to study the genomes of various legume types, including traditional landraces and their wild relatives, showing responsive considerable outputs under low inputs [66]. This thorough genomic study aids in the identification of genetic variants present in various genotypes. To determine relationships between certain genetic markers and phenotypic traits, mapping of quantitative trait locus (QTL) and genome-wide association studies (GWAS) are carried out [67]. These investigations entail the examination of vast populations of legumes in order to pinpoint specific genomic areas associated with desirable characteristics, such as resilience to disease or abiotic stresses. Further, this technique can mitigate crop losses and induce genetic gains by providing insights into alleles or QTLs showing the heterotic response [68]. Large-scale genetic data are created, and sophisticated bioinformatics techniques are used to process and analyze it [65]. These techniques aid in the identification of prospective alleles and the comprehension of their possible functions in the manifestation of traits. To find conserved areas and functional annotations for genes, comparative genomics is another tool used in bioinformatics analysis [65, 69].

12. Marker-assisted selection for legumes

Creating molecular markers connected to desirable qualities is the first stage in the MAS process. These markers may be simple sequence repeats (SSRs), single nucleotide polymorphisms (SNPs), or other genetic markers [70, 71]. The process of developing markers entails pinpointing genomic locations linked to particular characteristics and creating markers that can accurately identify these locations [72], which can further provide insights into trait-linked sequences throughout the whole genome of the crop. Favorable markers can be found in large breeding populations by screening them with high-throughput genotyping platforms. Selecting for multiple traits like high yields of pulses under low fertilizer inputs coupled with other stress-tolerant or quality parameters at once is possible [73–75], thanks to these systems, which enable the quick and affordable genotyping of thousands of plants. Breeding becomes easier and more accurate with high-throughput genotyping. By allowing breeders to screen plants early in growth on the basis of their genetic composition rather than just phenotypic features, MAS increases the efficiency of selection [70]. This method lessens the need for phenotypic evaluations, which are time-consuming and labor-intensive. Additionally, early selection makes it possible to identify favorable features that could be difficult to observed in the field, like those of some related to stress, which are expressed in stress condition only.

Backcrossing programs, which transfer particular alleles into elite breeding lines, rely heavily on MAS [76]. Pyramiding, through genomic level techniques like MABB, which combines several advantageous genes into a single variety, is another use for it, which is especially crucial for complicated traits that may involve numerous genes, such as disease resistance [77].

13. Yield stability in organic conditions

Yield is one of the most complex traits as not only multiple genes governing multiple pathways rather environmental influence is considerable. Novel plant nutrient sources, nutrient recycling, and renewable source-based fertilizer can assure higher yield including stability in organic system [78]. Under the organic system, soyabean yield remains the same as reported [79]. Intercropping of fababean and wheat, a good example of ecological intensification, enhances yield in organic crop production [80]; working in breeding for diversity and mixed cropping competition and nutrient acquisition mechanism may enhance yield stability and overall yield. Transcription regulator element FUN (function under nitrate) enhances nitrogen nutrition in legumes [81]. Breeding for compatibility in legumes is important for yield stability, whereas the genotypic variation is limited [82]. Enhancing growth under organic conditions is important by enhancing phosphate availability and introducing phosphate acquisition efficient genotype in the breeding program [83].

14. Challenges

Limited genetic diversity for combinability, mixed cropping, or intercropping in legumes causes slower genetic gain, which is contributing toward the gap between conventional and organic system. Major challenges in legume breeding remain untapped genomic resources which have potential genomic information to be utilized in breeding. Under low-input organic conditions, nutrient and nutrient remobilization, plant growth, disease, and resistance are traits of breeder's choice to address the cropping system. Understanding crop dynamics in multiple nutrient management conditions, exploiting easy methods of inducing genetic variation and screening them under actual field conditions rather than immediate prediction and modeling is to be emphasized. Crop with robust biomass and upward branching type, nonlodging type, moderate foliage intensity, robust rooting system, post-season rain tolerant genotypes, effective intercropping with short duration, and less perishable high-value crops can be taken up. Conserving agrobiodiversity for the identification of novel alleles remained untapped in local genetic stocks, which need to be nurtured for suitable legume crop and genotype choice under low input.

15. Conclusion

Legumes are a major group of crops in sustainable agriculture, promoting crop growth, beneficial soil microbes, nutrient mobilization, etc. Breeding legumes enhances soil carbon deposition and nutrient mobilization, considering root morphology, root biomass, etc., and may enhance soil health. Breeding for root-secreted biochemicals like acid phosphatase enzyme helps phosphate mobilization in phosphate-limited soil. Sustainable legume production in rice fallow needs more vigorous roots and enhanced nutrition acquisition ability along with heat and drought tolerance. Identification of major genetic loci, closely linked markers for organic, suitable traits, and identification of novel alleles may enhance organic, suitable variety development.

DOI: http://dx.doi.org/10.5772/intechopen.1008673

Author details

Anirban Roy*†, Suman Dutta†, Tapash Dasgupta, Payabrata Roy,
Satyaki Nandan Maity, Samita Saha, Sukamal Sarkar and Sourav Garai
Ramakrishna Mission Vivekananda Educational and Research Institute, Narendrapur, Kolkata, India

*Address all correspondence to: anirbanroygpb22@gmail.com

† These authors contributed equally.

References

[1] Singh A, Chahal HS. Organic grain legumes in India: Potential production strategies, perspective, and relevance. In: Legume Crops-Prospects, Production and Uses. London, UK: IntechOpen; 2020

[2] Nair A, Delate K. Composting, crop rotation, and cover crop practices in organic vegetable production. In: Organic Farming for Sustainable Agriculture. Cham, Switzerland: Springer; 2016. pp. 231-257

[3] Lal R. Improving soil health and human protein nutrition by pulses-based cropping systems. Advances in Agronomy. 2017;**145**:167-204

[4] Meena RS, Lal R. Legumes and sustainable use of soils. In: Legumes for Soil Health and Sustainable Management. Singapore: Springer; 2018. pp. 1-31

[5] Gamage A, Gangahagedara R, Gamage J, Jayasinghe N, Kodikara N, Suraweera P, et al. Role of organic farming for achieving sustainability in agriculture. Farming System. 2023;**1**(1):100005

[6] Kumar S, Meena RS, Datta R, Verma SK, Yadav GS, Pradhan G, et al. Legumes for carbon and nitrogen cycling: An organic approach. In: Carbon and Nitrogen Cycling in Soil. Singapore: Springer; 2020. pp. 337-375

[7] Kebede E. Contribution, utilization, and improvement of legumes-driven biological nitrogen fixation in agricultural systems. Frontiers in Sustainable Food Systems. 2021;**5**:767998

[8] Brandt K, Mølgaard JP. Organic agriculture: Does it enhance or reduce the nutritional value of plant foods? Journal of the Science of Food and Agriculture. 2001;**81**(9):924-931

[9] Fischer RA, Connor DJ. Issues for cropping and agricultural science in the next 20 years. Field Crops Research. 2018;**222**:121-142

[10] Goulding K, Jarvis S, Whitmore A. Optimizing nutrient management for farm systems. Philosophical Transactions of the Royal Society, B: Biological Sciences. 2008;**363**(1491):667-680

[11] Singh SP, Rani V, Sengar RS, Gupta S. Weed management in organic farming for better crop production. Progressive Agriculture. 2022;**22**(2):117-125

[12] Lang B, Conroy DM. Are trust and consumption values important for buyers of organic food? A comparison of regular buyers, occasional buyers, and non-buyers. Appetite. 2021;**161**:105123

[13] Foyer CH, Lam HM, Nguyen HT, Siddique KH, Varshney RK, Colmer TD, et al. Neglecting legumes has compromised human health and sustainable food production. Nature Plants. 2016;**2**(8):1-10

[14] Sanagi M, Aoyama S, Kubo A, Lu Y, Sato Y, Ito S, et al. Low nitrogen conditions accelerate flowering by modulating the phosphorylation state of FLOWERING BHLH 4 in Arabidopsis. Proceedings of the National Academy of Sciences of the United States of America. 2021;**118**(19):e2022942118. DOI: 10.1073/pnas.2022942118

[15] Guerra-Garcia A, Haile T, Ogutcen E, Bett KE, von Wettberg EJ. An evolutionary look into the history of lentil reveals unexpected diversity. Evolutionary Applications.

2022;**15**(8):1313-1325. DOI: 10.1111/eva.13467

[16] Roy A, Reddy MH, Sarkar M, Sagolsem D, Murmu SK, Das C, et al. A mis-splicing early flowering 3 (elf3) allele of lentil is associated with yield enhancement under terminal heat stress. Journal of Applied Genetics. 2023;**64**(2):265-273. DOI: 10.1007/s13353-023-00753-z

[17] Avasthe RK, Raghavendra S, Subhash B. Organic pulses production in India: Perspectives and opportunities. Indian Journal of Agronomy. 2016;**61**:144-152

[18] Veerappan V, Rangnathan U, Mannar J. Effect of organic foliar spray with pulse sprout extract on seed yield and quality of rice (*Oryza sativa*). Journal of Plant Nutrition. 2019;**42**(8):900-914

[19] Ramamoorthy P, Lakshmanan K, Upadhyaya HD, Vadez V, Varshney RK. Root traits confer grain yield advantages under terminal drought in chickpea (*Cicer arietinum* L.). Field Crops Research. 2017;**201**:146-161

[20] Reda T, Thavarajah P, Polomski R, Bridges W, Shipe E, Thavarajah D. Reaching the highest shelf: A review of organic production, nutritional quality, and shelf life of kale (*Brassica oleracea* var. *acephala*). Plants, People, Planet. 2021;**3**(4):308-318. DOI: 10.1002/ppp3.10183

[21] Thavarajah D, Lawrence T, Boatwright L, Windsor N, Johnson N, Kay J, et al. Organic dry pea (*Pisum sativum* L.): A sustainable alternative pulse-based protein for human health. PLoS ONE. 2023;**18**(4):e0284380

[22] Salaria S, Boatwright JL, Johnson N, Madurapperumage A, Joshi P, Thavarajah P, et al. Fatty acid composition and genome-wide associations of a chickpea (*Cicer arietinum* L.) diversity panel for biofortification efforts. Scientific Reports. 2023;**13**(1):14002

[23] Johnson N, Boatwright JL, Bridges W, Thavarajah P, Kumar S, Shipe E, et al. Genome-wide association mapping of lentil (*Lens culinaris* Medikus) prebiotic carbohydrates toward improved human health and crop stress tolerance. Scientific Reports. 2021;**11**(1):13926. DOI: 10.1038/s41598-021-93475-3

[24] Thavarajah D, Lawrence TJ, Powers SE, Kay J, Thavarajah P, Shipe E, et al. Organic dry pea (*Pisum sativum* L.) biofortification for better human health. PLoS ONE. 2022;**17**(1):e0261109. DOI: 10.1371/journal.pone.0261109

[25] Salaria S, Boatwright JL, Thavarajah P, Kumar S, Thavarajah D. Protein biofortification in lentils (*Lens culinaris* Medik.) toward human health. Frontiers in Plant Science. 2022;**13**:869713. DOI: 10.3389/fpls.2022.869713

[26] Reda T, Powers SE, Thavarajah D. Falling into line: Adaptation of organically grown kale (*Brassica oleracea* var. *acephala*) and kale relatives to fall planting. Scientia Horticulturae. 2022;**295**:110878

[27] Choukri H, El Haddad N, Aloui K, Hejjaoui K, El-Baouchi A, Smouni A, et al. Effect of high temperature stress during the reproductive stage on grain yield and nutritional quality of lentil (*Lens culinaris* Medikus). Frontiers in Nutrition. 2022;**9**:857469. DOI: 10.3389/fnut.2022.857469

[28] Lawrence T, Powers S, Jones B, Johnson N, Kay J, Bandaranayake A,

et al. Genetic variation in the prebiotic carbohydrate and mineral composition of kale (*Brassica oleracea* L. var. *acephala*) adapted to an organic cropping system. Journal of Food Composition and Analysis. 2021;**96**:103718

[29] Vijayakumar V, Sulthana S, Kannan B, Rathinavelu S, Krishnan A, Sudhagar R. Breeding for Macronutrient Use Efficiency (NTUE) in Legumes. London, UK: IntechOpen; 2023. DOI: 10.5772/intechopen.1002885

[30] De Santis MA, Rinaldi M, Menga V, Codianni P, Giuzio L, Fares C, et al. Influence of organic and conventional farming on grain yield and protein composition of chickpea genotypes. Agronomy. 2021;**11**(2):191

[31] Vandemark G, Thavarajah S, Siva N, Thavarajah D. Genotype and environment effects on prebiotic carbohydrate concentrations in Kabuli chickpea cultivars and breeding lines grown in the U.S. Pacific Northwest. Frontiers in Plant Science. 2020;**11**:112. DOI: 10.3389/fpls.2020.00112

[32] Kant S, Bi YM, Rothstein SJ. Understanding plant response to nitrogen limitation for the improvement of crop nitrogen use efficiency. Journal of Experimental Botany. 2011;**62**(4):1499-1509

[33] Ordóñez-Fernández R, Angel Repullo-Ruibérriz de Torres M, Márquez-García J, Moreno-García M, Carbonell-Bojollo R. Legumes used as cover crops to reduce fertilisation problems improving soil nitrate in an organic orchard. European Journal of Agronomy. 2018;**95**:1-13. DOI: 10.1016/j.eja.2018.02.001

[34] Blair MW. Breeding approaches to increasing nutrient-use efficiency: Examples from common bean. In: Rengel Z, editor. Improving Water and Nutrient-Use Efficiency in Food Production Systems. USA: Wiley Online Library; 2013. pp. 161-175

[35] Ganguly S, Roy A, Murmu SK, et al. Variation in P-acquisition ability and acid phosphatase activity at the early vegetative stage of lentil and their validation on P-deficiency field. Acta Physiologiae Plantarum. 2021;**43**:109. DOI: 10.1007/s11738-021-03280-8

[36] Powers S, Mirsky E, Bandaranayake A, et al. Field pea (*Pisum sativum* L.) shows genetic variation in phosphorus use efficiency in different P environments. Scientific Reports. 2020;**10**:18940. DOI: 10.1038/s41598-020-75804-0

[37] Carkner MK, Gao X, Entz MH. Ideotype breeding for crop adaptation to low phosphorus availability on extensive organic farms. Frontiers in Plant Science. 2023;**14**:1225174. DOI: 10.3389/fpls.2023.1225174

[38] Jha UC, Nayyar H, Parida SK, Beena R, Pang J, Siddique KHM. Breeding and genomics approaches for improving phosphorus-use efficiency in grain legumes. Environmental and Experimental Botany. 2023;**205**:105120. DOI: 10.1016/j.envexpbot.2022.105120

[39] Xu J, Li HD, Chen LQ, Wang Y, Liu LL, He L, et al. A protein kinase, interacting with two calcineurin B-like proteins, regulates K+ transporter AKT1 in Arabidopsis. Cell. 2018;**173**(7):1776-1787

[40] Garg R, Shankar R, Thakkar B, Kudapa H, Krishnamurthy L, Mantri N, et al. Transcriptome analyses reveal genotype- and developmental stage-specific molecular responses to drought and salinity stresses in chickpea. Scientific Reports. 2016;**6**:19228

[41] Kohli PS, Kumar Verma P, Verma R, Parida SK, Thakur JK, Giri J. Genome-wide association study for phosphate deficiency responsive root hair elongation in chickpea. Functional and Integrative Genomics. 2020;**20**:775-786

[42] Begum AA, Leibovitch S, Migner P, Zhang F. Specific flavonoids induced nod gene expression and pre-activated nod genes of rhizobium leguminosarum increased pea (*Pisum sativum* L.) and lentil (*Lens culinaris* L.) nodulation in controlled growth chamber environments. Journal of Experimental Botany. 2001;**52**(360):1537-1543

[43] Gamuyao R, Chin JH, Pariasca-Tanaka J, Pesaresi P, Catausan S, Dalid C, et al. The protein kinase PSTOL1 from traditional rice confers tolerance of phosphorus deficiency. Nature. 2012;**488**(7412):535-539

[44] Sulieman S, Schulze J, Tran LS. Comparative analysis of the symbiotic efficiency of *Medicago truncatula* and *Medicago sativa* under phosphorus deficiency. International Journal of Molecular Sciences. 2013;**14**(9):5198-5213

[45] Blair MW, Astudillo C, Grusak MA, Graham R, Beebe SE. Inheritance of seed iron and zinc concentrations in common bean (*Phaseolus vulgaris* L.). Molecular Breeding. 2009;**23**(2):197-207

[46] Wille L, Messmer MM, Studer B, Hohmann P. Insights to plant-microbe interactions provide opportunities to improve resistance breeding against root diseases in grain legumes. Plant, Cell & Environment. 2019;**42**(1):20-40. DOI: 10.1111/pce.13214

[47] Roy A, Sahu PK, Das C, Bhattacharyya S, Raina A, Mondal S. Conventional and newbreeding technologies for improving disease resistance in lentil (*Lens culinaris* Medik). Frontiers in Plant Science. 2023;**13**:1001682. DOI: 10.3389/fpls.2022.1001682

[48] Liu Z, Xing Y, Jin D, Liu Y, Lu Y, Chen Y, et al. Improved nitrogen utilization of faba bean (*Vicia faba* L.) roots and plant physiological characteristics under the combined application of organic and inorganic fertilizers. Agriculture. 2022;**12**:1999. DOI: 10.3390/agriculture12121999

[49] Aski M, Mehra R, Mishra GP, Singh D, Yadav P, Rai N, et al. Genotypic variation in root architectural traits under contrasting phosphorus levels in Mediterranean and Indian origin lentil genotypes. PeerJ. 2022;**10**:e12766. DOI: 10.7717/peerj.12766

[50] Jiao Z, Du H, Chen S, Huang W, Ge L. LAZY gene family in plant gravitropism. Frontiers in Plant Science. 2021;**11**:606241. DOI: 10.3389/fpls.2020.606241

[51] D'Agostino IB, Deruere J, Kieber JJ. Characterization of the response of the Arabidopsis response regulator gene family to cytokinin. Plant Physiology. 2000;**124**(4):1706-1717

[52] Rutschow HL, Baskin TI, Kramer EM. The carrier AUXIN RESISTANT (AUX1) dominates auxin flux into Arabidopsis protoplasts. New Phytologist. 2014;**204**(3):536-544

[53] Chen R, Hilson P, Sedbrook J, Rosen E, Caspar T, Masson PH. The *Arabidopsis thaliana* AGRAVITROPIC 1 gene encodes a component of the polar-auxin-transport efflux carrier. Proceedings of the National Academy of Sciences of the United States of America. 1998;**95**(25):15112-15117. DOI: 10.1073/pnas.95.25.15112

[54] Li X, Zhao J, Tan Z, Zeng R, Liao H. GmEXPB2, a cell wall β-expansin, affects

soybean nodulation through modifying root architecture and promoting nodule formation and development. Plant Physiology. 2015;**169**(4):2640-2653

[55] Khan MI, Afzal MJ, Bashir S, Naveed M, Anum S, Cheema SA, et al. Improving nutrient uptake, growth, yield and protein content in chickpea by the co-addition of phosphorus fertilizers, organic manures, and Bacillus sp. MN-54. Agronomy. 2021;**11**:436. DOI: 10.3390/agronomy11030436

[56] Wysokinski A, Lozak I. The dynamic of nitrogen uptake from different sources by pea (*Pisum sativum* L.). Agriculture. 2021;**11**:81. DOI: 10.3390/agriculture11010081

[57] Li M, Wang Y, Zhang P, et al. The role of GmSnRK1-GmNodH module in regulating soybean nodulation capacity. International Journal of Molecular Sciences. 2023;**24**(2):1225. DOI: 10.3390/ijms24021225

[58] Kumar J, Choudhary AK, Gupta DS, Kumar S. Towards exploitation of adaptive traits for climate-resilient smart pulses. International Journal of Molecular Sciences. 2019;**20**(12):2971. DOI: 10.3390/ijms20122971

[59] Balliu A, Zheng Y, Sallaku G, Fernández JA, Gruda NS, Tuzel Y. Environmental and cultivation factors affect the morphology, architecture and performance of root systems in soilless grown plants. Horticulturae. 2021;**7**(8):243. DOI: 10.3390/horticulturae7080243

[60] Kumar RS, Ganesh P, Tharmaraj K, Saranraj P. Growth and development of blackgram (*Vigna mungo*) under foliar application of Panchagavya as organic source of nutrient. Current Botany. 2011;**2**(3):09-11

[61] Reddy VRP, Aski M, Mishra GP, Dikshit HK, Singh A, Pandey R, et al. Genetic variation for root architectural traits in response to phosphorus deficiency in mungbean at the seedling stage. bioRxiv. 719773. DOI: 10.1101/719773

[62] Détain A, Bhowmik P, Leborgne-Castel N, Ochatt S. Latest biotechnology tools and targets for improving abiotic stress tolerance in protein legumes. Environmental and Experimental Botany. 2022;**197**:104824

[63] Ojiewo C, Monyo E, Desmae H, Boukar O, Mukankusi-Mugisha C, Thudi M, et al. Genomics, genetics and breeding of tropical legumes for better livelihoods of smallholder farmers. Plant Breeding. 2019;**138**(4):487-499

[64] Kumar GR, Sakthivel K, Sundaram RM, Neeraja CN, Balachandran SM, Rani NS, et al. Allele mining in crops: Prospects and potentials. Biotechnology Advances. 2010;**28**(4):451-461

[65] Salgotra RK, Stewart CN Jr. Genetic augmentation of legume crops using genomic resources and genotyping platforms for nutritional food security. Plants. 2022;**11**(14):1866

[66] Ali A, Altaf MT, Nadeem MA, Karaköy T, Shah AN, Azeem H, et al. Recent advancement in OMICS approaches to enhance abiotic stress tolerance in legumes. Frontiers in Plant Science. 2022;**13**:952759

[67] Ma Y, Coyne CJ, Grusak MA, Mazourek M, Cheng P, Main D, et al. Genome-wide SNP identification, linkage map construction and QTL mapping for seed mineral concentrations and contents in pea (*Pisum sativum* L.). BMC Plant Biology. 2017;**17**:1-17

[68] Gayacharan, Parida SK, Mondal N, Yadav R, Vishwakarma H, Rana JC. Mining legume germplasm for genetic gains: An Indian perspective. Frontiers in Genetics. 2023;**14**:996828

[69] Singh G, Gudi S, Amandeep, Upadhyay P, Shekhawat PK, Nayak G, et al. Unlocking the hidden variation from wild repository for accelerating genetic gain in legumes. Frontiers in Plant Science. 2022;**13**:1035878

[70] Bernardo R. Molecular markers and selection for complex traits in plants: Learning from the last 20 years. Crop Science. 2008;**48**(5):1649-1664

[71] Tanksley SD. Molecular markers in plant breeding. Plant Molecular Biology Reporter. 1983;**1**:3-8

[72] Nadeem MA, Nawaz MA, Shahid MQ, Doğan Y, Comertpay G, Yıldız M, et al. DNA molecular markers in plant breeding: Current status and recent advancements in genomic selection and genome editing. Biotechnology & Biotechnological Equipment. 2018;**32**(2):261-285

[73] Raju NL, Gnanesh BN, Lekha P, Jayashree B, Pande S, Hiremath PJ, et al. The first set of EST resource for gene discovery and marker development in pigeonpea (*Cajanus cajan* L.). BMC Plant Biology. 2010;**10**:1-22

[74] Dwivedi SL, Scheben A, Edwards D, Spillane C, Ortiz R. Assessing and exploiting functional diversity in germplasm pools to enhance abiotic stress adaptation and yield in cereals and food legumes. Frontiers in Plant Science. 2017;**8**:1461

[75] Tesfaye M, Liu J, Allan DL, Vance CP. Genomic and genetic control of phosphate stress in legumes. Plant Physiology. 2007;**144**(2):594-603

[76] Sheoran S, Ramtekey V, Kumar D, Kumar S, Meena RS, Kumawat A, et al. Grain legumes: Recent advances and technological interventions. In: Advances in Legumes for Sustainable Intensification. US: Academic Press; 2022. pp. 507-532

[77] Sahay G, Mehta BK, Dikshit N, Shashikumara P, Bhargavi HA, Priyadarshini P. Developing climate-resilient cowpea (*Vigna unguiculata* [L.] Walp.) through genomics-assisted breeding approaches. In: Developing Climate Resilient Grain and Forage Legumes. Singapore: Springer Nature Singapore; 2022. pp. 77-103

[78] Röös E, Mie A, Wivstad M, et al. Risks and opportunities of increasing yields in organic farming. A review. Agronomy for Sustainable Development. 2018;**38**:14. DOI: 10.1007/s13593-018-0489-3

[79] Knapp S, Gunst L, Mäder P, Ghiasi S, Mayer J. Organic cropping systems maintain yields but have lower yield levels and yield stability than conventional systems—Results from the DOK trial in Switzerland. Field Crops Research. 2023;**302**:109072. DOI: 10.1016/j.fcr.2023.109072

[80] Jensen E, Bedoussac L, Carlsson G, Journet E, Justes E, Hauggaard-Nielsen H. Enhancing yields in organic crop production by eco-functional intensification. Sustainable Agriculture Research. 2015;**4**. DOI: 10.5539/sar.v4n3p42

[81] Lin J, Bjørk PK, Kolte MV, Poulsen E, Dedic E, Drace T, et al. Zinc mediates control of nitrogen fixation via transcription factor filamentation.

Nature. 2024;**631**(8019):164-169. DOI: 10.1038/s41586-024-07607-6

[82] Annicchiarico P, Collins RP, De Ron AM, Firmat C, Litrico I, Hauggaard-Nielsen H. Chapter Three— Do we need specific breeding for legume-based mixtures? In: Sparks DL, editor. Advances in Agronomy. Vol. 157. Berlington, Canada: Academic Press; 2019

[83] Kadirimangalam SR, Jadhav Y, Nagamadhuri KV, et al. Genetic approaches for assessment of phosphorus use efficiency in groundnut (*Arachis hypogaea* L.). Scientific Reports. 2022;**12**:21552. DOI: 10.1038/s41598-022-24016-9

Chapter 8

Chickpea Breeding and Research Program in Tunisia

Mariem Bouhadida

Abstract

In Tunisia, chickpea cultivation is challenged by a range of both biotic and abiotic stresses, leading to decreased productivity and a waning interest among farmers. The chickpea breeding initiative has been focused on selecting and developing new varieties that exhibit tolerance or resistance to prevalent biotic threats, particularly the detrimental fungal diseases such as ascochyta blight and *Fusarium* wilt. Numerous winter chickpea varieties, predominantly sourced from germplasm supplied by ICARDA, have been introduced and evaluated across various environments. Results stemming from the breeding program highlight the promising prospects for chickpea cultivation, showcasing improved resistance to major diseases alongside enhanced yield potential. The identification of molecular markers closely associated with a gene governing a particular trait is essential for the effective application of molecular-assisted selection within a traditional breeding program. Consequently, integrating MAS for wilt and ascochyta blight resistances serves as a valuable tool, significantly expediting the process of developing new chickpea cultivars.

Keywords: chickpea, breeding, genetic resistance, ascochyta blight, fusarium wilt, Tunisia

1. Introduction

The only cultivated species from the genus Cicer is the chickpea (*Cicer arietinum* L.) [1]. It is a member of the Fabaceae family and can fix atmospheric nitrogen (N) to improve soil fertility, especially in dry environments. It also gets a large amount of its N needs through symbiotic N fixation [2]. It is mostly consumed by humans, and is a crucial part of the Mediterranean diet and a staple meal in Pakistan and India, [3]. The first evidence of grown chickpeas dates back to Syria about 7000 years ago. Chickpea's close wild relatives, *C. reticulatum* and *C. echinospermum*, are only found in southeast Turkey.

It is believed that domestication occurred maybe as early as 10.5 thousand years ago, perhaps in tandem with or shortly after the domestication of other crops found in the Fertile Crescent, including lentil, wheat, barley, and pea. There are few records regarding the dates of introduction into East Africa and the Indian subcontinent, although it is believed that domesticated chickpeas were carried to Syria some 7000 years ago [4].

Chickpea, is a self-pollinated, annual diploid (2n =2x = 16) species [5] with genome size of 740 Mb [6]. Desi and Kabuli are the two different varieties of cultivated chickpea. The Desi (microsperma) varieties have rich, colorful seed coats and pink blooms with anthocyanin coloration on stems. The macrosperma varieties known as Kabuli have smooth surfaces, thin coats, and ram's head-shaped, white flowers with no anthocyanin coloration on the stalks [7]. These two varieties have different geographic distributions: The Desi variety is grown in the semi-arid tropics, while the Kabuli variety is produced in temperate regions [8]. Because Kabuli-type chickpeas fetch a higher market price than Desi-type, they are seen as having greater economic significance. Nonetheless, the majority of genetic resources have been produced for chickpeas of the Desi variety [9].

2. Current status of chickpea production in Tunisia

Chickpeas rank third among the legumes cultivated in Tunisia, following beans and peas in terms of cultivated area. Over the past 5 years, chickpeas have been grown on around 7000 hectares of land, yielding an average of 1300 kilograms per hectare and totaling approximately 9130 tons in production [10]. Over the last decade, all stakeholders in the sector have raised the alarm on the situation of chickpea cultivation in Tunisia, which has shown a clear regression compared to the situation 30 years ago. The regression observed resulted in a drastic reduction in cultivated areas, mainly due to the lack of encouragement for this sector compared to the cereal sector. This clear drop in cultivated areas and chickpea production (**Figures 1** and **2**) has had repercussions on local supply, leading to an increase in imports. The use of chickpea imports to meet the needs of Tunisian consumers has had an impact on the increase in market prices (around 2.5$/kg of chickpeas). The high prices of imported products have also affected chickpea consumption, particularly for poor populations with low or limited income.

Chickpea cultivation also faces a number of technical and environmental problems. Indeed, the sector was dominated by small-scale farmers using largely rain-fed and traditional practices, which made the country vulnerable to climate variability. In addition, farmers are not sufficiently aware of the needs of the crop and are rarely supervised by extension services, which considerably affect the yield (**Figure 2**), production (**Figure 3**), and quality of the product. Over the past three decades, there

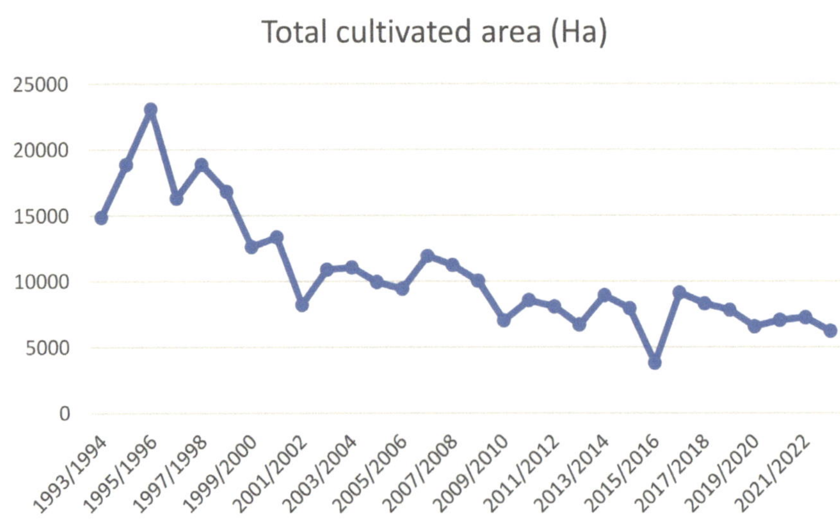

Figure 1.
Chickpea total cultivated areas in Tunisia (1000 ha) for the last 30 years (DGPA, Ministry of agriculture, 2024).

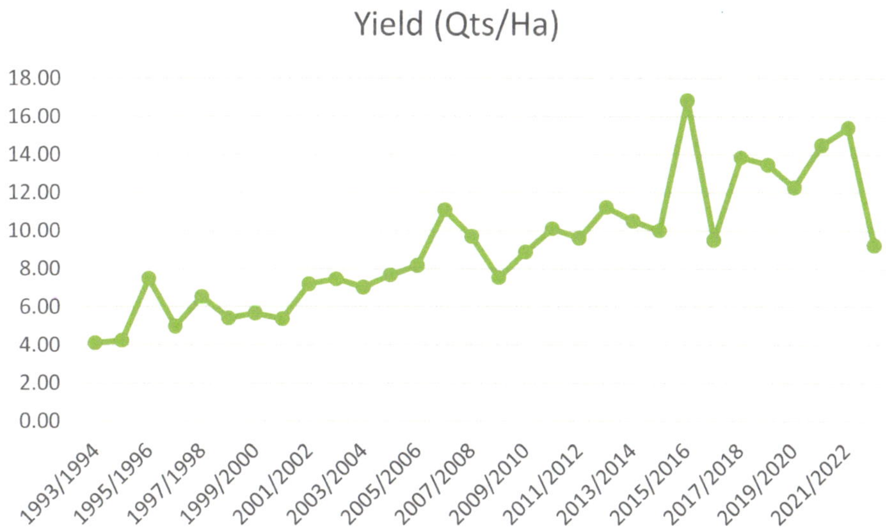

Figure 2.
Chickpea yield in Tunisia (Qts/ha) for the last 30 years (DGPA, Ministry of agriculture, 2024).

has been a notable improvement in chickpea yields, as evidenced by **Figure 2**. This increase can be attributed to the adoption of winter chickpea varieties developed through the Tunisian chickpea breeding program. These varieties are better suited to withstand the challenges posed by climate change and exhibit enhanced resistance to prevalent fungal diseases. Notably, in 2016, we observed a peak yield of 16 quintals per hectare (16 qts/Ha) (**Figure 2**). Despite these advancements, yield fluctuations have occurred due to various biotic factors, predominantly fungal diseases such as ascochyta blight, fusarium wilt, and an emerging concern: root rot, particularly prevalent during wet seasons. Environmental constraints such as drought and irregular distribution of precipitation are pronounced more and more these last years

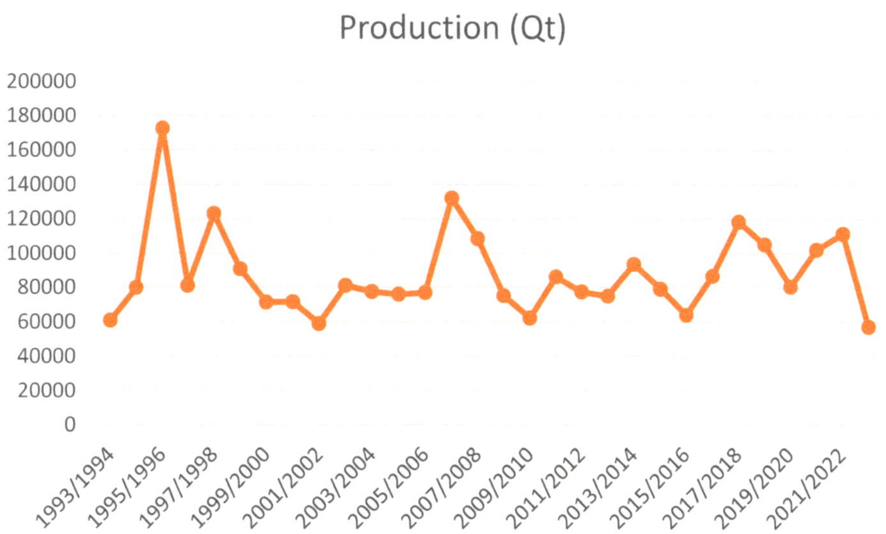

Figure 3.
Chickpea production in Tunisia (Qt) for the last 30 years (DGPA, Ministry of agriculture, 2024).

by climate changes and affect significantly chickpea yields that discourage farmers to develop this crop. Huge efforts are made by research for decades to select new varieties with high yield potential and good resistance to major diseases. Indeed, the research program conducted at the National Institute of Agricultural Research of Tunisia (INRAT) are covering the breeding aspects for legumes (faba bean, chickpea, and lentil) with the collaboration of International Center for Agricultural Research in the Dry Areas (ICARDA). The chickpea breeding programs starts since 1980 at INRAT and released several varieties. However, in order to be most appropriate for the system approach, research on chickpeas must focus on issues brought on by climate change in order to create climate-resilient chickpea varieties. Indeed, for sustainable food security, and soil health, the cereal-based systems in Tunisia need to be diversified with best resilient crops such as chickpea that provide these benefits.

3. Key constraints for chickpea production in Tunisia

The crop faces significant limitations in Tunisia, notably the challenge of small seed size, hindering export potential. Additionally, Tunisia's Mediterranean climate, characterized by modest rainfall in spring and early summer, restricts chickpea growth to residual moisture from winter rains. Moreover, the cultivation is predominantly undertaken by small-scale farmers with limited access to technology, often relying on self-sourced seeds, manual sowing, and minimal fertilizer application. Nitrogen-fixing bacteria inoculation of seeds is uncommon, resulting in ineffective nodulation. Weed management relies heavily on manual methods, with herbicides proving inefficient due to low humidity and inadequate soil preparation.

The yield potential of chickpeas has been greatly enhanced through breeding efforts; nonetheless, farmers' adoption of this crop is still greatly hampered by the crop's unstable production [11]. The vulnerability of cultivars to several biotic and abiotic factors that negatively impacts yield, such as fungal diseases (ascochyta blight and fusarium wilt), pests (pod borer), and drought, has been identified as a significant limiting factor affecting yield and production. Indeed, drought is considered as a serious situation for Tunisian agriculture in the context of climate change.

Parasitic plants could also be a big problem in such particular environments such as Mediterranean conditions [11–13].

Further, the presence of foliar and soil-borne diseases, particularly ascochyta blight and fusarium wilt, poses additional challenges to chickpea cultivation in Tunisia. The lack of government policies that promote chickpea cultivation, research, and investment in plant breeding and sustainable resource management exacerbates the crop's underdevelopment.

One of the main goals of chickpea breeding projects is to create cultivars that are resistant to the main diseases. In Tunisia and many other countries, cultivars resistant to fusarium wilt and ascochyta blight have been released and registered in recent years [14].

4. Main fungal diseases of chickpea in Tunisia

4.1 Ascochyta blight

Ascochyta blight (AB), caused by *Ascochyta rabiei* (Pass.) Labr., is a devastating foliar disease prevalent in Tunisia, capable of causing complete crop loss in numerous chickpea-growing regions worldwide [15]. Various terms such as pathogenic groups, races, and virulence forms have been proposed to classify the pathogenic variation of AB isolates, with different pathotypes documented in various countries, ranging from 2 to 14 [16–18].

The yield losses attributed to AB can vary from 10 to 100 percent during severe natural epidemics [1], leading to diminished yield and product quality while escalating production costs due to the necessity for multiple fungicide applications [19]. Consequently, effective AB control is crucial for enhancing chickpea productivity in both spring and winter seasons.

Cultivating resistant cultivars is recognized as the most economical and practical long-term solution for AB management [20]. Although chickpea cultivars with improved resistance levels to AB have been developed and commercialized, only partial resistance is present within cultivated chickpea germplasm, with the improved cultivars demonstrating moderate resistance to AB. Pyramiding resistance genes from diverse sources in cultivated chickpea stands out as the optimal strategy for enhancing resistance levels [19] and ensuring durability in disease management.

Previous studies have endeavored to unravel the genetic underpinnings of AB resistance [21]. However, resistance to AB appears to be multifaceted; it can be governed by one gene [22], two genes [23], and one recessive and two dominant genes [22], or exhibit polygenic inheritance [24, 25].

Several research teams have identified a shared set of ascochyta blight resistance quantitative trait loci (QTL) across different crosses and genetic backgrounds [19]. The primary AB resistance QTLs are situated on Linkage Groups 2, 3, 4, 6, and 8 [3, 17, 24–27].

4.2 Fusarium wilt (FW)

Chickpea wilt, prevalent in 32 countries across six continents [12], was first noted in India by Butler in 1918, with its etiology elucidated by Padwick in 1940 [28]. Fusarium wilt epidemics, as highlighted by Refs. [11, 29], can wreak havoc on individual crops, leading to up to 100% loss under favorable conditions. The pathogen, *Fusarium oxysporum* f. sp. *ciceris* Schlechtend: Fr. f Sp. *ciceris* (Padwick) T. Matuo and K. Satô [30–32], can survive in soil for over 6 years even without a host [1, 33–35].

Wilting in chickpeas is primarily instigated by the invasion of *Fusarium* species, notably *F. oxysporum*, with similar symptoms observed in other *Fusarium* species such as *F. redolens* [36]. This wilt results from a combination of pathogenic activities, including fungal mycelium accumulation in the xylem, toxin production, and host defense responses such as gels, gums, and tyloses production, alongside vessel crushing by adjacent parenchyma cell proliferation [37].

The fungus infiltrates the plant's vascular system via the roots, facilitated by cell wall degrading enzymes that break down host cell walls, leading to gel formation, blockage of the plant's transport systems, and subsequent wilting [38]. Disease symptoms can manifest at any growth stage, with affected plants appearing in patches or scattered across fields [39].

Resistance to fusarium wilt in chickpeas has been reported as either monogenic or oligogenic, contingent upon the race or resistance source [40–42].

Managing fusarium wilt of chickpeas is challenging, with no single control measure proving entirely effective [35]. As fusarium wilt is a monocyclic disease, its management focuses on excluding the pathogen and reducing the initial inoculum's amount and efficiency [39]. Strategies for control encompass using pathogen-free planting material, avoiding high-risk soils, reducing or eliminating *F. oxysporum* inoculum through chemical, physical, or biological means, employing resistant cultivars, and implementing optimal cropping practices to deter plant-pathogen interactions.

5. Main goals of the Tunisian breeding program

Breeding goals include:

- Large seed size cultivars that match or surpass the agronomic traits of the commonly cultivated varieties in the country (100 grain height $\geq$ 42 grams)
- Disease-resistant genotypes, especially for ascochyta blight and fusarium wilt
- Genotypes with adapted growth habit to the mechanical harvesting
- Genotypes with high-yield potential and high technological quality

6. Breeding method for dual resistance to ascochyta blight and fusarium wilt

In breeding programs, it is imperative to incorporate diverse sources of resistance to broaden the genetic basis of resistance in advanced chickpea lines and elevate resistance levels through gene pyramiding.

Developing resistant cultivars to combat major diseases stands as a paramount objective within the Tunisian chickpea breeding programs. Recent years have seen the breeding and release of cultivars resistant to ascochyta blight (*Ascochyta rabiei*) (AB) and fusarium wilt (*Fusarium* spp.) by these programs.

Traditionally, chickpea in Tunisia was predominantly grown as a spring crop. However, challenges like spring drought and low production prompted a strategic shift toward the development of high-yielding winter chickpea varieties. Winter cropping is deemed crucial for maximizing yield in Tunisia [43], as well as in other Mediterranean countries. The transition to winter sowing necessitated the breeding of cultivars with enhanced resistance to AB, thereby facilitating a substantial increase in production [20].

Annually, numerous nurseries and trials, including material in F3:4 generation or advanced lines, are received mainly from ICARDA and subjected to field evaluations for resistance to *A. rabiei* and *Fusarium* spp. under varying environmental conditions and exposure to different disease pathotypes. This evaluation is conducted across different experimental research stations (**Table 1**). For assessing resistance to ascochyta blight, chickpea genotypes are inoculated with infested debris collected from the previous year at the seedling stage in the field and then scored using the 1–9 scale based on the severity of infection on leaves, stems, and pods as per [44]. The most resistant genotypes exhibiting favorable agronomic traits are selected for generation advancement.

Similarly, resistance to Fusarium wilt complex is evaluated during the same cropping season in a wilt-sick plot at the experimental station of Beja (**Table 1**). The main *Fusarium* spp. involved include *Fusarium oxysporum* f. sp. *ciceris* and other species

Experimental station	Bioclimatic stage	Evaluation and screening
Oued Béja (3644′03.53"N; 913′37.73″E)	Sub-humid (580–600 mm)	Ascochyta blight Fusarium wilt Root rot Orobanche Agronomic traits
Oued Mliz (3628′44.58"N; 829′37.87″E)	Semi-arid (450 mm)	Ascochyta blight Agronomic traits
Kef (3607′31.26"N; 843′19.61″E)	Semi-arid (350–400 mm)	Ascochyta blight Drought, cold Agronomic traits

Table 1.
Experimental stations for chickpea breeding program is Tunisia.

such as *F. redolens*. Using a 0–100% scale for dead plants per plot, resistant lines are identified. Only genotypes demonstrating resistance to both ascochyta blight and fusarium wilt are retained and progressed to the next generation to select for dual resistance. Field evaluations are corroborated by laboratory screening and molecular studies for confirmation.

7. Marker-assisted selection

Integration of genomics tools, such as marker-assisted selection (MAS), holds immense potential for enhancing chickpea breeding programs [13]. The efficacy of MAS hinges on the strength of linkage between markers and gene loci-controlling traits, situated in a densely saturated genomic region with highly polymorphic and easily interpretable markers [1, 11]. MAS offers several advantages over conventional selection methods, including the ability to select for traits that are challenging to score directly, streamlining the process for breeding disease resistance, and facilitating the pyramiding of multiple resistance genes in a single genotype, thereby expediting line development and variety release [11, 45].

Despite the availability of a saturated genetic map and markers closely linked to genes or QTLs, the adoption of MAS in chickpea breeding programs has not been widespread. However, it has been successfully shown that marker-assisted backcrossing can be used to introduce ascochyta blight (AB) resistance with double-podding features in chickpea cultivars. Every backcross generation has used SSR markers to help select for resistance to Fusarium wilt race 1 and blight [19, 46].

AB resistance is closely linked to QTLAR2, and the sequence characteristic-amplified regions (SCARs) SCY17590 and SCAE19336 have been used to tag resistant lines in a collection of chickpea germplasm [47]. When combined with the co-dominant marker CaETR, the SCAR marker SCY17590 located in the genomic region of QTLAR1 for AB resistance enabled the detection of resistance alleles in 90% of resistant accessions in a landrace collection [48]. Recent studies have reported the efficient utilization of both markers in MAS of blight-resistant genotypes in the Tunisian and the Spanish breeding programs [49, 50].

Additionally, in LG2 and LG5, STMS markers that are closely associated with foc01 (TR59, TS35) and foc5 (TA59) have been utilized to help select resistant and susceptible genotypes for the creation of near-isogenic lines (NILs) [51]. A subset of markers

(TA186, TA200, TA106, TA113, TA117, and TA30) identified by Ref. [52] could serve as a reference set for chickpea cultivar identification.

Despite these efforts in utilizing molecular markers across different breeding programs and for various purposes, there is a pressing need to develop useful molecular markers for MAS targeting genes or QTLs associated with key traits such as fusarium and ascochyta resistance, yield components, growth habit, flowering time, and nodulation.

8. Outstanding achievements

The Tunisian breeding program has heavily relied on introduced germplasm, with many crucial traits now firmly established within the program originating from international sources. Traits such as growth habit suited for mechanical harvesting, large seed size, and resistance to ascochyta blight and fusarium wilt have been successfully introgressed from international germplasm.

Ongoing efforts within the program are directed toward enhancing agronomic characteristics such as flowering and early maturation, as well as improving overall yield. Notably, there has been a concerted endeavor to enhance the seed quality of Tunisian chickpea varieties.

During the last two decades, the breeding program achieved significant milestones with the selection, release, and registration of numerous winter chickpea varieties (**Table 2**). These newly developed high-yielding varieties exhibit tolerance or resistance to the most detrimental diseases affecting this crop, reflecting the program's dedication to continual improvement and adaptation to changing agricultural conditions.

9. Conclusion

Remarkable strides have been achieved in chickpea production over the past few decades, largely attributed to sustained research efforts by international centers such as ICARDA and ICRISAT, as well as national research and breeding programs. However, with the increasing impact of rapid climate change in recent years, the primary focus of breeding programs must shift toward adapting germplasm to withstand drought and heat stress, particularly during the critical beginning of flowering.

It is imperative that the state commits to implementing this strategy to bolster chickpea cultivation. Research efforts should prioritize the development of high-performance varieties with enhanced yield potential and technological quality, ensuring farmers have access to resilient options. Additionally, there should be a concerted effort to enhance agronomic practices and facilitate the efficient transfer of appropriate techniques by extension workers.

Moreover, heightened awareness and oversight from industrialists regarding the processing operations of raw chickpea products (including storage, processing, and packaging) are vital. Furthermore, a better understanding and selection of appropriate varieties for product development, alongside increased consumer awareness regarding the nutritional value of legumes like chickpeas, are critical components for promoting the chickpea sector.

Collectively, these elements constitute the cornerstone of success for chickpea cultivation, addressing challenges and contributing to the establishment of a

Variety	Pedigree	Date of registration	Characteristics
Chetoui	ILC3279	1987	• small seed size, • adapted for MH* • tolerance to *Ar** • adapted to winter sowing
Kasseb	FLIP83-46C	1987	• small seed size, • suited for mechanical harvesting • tolerance to *Ar** • adapted to winter sowing
Amdoun 1	—	1987	• large seed size, • adapted for MH* • resistance to Foc* race 0 • adapted to spring sowing
Bochra	FLIP84-79C (X80TH176/ ILC72x ILC215)	2003	• small seed size, • tolerance to *Ar**
Nayer	FLIP84-92C (X80TH176/ ILC72 x ILC215)	2003	• small seed size, • tolerance to *Ar** • adapted to winter sowing
Béja 1	(Amdoun1x ILC3279)x ILC200	2003	• small seed size, • tolerance to *Ar** • resistant to race 0 of Foc* • adapted to winter sowing
Nour	X96TH61-A3-W1-A2-W1-A1-W1-W1	2011	• medium seed size, • tolerance to *Ar** • resistance to Foc* race 0 • adapted to winter sowing
Rebha	FLIP03-27C-A1	2015	• large seed size, • tolerance to *Ar** • resistance to Foc* race 0 • adapted to winter sowing
Joud	FLIP05-58C	2018	• large seed size, • tolerance to *Ar** • resistance to Foc* race 0 • adapted to winter sowing
Zina	(S00787Xsel01ter73616)x FLIP98-22C	2023	• large seed size, • tolerance to *Ar** • resistance to Foc* race 0 • adapted to winter sowing

**Foc: Fusarium oxysporum f. sp. ciceris; Ar: Ascochyta rabiei; MH: mechanical harvesting.*

Table 2.
Chickpea varieties released by the Tunisian legume breeding program.

well-organized food value chain for chickpeas. By addressing these aspects comprehensively, the chickpea sector can thrive amidst evolving agricultural landscapes and consumer demands.

Acknowledgements

This chapter was written by Mariem Bouhadida in the framework of the National project PROPOS on the enhancement of chickpea crop in Tunisia financed by the IRESA (Institution de la Recherche et de l'Enseignement Supérieur Agricole) Ministry of Agriculture of Tunisia.

Conflicts of interest

The author declares no conflict of interest.

Author details

Mariem Bouhadida
Field Crop Laboratory, National Institute of Agricultural Research of Tunisia, Carthage University, Tunis, Tunisia

*Address all correspondence to: mariembouhadida76@gmail.com

References

[1] Singh R, Sharma P, Varshney RK, Sharma SK, Singh NK. Chickpea improvement: Role of wild species and genetic markers. Biotechnology and Genetic Engineering Reviews. 2008;**25**(1):267-314

[2] Kumar J, Abbo S. Genetics of flowering time in chickpea and its bearing on productivity in semiarid environments; 2001

[3] Millan T, Winter P, Jüngling R, Gil J, Rubio J, Cho S, et al. A consensus genetic map of chickpea (*Cicer arietinum* L.) based on 10 mapping populations. Euphytica. 2010;**175**: 175-189

[4] Roorkiwal M, Von Wettberg EJ, Upadhyaya HD, Warschefsky E, Rathore A, Varshney RK. Exploring germplasm diversity to understand the domestication process in Cicer spp. using SNP and DArT markers. PLoS One. 2014;**9**(7):e102016

[5] Cobos MJ, Rubio J, Fernández-Romero MD, Garza R, Moreno MT, Millán T, et al. Genetic analysis of seed size, yield and days to flowering in a chickpea recombinant inbred line population derived from a Kabuli× desi cross. Annals of Applied Biology. 2007;**151**(1):33-42

[6] Varshney RK, Mohan SM, Gaur PM, Gangarao NVPR, Pandey MK, Bohra A, et al. Achievements and prospects of genomics-assisted breeding in three legume crops of the semi-arid tropics. Biotechnology Advances. 2013;**31**(8):1120-1134

[7] Jukanti AK, Gaur PM, Gowda CLL, Chibbar RN. Nutritional quality and health benefits of chickpea (*Cicer arietinum* L.): A review. British Journal of Nutrition. 2012;**108**(S1):S11-S26

[8] Maheri-Sis N, Chamani M, Ali-Asghar S, Mirza-Aghazadeh A, Aghajanzadeh-Golshani A. Nutritional evaluation of kabuli and desi type chickpeas (*Cicer arietinum* L.) for ruminants using *in vitro* gas production technique. African Journal of Biotechnology. 2008;**7**(16):2946-2951

[9] Agarwal G, Jhanwar S, Priya P, Singh VK, Saxena MS, Parida SK, et al. Comparative analysis of kabuli chickpea transcriptome with desi and wild chickpea provides a rich resource for development of functional markers. PLoS One. 2012;**7**(12):e52443

[10] DGPA. Statistics from the General Directorate of Agricultural Production, Ministry of Agriculture, Water Resources and Fisheries (Tunisia). 2024

[11] Millán T, Madrid E, Cubero JI, Amri M, Castro P, Rubio J. Chickpea. In: Grain Legumes. Springer.; 2015. pp. 85-109

[12] Singh M, Bisht IS, Dutta M, editors. Broadening the Genetic Base of Grain Legumes. Springer; 2014

[13] Gaur PM, Jukanti AK, Varshney RK. Impact of genomic technologies on chickpea breeding strategies. Agronomy. 2012;**2**(3):199-221

[14] Singh KB, Malhotra RS, Halila MH, Knights EJ, Verma MM. Current status and future strategy in breeding chickpea for resistance to biotic and abiotic stresses. Euphytica. 1993;**73**:137-149

[15] Pande SKHMS, Siddique KHM, Kishore GK, Bayaa B, Gaur PM,

Gowda CLL, et al. Ascochyta blight of chickpea (*Cicer arietinum* L.): A review of biology, pathogenicity, and disease management. Australian Journal of Agricultural Research. 2005;**56**(4):317-332

[16] Chen W, Coyne CJ, Peever TL, Muehlbauer J, F. Characterization of chickpea differentials for pathogenicity assay of ascochyta blight and identification of chickpea accessions resistant to *Didymella rabiei*. Plant Pathology. 2004;**53**(6):759-769

[17] Udupa SM, Baum M. Genetic dissection of pathotype-specific resistance to ascochyta blight disease in chickpea (*Cicer arietinum* L.) using microsatellite markers. Theoretical and Applied Genetics. 2003;**106**(7):1196-1202

[18] Udupa SM, Weigand F, Saxena MC, Kahl G. Genotyping with RAPD and microsatellite markers resolves pathotype diversity in the ascochyta blight pathogen of chickpea. Theoretical and Applied Genetics. 1998;**97**:299-307

[19] Taran B, Warkentin TD, Vandenberg A. Fast track genetic improvement of ascochyta blight resistance and double podding in chickpea by marker-assisted backcrossing. Theoretical and Applied Genetics. 2013;**126**:1639-1647

[20] Singh KB, Reddy MV. Improving chickpea yield by incorporating resistance to ascochyta blight. Theoretical and Applied Genetics. 1996;**92**:509-515

[21] Coram TE, Pang EC. Isolation and analysis of candidate ascochyta blight defence genes in chickpea. Part I. Generation and analysis of an expressed sequence tag (EST) library. Physiological and Molecular Plant Pathology. 2005;**66**(5):192-200

[22] Tewari SK, Pandey MP. Genetics of resistance to ascochyta blight in chickpea (*Cicer arietinum* L.). Euphytica. 1986;**35**(1):211-215

[23] Dey SK, Singh G. Resistance to Ascochyta blight in chickpea-genetic basis. Euphytica. 1993;**68**:147-153

[24] Flandez-Galvez H, Ford R, Pang ECK, Taylor PWJ. An intraspecific linkage map of the chickpea (*Cicer arietinum* L.) genome based on sequence tagged microsatellite site and resistance gene analog markers. Theoretical and Applied Genetics. 2003;**106**(8):1447-1456

[25] Cho S, Chen W, Muehlbauer FJ. Pathotype-specific genetic factors in chickpea (Cicer arietinum L.) for quantitative resistance to ascochyta blight. Theoretical and Applied Genetics. 2004;**109**:733-739

[26] Anbessa Y, Taran B, Warkentin TD, Tullu A, Vandenberg A. Genetic analyses and conservation of QTL for ascochyta blight resistance in chickpea (*Cicer arietinum* L.). Theoretical and Applied Genetics. 2009;**119**:757-765

[27] Tar'an B, Warkentin TD, Tullu A, Vandenberg A. Genetic mapping of ascochyta blight resistance in chickpea (*Cicer arietinum* L.) using a simple sequence repeat linkage map. Genome. 2007;**50**(1):26-34

[28] Cunnington J, Lindbeck K, Jones RH. National Diagnostic Protocol for the Detection of Fusarium Wilt of Chickpea (Fusarium Oxysporum f. sp. Ciceris). Camberra, Australia: Plant Health Australia; 2007

[29] Ali H, Ul Haq MA, Shah TM, Ur Rahman M, Chen W. Validation of molecular markers for resistance among Pakistani chickpea germplasm to races of *Fusarium oxysporum* f. sp. ciceris. European Journal of Plant Pathology. 2012;**132**:237-244

[30] Jimenez Diaz RM, Jimenez Gasco MM, Landa BB, Castillo P, Navas-Cortes JA. Fusarium wilt of chikpea. In: Chen W, Sharma HC, editors. Compendium of Chikpea and Lentil Diseases and Pests. USA: The American Phytopathological Society; 2011

[31] Landa BB, Navas-Cortés JA, del Mar Jimenez-Gasco M, Katan J, Retig B, Jiménez-Díaz RM. Temperature response of chickpea cultivars to races of *Fusarium oxysporum* f. sp. ciceris, causal agent of Fusarium wilt. Plant Disease. 2006;**90**(3):365-374

[32] Haware MP. Fusarium wilt and other important diseases of chickpea in the mediterranean area. Agricultural and Food Sciences. 1990;**9**:61-64

[33] Halila MH, Strange RN. Identification of the causal agent of wilt of chickpea in Tunisia as *Fusarium oxysporum* f. sp. ciceri race 0. Phytopathologia Mediterranea. 1996;**35**:67-74

[34] Haware MP, Nene YL, Pundir RPS, Rao JN. Screening of world chickpea germplasm for resistance to fusarium wilt. Field Crops Research. 1992;**30**(1-2):147-154

[35] Landa BB, Navas-Cortés JA, Jiménez-Díaz RM. Integrated management of Fusarium wilt of chickpea with sowing date, host resistance, and biological control. Phytopathology. 2004;**94**(9):946-960

[36] Bouhadida M, Jendoubi W, Gargouri S, Beji M, Kharrat M, Chen W. First report of *Fusarium redolens* causing Fusarium yellowing and wilt of chickpea in Tunisia. Plant Disease. 2017;**101**(6):1038-1038

[37] Beckman CH. The Nature of Wilt Diseases of Plants. St Paul, Minnesota: The American Phytopathological Society; 1987

[38] Li H, Rodda M, Gnanasambandam A, Aftab M, Redden R, Hobson K, et al. Breeding for biotic stress resistance in chickpea: Progress and prospects. Euphytica. 2015;**204**:257-288

[39] Jiménez-Díaz RM, Castillo P, del Mar Jiménez-Gasco M, Landa BB, Navas-Cortés JA. Fusarium wilt of chickpeas: Biology, ecology and management. Crop Protection. 2015;**73**:16-27

[40] Sharma KD, Muehlbauer FJ. Fusarium wilt of chickpea: Physiological specialization, genetics of resistance and resistance gene tagging. Euphytica. 2007;**157**:1-14

[41] Sharma KD, Chen W, Muehlbauer FJ. Genetics of chickpea resistance to five races of Fusarium wilt and a concise set of race differentials for *Fusarium oxysporum* f. sp. ciceris. Plant Disease. 2005;**89**(4):385-390

[42] Upadhyaya HD, Haware MP, Kumar J, Smithson JB. Resistance to wilt in chickpea. I. Inheritance of late-wilting in response to race 1. Euphytica. 1983;**32**:447-452

[43] Halila MH, Harrabi MM. Breeding for dual resistance to Ascochyta and wilt diseases in chickpea. Options Méditerranéennes-Série Séminaires. 1990;**9**:163-166

[44] Reddy MV, Singh KB. Evaluation of a world collection of chickpea

germ plasm accessions for resistance to Ascochyta blight. Plant Disease. 1984;**68**(10):900-901

[45] Collard BC, Mackill DJ. Marker-assisted selection: An approach for precision plant breeding in the twenty-first century. Philosophical Transactions of the Royal Society B: Biological Sciences. 2008;**363**(1491):557-572

[46] Varshney RK, Mohan SM, Gaur PM, Chamarthi SK, Singh VK, Srinivasan S, et al. Marker-assisted backcrossing to introgress resistance to Fusarium wilt race 1 and ascochyta blight in C 214, an elite cultivar of chickpea. The Plant Genome. 2014;**7**(1):plantgenome 2013-10

[47] Imtiaz M, Materne M, Hobson K, Van Ginkel M, Malhotra RS. Molecular genetic diversity and linked resistance to ascochyta blight in Australian chickpea breeding materials and their wild relatives. Australian Journal of Agricultural Research. 2008;**59**(6):554-560

[48] Madrid E, Chen W, Rajesh PN, Castro P, Millán T, Gil J. Allele-specific amplification for the detection of ascochyta blight resistance in chickpea. Euphytica. 2013;**189**:183-190

[49] Bouhadida M, Benjannet R, Madrid E, Amri M, Kharrat M. Efficiency of marker-assisted selection in detection of ascochyta blight resistance in Tunisian chickpea breeding lines. Phytopathologia Mediterranea. 2013;**52**(1):202-211

[50] Castro P, Rubio J, Madrid E, Fernández-Romero MD, Millán T, Gil J. Efficiency of marker-assisted selection for ascochyta blight in chickpea. The Journal of Agricultural Science. 2015;**153**(1):56-67

[51] Castro P, Pistón F, Madrid E, Millan T, Gil J, Rubio J. Development of chickpea near-isogenic lines for fusarium wilt. Theoretical and Applied Genetics. 2010;**121**:1519-1526

[52] Castro P, Millan T, Gil J, Merida J, Garcia ML, Rubio J, et al. Identification of chickpea cultivars by microsatellite markers. The Journal of Agricultural Science. 2011;**149**(4):451-460

Chapter 9

Peruvian Popping Beans: Breeding and Performance Data of Beans Cultivated in the U.S.

Jill F. Keith, Jim Heitholt and Donna Harris

Abstract

Nuña beans (*Phaseolus vulgaris* L.) are a class of common beans found in South America. They are grown in the mountainous areas of Peru, Ecuador and Bolivia. Nuña beans are unique because of their ability to "pop" after being exposed to heat. This produces a toasted edible snack that is soft-textured. Nuña bean preparation has reduced requirements for cook time and for fuel making them more energy-efficient compared to other types of dry beans. Scientists have been working to remove the photoperiod-sensitive gene that is found in traditional nuña beans, while at the same time retaining the popping trait for cultivation in the U.S. to be possible. Evaluation of popping bean field performance data, greenhouse performance data, and popping protocols and characteristics of beans cultivated in the U.S. will (1) increase awareness of popping beans as a nutritious source of both protein and dietary fiber as well as an alternative dry bean market class and (2) contribute to knowledge about factors that impact consumption of dry edible beans.

Keywords: popping beans, dry beans, legumes, dietary fiber, plant breeding, protein

1. Introduction

1.1 History and use

Dietary fiber and consumption of bean and legumes among United States (U.S.) consumers is well below recommendations from the U.S. Department of Agriculture (USDA) 2020–2025 Dietary Guidelines. U.S. adults consume half of the recommended amount of fiber per day and individuals aged 19 and 70 fail to meet bean, pea, and lentil consumption recommendations [1, 2]. Dry beans are consumed worldwide as an affordable source of protein and rich source of fiber, starch and vitamins/minerals such as folate, zinc, iron, and magnesium [3]. Intake of dietary fiber, as part of a plant-based diet, including dry beans, promotes healthy weight and prevents the development of chronic diseases such as diabetes, cardiovascular disease, and cancer [3–5]. Incorporating more beans into the diets of both adults and teenagers could provide significant nutritional advantages. A bean variety that cooks quickly and offers an appealing, distinctive taste might encourage greater bean intake.

Important characteristics for bean consumption and production include fast cooking, good flavor, and shelf-stability [6, 7]. Nuña beans (*Phaseolus vulgaris* L.) are a type of common beans found in the highland areas of South America at elevations from 1500 to 3000 m above sea level [8, 9]. Nuna beans are classified within the Andean gene pool of wild beans. The Andean gene pool can be further divided into three subraces: Nueva Granada (N), Chile (C), and Peru (P). Race Peru (P) encompasses the popping Andean nuña beans [10]. Nuna beans from Race P are classified into the Andean Center of Domestication, and are characterized by their medium to large seed, indeterminate strong climbing growth habitat, very late maturity, and short-daylength photoperiod sensitivity [8]. Seed shape can be rounded to elongated with a wide range of colors and sometimes a spotted or mottled surface. Seed size can range from 0.3 g to 1.0 g per seed and it is related to expansion during cooking with the larger seeds having a greater expansion ability [11–14]. Currently, production of nuña beans is mostly found in its native range of the mountainous areas from Bolivia to Peru where there are more than 100 different varieties cultivated with varying yields, colors, and maturities [13, 15–17]. However, nuña beans have been bred and grown in North America by plant scientists in Colorado and Wisconsin [8]. They are a unique edible dry bean because of the "popping" characteristics and ability to expand when being exposed to heat, producing a soft-textures, toasted snack [8, 18, 19]. The popping is a result of rapid expansion of the cotyledons, producing a nut-like flavor like that of roasted peanuts [19]. The expansion of the cotyledon mesophyll tissue in the beans results in a softening and expansion of the entire seed [8]. Studies also indicate that a greater percentage of popping frequency (>90%) can be achieved through reducing moisture content (<5%) in the beans [19].

Nuña beans have a unique potential as a snack food due to their beneficial nutritional profile, including higher starch, amylose, and copper when compared to other dry bean classes [8, 14]. Also included in the beans' nutrients are dietary fiber, minerals, vitamins, and assorted phytonutrients [14, 20]. Nuña beans also have a distinct advantage for consumption due to reduced requirements for fuel and cook time making them energy-efficient compared to other dry beans [21]. They are versatile in that they can be popped in hot oil, hot air, or a microwave oven [19]. Nuña beans offer a high potential for marketing for both producers and consumers, as a niche for small-scale growers, and to be marketed as a healthy snack for consumers of all age ranges [20]. Additionally, nuña bean flour could be used as an alternative source of protein and fiber for extruded nutritious food products as the flour has been shown to have good expansion properties [22]. The unique popping characteristics, preparation method, and flavor profile of popping beans could appeal to consumers. Popping beans cultivated in the U.S. are displayed in **Figure 1**.

1.2 Germplasm resources

The U.S. Patent Office allows the claim of accessions through the USDA germplasm bank, available through the National Plant Germplasm System, and the Germplasm Resources Information Network (GRIN) [23, 24]. Phaseolus genetic material is very diverse, and includes the major cultivated species, as well as their respective conspecific ancestral forms, and many wild species very distant from these gene pools [25]. The Western Regional Plant Introduction Station holds more than 20,000 *Phaseolus* accessions, the most abundant of which is *P. vulgaris* L., a collection which includes 90 accessions classified as nuña beans. The beans were sourced from populations raised among Andean natives in Peru [24]. The CIAT (Centro Internacional de

(a) (b)

Figure 1.
CO 49957 beans dry (a) and popped with oil (b) in the HNF lab at the University of Wyoming in Laramie, WY (2195 m above sea level).

Agricultura Tropical) bean core collection makes available one hundred eighty-two landraces originating in the Andean countries and secondary centers of diversity. Distinction between landraces was classified within agroecological bases of precipitation, temperature, day length at flowering, and soil type [10]. Given that the available germplasm sources of nuña bean exist along a gradient of elevation and precipitation, there is likely enough genetic variation between both nuña beans and the common dry bean to adapt to the photoperiod and climate in North America [24].

1.3 Breeding goals

Although nuña beans have remained a staple crop in their South American center of origin for centuries, efforts to introduce the crop into the North American continent have been largely inhibited by the plants' photoperiod sensitivity, which prevents flowering and seed production in some varieties if the daylight hours deviate from their adapted range [20]. Plant scientists aimed to develop a population of nuña bean crosses which not only expressed consistently high popping ability, but also eliminate photoperiod sensitivity and indeterminate growth habit [19]. Incorporating these traits into a single variety would aid in the adaptation to North American environments. Their study initially used six nuña bean landraces to develop hybrid crosses with temperate-adapted populations through the F_3 generation. The results of the study indicate that the heritability of the popping trait in temperate adapted varieties was relatively high, and selection among family means can be effective for increasing popping percentage [19]. A recombinant inbred line population of 185 individual lines was developed for mapping purposes between two Andean genotypes. The results from this study identified 19 quantitative trait loci (QTL) using three popping traits which included unpopped seeds (PUS), popping dimension index (PDI), and expansion coefficient (EC). QTL that co-localized for popping ability traits were identified on linkage groups 3, 5, 6, and 7. For each of the three traits, PUS, PDI, and ED, 25.3, 24.9., and 14.5% of the phenotypic variation was explained respectively. Conclusions from this study indicated that popping ability is a quantitative trait and epistatic interactions could play a role in its genetic control [26].

Additional breeding efforts set out to combine the popping ability of a nuña bean with the photoperiod insensitivity of a commercial light red kidney bean, by evaluating the plant responses of ten advanced backcross lines of nuña bean at three locations across Colorado [8]. The main effects for nuña lines and location were for seed yield, popping frequency, and seed size. The heritability of the popping frequency was shown to be high among temperate-adapted nuña bean populations. However, there was also a statistical significance for line x location interaction [8]. Lines from this work were used in greenhouse and field evaluations conducted at the University of Wyoming.

2. Nutrient and popping characteristics

2.1 Preparation and popping of popped beans

There is no published standard for popping beans although different methods have been utilized including popping in a small amount of hot oil on a stovetop pan, commercial air poppers, and microwaves [8, 14, 19]. As there was no identified standard, our pilot experiment was developed to define and establish parameters for recommended time and temperature of popping beans that resulted in max popping percentage. Popping analysis was performed in the Human Nutrition and Food (HNF) lab at the University of Wyoming campus in Laramie (2195 m above sea level). Initial testing conducted in the HNF Lab for quality of product and popping percentage included heating beans from six different nuña lines (CO 49957, CO 49956, Wisc 19, Wisc 21, PI 298824, and PI 577678) acquired from the USDA dry bean collection for 1 to 4 min to initiate popping using four different methods: (1) commercially available air popper (without a fat source), (2) aluminum pan placed on an electric range (using canola oil), (3) microwave (without a fat source), and (4) cast-iron pan placed on a gas range (using canola oil). Initial testing using the methods of the microwave and aluminum pan produced inconsistent popping and quality of beans. The second round of pilot pop testing was conducted on CO 49957, CO 49956, Wisc 19, Wisc 21, PI 298824, and PI 577678 which were also obtained from the USDA dry bean collection and grown at the University of Wyoming's Sustainable Agriculture Research and Extension Center (SAREC) located near Lingle, WY. Pilot testing on these six lines used three methods of heating to identify a finalized method of popping: (1) a cast iron skillet placed on a gas range using canola oil, (2) a cast iron skillet placed on a gas range using lard, and (3) air popper. Beans that were popped using fat source were placed in a cast iron skillet with 5 ml of canola oil or lard and heated to induce popping. Researchers selected two distinct fat sources for popping nuña seeds in a cast-iron skillet: canola oil, which is widely consumed in the U.S., and lard, an animal fat similar to lipids traditionally used in the Peruvian highlands. To assess popping performance, researchers tested seeds from six different nuña lines (CO 49956, CO 49957, Wisc 19, Wisc 21, PI 298824, and PI 577678) using three heating methods. For each line, ten seeds were heated for one to four minutes, with care taken to ensure complete heating without burning. Beans that were popped using the air popper (Presto Poplite Hot Air Popcorn Popper) were placed into the heating chamber to induce popping. There was no significant difference in popping percentages for each method during the pilot testing. Although popping percentage was consistent between the air popper and cast-iron pan, the risk of burning the beans was higher using the air popper which negatively impacted sensory qualities. The outcome of the pilot study identified that the method of heating the popping beans in 5 ml of oil in a cast-iron pan over medium heat was optimal. This method of heating would also be more similar with traditional use of the beans in an environment where electricity and fuel may be limited. An infrared thermometer (ELECALL EIRT550E) was used for additional analysis of the cooking temperature which was held 25.4 mm above the skillet surface. The final method used to prepare the beans for subsequent popping included preheating the cast-iron pan for 5 min over medium heat, rinsing and drying the beans, heating the beans in the pan for one to four min, then removing them from the pan and allowing the beans to cool on a paper towel. The popping protocol is displayed in **Figure 2**.

2.2 Analysis of popping percentage

Important characteristics for bean consumption and production include physical properties (such as seed size and color) and fast cooking [6, 27]. Standardizing the popping method is crucial for achieving consistent popped beans in both home and commercial environments, preventing burning and uneven results.

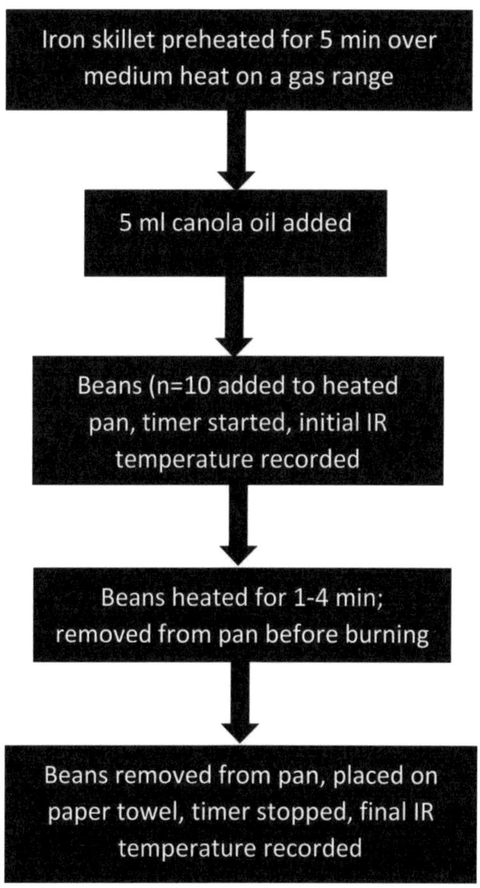

Figure 2.
Finalized popping protocol at the University of Wyoming (2195 m above sea level) for nuña beans cultivated in the U.S.

2.2.1 Pop testing: Greenhouse cultivation

In the fall of 2018, popping percentage was evaluated on 6 lines of nuña beans grown at the Laramie Research & Extension Center (LREC), which included 2 day-neutral lines. Four day-neutral nuña lines (CO 49957, CO 49956, Wisc 19, Wisc 21) and two short-day nuña PI's (PI 298824, PI 577678) from the USDA dry bean collection were planted in the greenhouse in 11.4 L pots using a mixture of LREC site soil, pine bark mix, and sand (1:1:1; v,v,v). No supplemental lighting was used for this experiment. CO 49957, CO 49956, Wisc 19, and Wisc 21 seeds were harvested 74 d after planting. PI 298824 and PI 577678 seeds were harvested 135 d after planting. Pods were shelled, seeds weighed, and kept dry, room temperature storage for 25–30 d before analysis of popping percentage was conducted. For analysis of the popping percentage the final popping protocol method was used (**Figure 2**). There was no significant difference in popping percentage by nuña line, however the day-neutral lines had slightly higher performance. Popping percentage was not significantly impacted by seed size. An additional evaluation of popping percentage on one line grown in the field in Lingle, WY in 2018 and 2019 was conducted. Although initial analysis indicated a decrease in popping percentage the longer the beans were stored, the popping percentage averaged 90–100% in the study. Additional analysis on the impact of storage on popping percentage is described in Section 2.2.3.

2.2.2 Pop testing: Influence of maturation stage and pod color at harvest

In 2019 an experiment was conducted at the Powell Research and Extension Center as well as in Lingle (SAREC) to evaluate the impact of location and harvesting date on popping percentage. Five day-neutral popping bean lines (CO 49956, CO 49957, CO 50004, Wisc 19, Wisc 21) from the USDA dry bean collection were used. Beans were cultivated at both locations using conventional cultural practices (planting date 3 June 2019). During July and August, plots were surface irrigated. During mid-September pods were harvested periodically as they matured. In late September 2019, a – 3.89°C forecast was predicted for early October and the temperature reached −5°C on 3 Oct 2019. Based on the weather prediction, pods were harvested, categorized by their hues (brown, yellow, green), and then air-dried inside for six weeks before manual removal of the shells. Testing of the different seed batches was conducted for seed size, germination percent (5 Feb 2020) and popping percent (13 Feb 2020). Ten seeds from each line and pod color were popped using the final popping protocol to assess popping percentage (**Figure 2**). Mature pods harvested from SAREC from CO 49957 had a popping percentage of 90–100% [28]. Brown and yellow pods from all nuña lines grown at PREC, except CO 49956, also achieved popping percentages of 90–100%. For CO 49957, pod color did impact the percentage of popped beans with green pods having a lower percentage of seeds that popped. However, the pod color had no impact on popping ability of other nuña lines. Popping percentage was not significantly impacted by seed size [28–31].

2.2.3 Pop testing: Influence of storage

CO 49957 was evaluated from 2019 to 2021, to examine the impact of storage on popping percentage [32] The popping line was grown at SAREC near Lingle, WY. The line was grown under conventional practices, harvested by hand in the fall of 2019, and transported to the HNF Lab at UW. Beans of CO 49957 were air dried, hand shelled, and stored in paper bags. They were then popped using the final popping protocol (**Figure 2**). Assessment of popping percentage was conducted at baseline (0 months) as well as after 6 months, 12 months, and 15 months of storage. The calculation for popping percentage was popped beans/total beans assessed x 100. An analysis of the comparison of the difference of baseline (0 months) popping and testing over a period of 6, 12, and 15 months using repeated measures ANOVA with post hoc Tukey correction to account for family-wise error was conducted. Popping onset ranged from 230 to 240°C at baseline, 213 to 240°C at 6 months, 199 to 231°C at 12 months, and 190 to 231°C at 15 months [32]. There was no significant difference observed in onset of popping temperature between the trials. Popping percentage was 90% at baseline, 100% at 6 months, 87% at 12 months, and 80% at 15 months. There was no significant difference in popping percentage at 6, 12, or 15 months when compared to the baseline [32]. Additionally, popping percentage was not reduced significantly at any time point when compared to immediately post-harvest or baseline. A popping percentage increase was observed at 6 months (100%) when compared to right after harvest (90%). The rise in popping percentage might be attributed to reduced moisture content in the beans over time, though this study did not measure the beans' moisture levels. Scientists that study nuña beans have shown that with a seed moisture content lower than 5% maximizes popping percentage [19]. Although not significant, a numeric reduction in popping percentage was observed at 12 and 15 months. Other studies have shown popping percentages between 17 and 81% within 3 months of harvest but were not evaluated further for popping ability after 3 months [8]. Future research is needed where the duration of shelf storage is extended beyond 15 months to determine when a significant reduction in popping percentage might begin. Onset of popping averaged 233°C at baseline, 228°C at 6 months, 228°C at 12 months, and 221°C at 15 months. Other reports show that the ideal temperature was above 244°C which is slightly higher than what was found in this study [19]. Difference in popping methods might help explain the difference in popping temperature as other studies used an air popper that heats popping beans through convection [19]. In this study, the nuña beans were heated through a conduction method. Nuña bean trials in this study were conducted at an altitude of 2195 m while Vorwald and Nienhuis conducted their popping bean trials at an altitude of approximately 267 m [19]. Pressurization within the cotyledon could have been affected because of the high altitude which decreases atmospheric pressure. This is another possible explanation for the difference in ideal popping temperature found in this study when compared to the work published by Vorwald and Nienhuis study [19].

2.2.4 Pop testing: Comparison of popping in oil versus no oil

In the fall of 2022, four lines (CO 49957, CO 50004, Wisc 19, Wisc 21) were evaluated for the impact of oil on popping percentage. Seed was obtained from the USDA dry bean collection and grown in the field in Powell and at the Sheridan Research & Extension Center. CO 49957 was additionally cultivated at SAREC in Lingle, WY. Harvest, storage, and popping protocol are described in 2.1. For each replication of the experiment, ten beans were used. Experiment was replicated at least ten times (n = 100 beans) for each line from each field. The popping percentage for CO 49957 with oil was 82% and 43% without oil, for CO 50004 with oil was 87% and 35% without oil, for Wisc 19 with oil was 71% and 11% without oil, for Wisc 21 with oil was 70% and 26% without oil. With using oil, the highest popping percentage was found with CO 50004. Without oil, the highest popping percentage was found with

CO 49957. The overall 107 replicates of popping trials (n = 10 in each trial) with oil (M = 78.13, SD = 17.00) compared to the 130 trials without oil (M = 33.54, SD = 21.24) demonstrated significantly higher popping percentage with oil ($p < .001$). Popping percentage was significantly higher for every line when oil was used compared to when oil was not used (CO 499957, CO 50004, Wisc 19, Wisc 21, $p < .001$, with 95% confidence level). Results of popping with and without oil are displayed in **Table 1**.

2.3 Sensory analysis of popping beans

Sensory analysis and consumer acceptance studies offer valuable insights into the factors that impact the consumption of dry edible beans and bean-based extruded products. Through these evaluations, key sensory attributes such as texture, taste (encompassing potential bitter, beany, or lingering flavors), and mouthfeel can be thoroughly examined and understood [33–35]. Additionally, physical properties such as seed size and color as well as fast cooking are other important characteristics for bean consumption and production [6, 27]. Willingness to pay and purchasing habits are also included in consumer testing of food products. Price, taste, quality, and nutritional value are factors that have been shown to have an influence on a consumer's willingness to pay [36–38]. There is limited literature related to research on sensory characteristics and consumer feedback of consumer feedback of popping beans. In Wyoming in the spring of 2019, consumer testing resulted in initial feedback from consumers in a university setting (University of Wyoming) using one line of popping bean.

Consumers were introduced to nuña beans, and a sensory analysis was performed with a protocol that was approved by the University of Wyoming's Institutional Review Board (IRB). E-mail and personal invitation were used to recruit participants for on-campus consumer testing. Faculty, staff, students, and community members all 18 or older participated in this study. Those with food allergies including nuts or legumes were excluded from the study. Participants were given an overview of taste testing procedures and instructions prior to sampling which included informed consent before filling out the sensory evaluation form. The panelists evaluated samples using a survey that included a 9-point hedonic scale based on flavor, appearance, aroma, texture, and acceptability. The range for the scale was from "like extremely"

Line	Oil	Mean time (min:sec)	Mean weight of a bean (grams)	Percent popped (%)	Mean initial temperature (°C)	Mean end temperature (°C)
CO49957	No	04:21	4.72	43.17	180.76	270.96
CO49957	Yes	02:26	4.9	81.71	184.63	259.59
CO50004	No	03:37	4.78	34.67	182.87	291.48
CO50004	Yes	02:20	4.63	87.27	181.19	257.67
Wisc 19	No	03:28	4.26	11.00	184.74	289.83
Wisc 19	Yes	02:30	4.23	70.95	186.12	278.65
Wisc 21	No	03:48	4.3	25.50	182.65	286.76
Wisc 21	Yes	03:13	4.21	69.57	180.12	263.48

Table 1.
Results of popping with and without oil in the HNF lab.

(9) to "dislike extremely" (1). Comments could also be added to the evaluation sheets by the panelists. Individual plates were used to serve samples, and the analysis took place in a brightly lit room with white fluorescent lighting. Questions were also asked to the panelists regarding intent to add popping beans to their diet and previous knowledge about the beans. These additional questions were given using a 7-point Likert scale which ranged from "strongly disagree" (1) to "strongly agree" (7). Lastly, participants were asked to respond to questions regarding a willingness to pay. During the evaluation, participants were asked not to interact with each other, and they were seated at separate stations. Beans from Line CO 49957 grown at SAREC in Lingle, WY in 2018 were popped using the final popping protocol (**Figure 2**). No flavors or seasoning were added. Consumer acceptability, sensory attributes, and a willingness to pay mean response main effects were calculated using t-test and analysis of variance (SPSS 26). P values of $p < 0.05$ were considered as significantly different.

Characteristic	n (%)
Sex	
Male	29 (22.3)
Female	100 (76.9)
Age	
18–25	97 (74.6)
26–34	10 (7.7)
35–44	6 (4.6)
45–54	10 (7.7)
55–64	7 (5.4)
65+	0
Ethnicity	
White	101 (77.7)
Hispanic or Latino	16 (12.3)
Black or African American	3 (2.3)
Native American or American Indian	2 (1.5)
Asian/Pacific Islander	7 (5.4)
Other	1 (0.8)
Personal	
Community member	1 (0.8)
Undergraduate student	98 (75.4)
Graduate student	9 (6.9)
Staff	11 (8.5)
Faculty	10 (7.7)
Other	1 (0.8)

Table 2.
Demographics of a sample of 130 consumers evaluating popped beans at a Mountain West university community in Wyoming.

One hundred and thirty faculty, students, and staff panelists evaluated the nuña bean samples. It was a diverse group of panelists which included both women and men of various ages ranging from 18 to 65. There was no clear preference for the popped beans by the panelists. The overall perception was that panelists "dislike slightly" the popped nuña beans. Participant demographics are presented in **Table 2**. Sensory characteristics are shown in **Table 3**.

The study revealed notable variations in "overall acceptability" of popped beans across different demographic groups. Participants aged 18–25 rated the beans significantly lower compared to those aged 35–44 ($p = 0.023$). Similarly, significant differences emerged among racial and ethnic groups, with Native American, Hispanic/Latino, and Asian/Pacific Islander participants scoring the beans higher ($p = 0.008$). Differences in overall acceptability by race/ethnicity are shown in **Table 4**.

When comparing panelists based on their bean consumption habits, interesting insights emerged. Those who regularly included beans in their diet demonstrated a more positive attitude towards bean-related knowledge. They expressed greater interest in learning about different bean and legume varieties and reported higher awareness of preparation methods compared to those who do not frequently consume beans. Furthermore, bean consumers showed a comparable valuation of popped nuña beans, indicating they would be willing to pay prices similar to those of peanuts and kidney beans. This suggests a potential market acceptance for this novel bean product among consumers who are already familiar with and appreciate beans in their diet.

Overall, willingness to pay and consumer acceptance did not show a positive perception of the popped nuña beans. This could be due to a lower average age among

Sensory criteria	**Popped beans**
Appearance	4.38 ± 1.86
Aroma	3.85 ± 2.11
Flavor	4.00 ± 2.29
Mouthfeel	4.37 ± 2.28
Texture	4.32 ± 3.0
Overall acceptability	3.98 ± 2.18

Table 3.
Sensory evaluation of popped nuña beans on a 9-point scale (1 = dislike extremely, 9 = like extremely) with mean attributes and standard deviations reported.

Race/Ethnicity	**Overall acceptability**
White	3.74 ± 1.95
Hispanic or Latino	5.38 ± 2.63
Black or African American	1.67 ± 0.58
Native American or American Indian	5.00 ± 2.83
Asian/Pacific Islander	4.43 ± 2.88

Table 4.
Overall acceptability of popped nuña beans on a 9-point scale (1 = dislike extremely, 9 = like extremely) by race/ethnicity with mean attributes and standard deviations reported.

panelists. It could also reflect the national trend of middle-aged adults and ethnic minority populations which consume beans in higher quantities. Those consumers that already eat beans may be more open and willing to buy and try new types of bean products such as popping beans. Further analysis of sensory attributes using a more diverse panel of consumers could provide a clearer understanding of the perception of popping beans among potential U.S. consumers [39].

2.4 Nutrient characteristics of popped beans

In the fall of 2019 and the spring of 2020, popping beans were evaluated for fat content and protein [32]. Popping beans used for crude protein analysis were obtained from the USDA dry bean collection (CO 49957, CO49956, PI577678, Wisc 19, Wisc 21) and grown in UW greenhouses in 2018 in 11.4 L pots which contained a mixture of pine bark mix, soil, and sand (1:1:1; v,v,v). CO 49956, Wisc 19, and Wisc 21 popping lines were harvested 74 d after planting. PI 577678, a line that is photoperiod sensitive, was harvested 135 d after planting. Pods were shelled, and the beans stored, and popped according to the final popping protocol method (**Figure 2**). Five lines grown in Wyoming were used to measure crude protein content utilizing the Dumas method for nitrogen quantification. The WY State Analytical Lab performed the analysis on crude protein content. To achieve combustion using the Dumas method, A LECO TruMacN was used for the quantitative determination of nitrogen.

Gas-liquid chromatography was used to determine the total fatty acid profile and fatty acid content for line CO 49957. Pods were shelled, beans were stored, and popped according to the protocol described in 2.1 using air popped beans for comparison of fat. A paper towel was used to manually remove excess oil from beans popped in oil, and the popping beans were ground to a fine powder using a food processor. Air popped beans were directly placed into a food processor where they were ground into a fine powder. Samples were partitioned out into 0.5 g amounts and put into freezer bags, treated with nitrogen gas, sealed, and then stored at −80°C until time for fatty acid analysis [32]. Gas-liquid chromatography was used to conduct fatty acid analysis and the fatty acid methyl esters were prepared using the method as described by Weston et al. [40].

A one-way ANOVA was used for analysis of crude protein percentage among the different lines (CO 49957, CO 49956, Wisc 19, Wisc 21, and PI 78767). Mean value for each nuña line was compared after analysis was repeated in sets of three. To account for family-wise error, post hoc Turkey correction was performed. The mean values for total fat content and fatty acid composition were determined through triplicate analyses. Statistical analysis of the total fat content and fatty acid profile was conducted using two-way ANOVA [32].

There was a significant difference ($p < 0.001$) in crude protein content between all five lines. Crude protein content for nuña lines was 22.1% (Wisc 19), 20.9% (CO 49957), 20.1% (Wisc 21), 18.9% (PI 787678), and 17.3% (CO 49956). The total fatty acid content of line CO 49957 averaged 2.90 g/100 g wet weight. The average fatty acid profile of line CO 49957 which was popped in canola oil was comprised of oleic acid (41.4%), linoleic acid (20.4%), α-linolenic acid (18.6%), palmitic acid (10.4%), and stearic acid (2.23%). Preparation of popping beans using the canola oil method, as compared to air popping significantly increased stearic acid (86%, $p < 0.001$), oleic acid (715%, $p < 0.001$), and linoleic acid content (76%, $p < 0.001$). Additionally, a significant increase (115%, $p < 0.001$) in total fatty acid content was identified when popping the nuña lines in canola oil as compared to air popping [32].

Oil popped beans contained 2903 mg/100 g bean of total fat while the air popped beans contained 1348 mg/100 g bean of total fat. Saturated fat content (27%, $p = 0.022$), monounsaturated fat content (715%, $p < 0.001$), and polyunsaturated fat content (33%, $p < 0.001$) significantly increased when popping beans in oil as compared to the air popper [32].

The high oleic acid content in canola oil significantly contributed to the increase in oleic acid levels, with oleic acid comprising 59% of the oil. This insight can help consumers make informed choices about popping methods and oil selection. A standard serving of 100 g of popping beans is equivalent to 118.3 ml, providing a useful reference for portion sizes. An air-popped bean serving of 118.3 ml contains 1.30 g of total fat, while the same serving of nuña beans prepared with canola oil increases to 2.64 g of total fat.

Comparing fat content, air-popped popcorn has a notably higher crude fat content of 3.9 g per 100 g [41]. In response to consumer preferences, low-fat microwavable popcorn options have been developed, with fat content as low as 5.5 g per 100 g [42]. Conversely, stovetop-prepared popcorn can contain up to 26.0 g of fat per 100 g [41]. In this context, a 118.3 ml serving of oil-popped nuña beans with 2.9 g of fat remains well within current guidelines for low-fat snack alternatives.

3. Breeding and cultivation of popping beans

3.1 Greenhouse trials of popping beans

Agronomic traits of existing and advanced breeding lines of nuña beans were evaluated in 2018 for for popping characteristics. The objective of this study was to compare six existing nuña lines which included four day-neutral nuña bean lines (CO 49957, CO 49956, Wisc 19, Wisc 21) and two short-day nuña PI's (PI 298824, PI 577678) from the USDA dry bean collection. The lines were sown in the greenhouse on 30 July 2018 in 11.4 L pots with a mixture of LREC soil, pine bark mix, and sand (1:1:1; v,v,v). Lines were inoculated with a commercial rhizobia specific for *Phaseolus*, and seedlings were thinned to three per pot. Lines were advanced in the greenhouse during the winter of 2017 to 2018 to generate enough seed for this study. Sixteen days after planting, three pots of each entry were fertilized with a common fertilizer (high N) or with an organic fertilizer (low N). Both fertilizers were slow-release. Therefore, the fertilizer source was confounded with rate. Maturity (65% of pods mature) and flowering dates were recorded for each pot and mature pods were harvested as they matured. Air-dried seed was used to calculate yield and yield components. The experimental design used in this study was a completely randomized design and the sources of variation included entry (5 df), fertilizer (1 df), and entry-by-fertilizer (5 df). Due to four missing pots, the error df was 20 instead of 24. Although there was not a significant difference in seed yield among the entries, the common fertilizer (high N) yielded higher than the organic treatment (low N) by 89%. Pod number per pot was 58% greater in the high N treatment, but seed size was not affected by fertilizer. The remaining yield difference can be explained by a 21% greater seed per pod with the high N treatment. One of the PIs had the largest seed (458 mg) while the line, Wisc 21, had the smallest seed size (328 mg). The four adapted lines flowered at around 30 dap (data not shown), and PI 298824 and PI 577678 flowered at 95 and 96 dap, respectively. There was no entry by fertilizer interactions detected [29].

3.2 Field evaluations of popping beans

Wyoming has a unique selection environment for nuña beans due to the greater elevation and shorter growing season compared to the four surrounding states where much of the US dry bean acreage exists. To evaluate the agronomic potential of the aforementioned day-neutral popping bean lines, field trials were established at the UW, Powell REC in 2020 and 2021. In 2020, varieties included five popping bean types, three commercial pinto types commonly grown near Powell, and five novelty/heirloom types not typically grown near Powell. In 2021, varieties included eleven day-neutral popping types, three commercial pinto types, and nine novelty/heirloom types. Seed were sown on 29 May 2020 and 1 June 2021. Plots were six-rows wide and 15-feet long and 22-inch rows were used. Seeding rate was approximately 222,395 seed per hectare and furrow irrigation was employed approximately every ten d during July and August. Variables collected included flowering date, upright stature (0 = prostrate and 10 = fully upright), plant height, maturity date, grain yield per ha, and average seed size. The experimental design used in this study was a completely randomized design with two replications for some varieties that were in short seed supply and three replications for the other varieties. Seed yield was determined after threshing 3-m of row that was harvested from the two center rows of each plot. The performance results indicated that

Entry	Market class	Yield† (kg/ha)	Seed Size (mg)	Flowering (dap)	Maturity (dap)	Upright‡	Height (cm)
CO 49956	Nuña ††	2905	367	42	94	1.7	47
CO 49957	Nuña ††	2726	455	44	99	3.0	44
CO 50004	Nuña ††	3235	402	48	103	3.7	49
Cayenne	Small Red	2369	277	50	87	9.0	55
La Paz	Pinto	1417	339	55	87	9.3	46
Poncho	Pinto	2107	343	47	79	4.0	67
UC-SoWst Gold	Gold Cow	1571	251	48	76	8.0	32
UC-SoWst Red ¶	Red Cow	1610	245	48	73	8.3	35
UC Tiger's Eye ¶	Gold w/Stripe	1045	383	42	74	9.0	41
Wisc 19	Nuña ††	1678	354	45	100	6.0	41
Wisc 21	Nuña ††	1663	350	43	101	6.7	39
Windbreaker	Pinto	1735	364	51	86	7.6	51
Yellowstone	Yellow	2235	331	47	92	7.7	39
LSD (0.05)		809	41	3	5	1.2	13

†*Grain moisture was approximately 10% and no corrections or adjustments were made.*
‡*Upright rating was assessed visually on 18 Aug 2020 and is analogous, but inversely related to, lodging with the higher value representing more upright stature and low values representing prostrate plants.*
¶*UC-SoWst Gold has a formal name of UC Southwest Gold (doi: 10.1002/plr2.20117). UC-SoWst Red has a formal name of UC Southwest Red (doi: 10.1002/plr2.20092).*
††*The five nuña lines were tested for percent popping. Values were: CO 49956, 78%; CO 49957, 93%, CO 50004, 100%; Wisc 19, 67%; Wisc 21, 87% (P = 0.069).*

Table 5.
Results of the field trial at Powell REC in 2020.

several of the day-neutral popping bean types had yield potential equal to or better than the commercial pinto types (**Tables 5** and **6**). The popping bean types were clearly later in maturity than the commercial pinto types. Popping percentages were highest in the CO 49957 and CO 50004 lines for 2020 and for PO-2, PO-4, and PO-5 lines in 2021.

Entry	Market class	Yield† (kg/ha)	Seed Size (mg)	Flowering (dap)	Maturity (dap)	Upright‡	Height (cm)
CO 49956	Nuña ††	4460	422	43	106	7.7	49
CO 49957	Nuña ††	4450	540	45	109	8.3	41
CO 50004	Nuña ††	4519	489	48	115	8.0	55
CO 39704–8	Nuña ††	—	—	53	frost	7.0	93
CO 39706–20	Nuña ††	4251	531	46	113	6.5	81
PI 136692	Heirloom	3505	373	44	97	3.5	48
PI 136712	Heirloom	2019	309	40	75	9.5	37
PI 353548	Heirloom	3174	258	29	87	8.5	38
PI 433617	Heirloom	2078	480	39	82	8.5	39
PI 598303	Heirloom	3556	547	43	109	6.5	93
Othello	Pinto	3982	371	42	81	4.7	57
PO-2	Nuña ††	3643	418	45	101	9.5	42
PO-3	Nuña ††	3449	391	48	105	8.5	51
PO-4	Nuña ††	3404	395	46	101	8.5	48
PO-5	Nuña ††	3514	416	44	97	8.5	47
Poncho	Pinto	3845	406	43	81	5.3	79
UC-SoWst Gold	Gold Cow	2207	287	43	77	8.3	38
UC-SoWst Red ¶	Red Cow	2736	288	43	74	7.5	41
UC Tiger's Eye ¶	Gold w/Stripe	1525	446	39	72	9.1	46
Wisc 19	Nuña ††	4148	446	44	113	8.0	48
Wisc 21	Nuña ††	4072	450	40	115	9.0	44
Windbreaker	Pinto	3726	439	47	96	7.3	64
Yellowstone	Yellow	3225	384	43	93	9.0	39
LSD (0.05)		749	21	3	5	1.3	10

†*Grain yields were corrected to 14% moisture. Moisture at the time of weighing ranged from 8.9% to 10.5%.*
‡*Upright rating was assessed visually at or about the time of maturity; this value is inversely related to lodging; higher values represent more upright stature; low values represent prostrate plants.*
¶*UC-SoWst Gold has a formal name of UC Southwest Gold (doi: 10.1002/plr2.20117). UC-SoWst Red has a formal name of UC Southwest Red (doi: 10.1002/plr2.20092).*
††*The eleven nuña lines were tested for percent popping. Values were: CO 49956, 83%; CO 49957, 95%; CO 50004, 98%; CO 39704–8, na; CO 39706–20, 66%; PO-2, 99%; PO-3, 98%; PO-4, 100%; PO-5, 100%; Wisc 19, 82%; Wisc 21, 87% (P = 0.0005; LSD = 12).*

Table 6.
Results of the field trial at Powell REC in 2021.

4. Conclusions and next steps

Individuals of all ages are unable to meet the USDA's recommendations for pea, bean, and lentil consumption, even though the nutritional and health benefits of regular bean consumption are known. The popping bean, with it's unique characteristic, combined with its rapid preparation time relative to other dry bean varieties, may appeal to some consumers. The results form these studies show that various nuña bean lines can be consumed as a nutritious snack with a short preparation time. The fast-cooking method used in this study has the advantage of both a decreased preparation time and a lesser amount of fuel usage. As a snack, the popping beans could be more appealing to consumers when popped with oil, as it would increase the popping percentage of the beans. Popping beans are an excellent source of plant-based proteins, essential minerals, and fiber, as part of a healthy diet. Intake of dietary fiber, as part of a plant-based diet, which includes dry beans such as popping beans, has been shown to promote healthy weight and chronic disease prevention. Options like that of the nuña popping bean may help contribute to an increase in nutrient dense options for consumers. As human consumption does not currently meet the USDA guidelines, there should be efforts made to increase consumer consumption of dry beans as well as increase consumer education on how to prepare and use them meal planning.

Acknowledgements

This project was supported by funds from the USDA AMS, Specialty Crop Block Grant Program and the Wyoming Department of Agriculture under grant #178064 and by USDA-ARS Hatch project WYO-558-15, Multistate Bean Breeding project W-3150 (i.e., WYO 562 16), and the Wyoming Department of Agriculture under grant #1003874. The authors thank Mark Brick, Barry Ogg, Jim Nienhuis, and the USDA-ARS Plant Germplasm Collection in Pullman, WA for providing the stock seeds. The authors thank graduate students Adrienne Griebel-Thompson, Omer Faruq, Justin Bolak, Devini Wettimuny, Zoe Wall, and Price Akiina for their contributions to the popping bean work at the University of Wyoming.

Conflict of interest

The authors declare that they do not have any conflict of interest.

Author details

Jill F. Keith*, Jim Heitholt and Donna Harris
University of Wyoming, Laramie, Wyoming, USA

*Address all correspondence to: jkeith5@uwyo.edu

References

[1] United States Department of Agriculture. 2020-2025 dietary guidelines [Internet]. 2020. Available from: https://www.dietaryguidelines.gov/sites/default/files/2020-12/Dietary_Guidelines_for_Americans_2020-2025.pdf

[2] Hoy MK, Goldman JD. Fiber Intake of the U.S. Population; What We Eat in American, NHANES 2009-2010 [Internet]. 2014. Available from: https://www.ars.usda.gov/research/publications/publication/?seqNo115=309591

[3] Mitchell D, Lawrence F, Hartman T, Curran J. Consumption of dry beans, peas, and lentils could improve diet quality in the US population. Journal of the American Dietetic Association. 2009;**109**(5):909-912. DOI: 10.1016/j.jada.2009.02.029

[4] Aune D, Chan DSM, Lau R, Vieira R, Greenwood DC, Kampman E, et al. Dietary fibre, whole grains, and risk of colorectal cancer: Systematic review and dose-response meta-analysis of prospective studies. BMJ. 2011;**343**:d6617. DOI: 10.1136/bmj.d6617

[5] Brownlee IA, Chater PJ, Pearson JP, Wilcox MD. Dietary fibre and weight loss: Where are we now? Food Hydrocolloids. 2016;**68**:186-191. DOI: 10.1016/j.foodhyd.2016.08.029

[6] Scott J, Maiden M. Socio-economic survey of three bean-growing areas in Malawi. In: Centro Internacional de Agricultura Tropical (CIAT), Network on Bean Research in Africa. Kampala, UG; 1998;39 p. (Working documents no. 172). Available from: https://alliancebioversityciat.org/publications-data/socio-economic-survey-three-bean-growing-areas-malawi

[7] Reid I, Alberta Agriculture and Rural Development. Factors Influencing Pulse Consumption in Canada [Internet]. 2010. Available from: https://www1.agriculture.alberta.ca/$Department/deptdocs.nsf/all/sis13117/$FILE/v3_factors_influencing_pulse_consumption_final_report_feb24_2010.pdf

[8] Pearson CH, Ogg JB, Brick MA, Berrada A. Popping and yield characteristics of nuña bean lines developed for temperate climates. Agronomy Journal. 2012;**104**:1574-1578

[9] Ogg JB, Brick MA, Pearson C. Introgression of popping ability into dry beans adapted to Colorado. Annual Report of the Bean Improvement Cooperative. 1998;**41**:35-36

[10] Beebe S, Rengifo J, Gaitan E, Duque MC, Tohme J. Diversity and origin of Andean landraces of common bean. Crop Science. 2001;**41**(3):854-862. DOI: 10.2135/cropsci2001.413854x

[11] Cruz Balarezo J, Camarena Mayta F, Baudoin JP, Huaringa Joaquín AH, Sevillano RB. Evaluación agromorfológica y caracterización molecular de la ñuña (*Phaseolus vulgaris* L.). IDESIA (Chile). 2009;**27**(1):29-40

[12] Pesantes Vera MF, Rodríguez Soto J. Caracterización y evaluación de la variabilidad de doce colectas de *Phaseolus vulgaris* tipo reventón procedente de la provincia de Santiago de Chuco. REBIOL. 2013;**33**(2):23-33

[13] Santa Cruz Padilla AE, Vásquez Orrillo JL. Catálogo de Ñuña (*Phaseolus vulgaris* L.) del Banco de Germoplasma del INIA. Lima: Instituto Nacional de Innovación Agraria; 2021

[14] Van Beem J, Kornegay J, Lareo L. Nutritive value of the Nuna popping bean. Economic Botany. 1992;**46**(2):164-170. Available from: https://www.jstor.org/stable/4255422

[15] Gutierrez FL. Selección y caracterización de 21 cultivares de frijol ñuña (*Phaseolus vulgaris* L.) Canaán a 2735 msnm, Ayacucho. Thesis (Bachelor) – Facultad de Sciencias Agrarias. Ayacucho, Peru: Universidad Nacional de San Cristóbal de Huamanga; 2011

[16] Tohme J, Toro Ch O, Vargas J, Debouck DG. Variability in Andean nuña common beans (*Phaseolus vulgaris*, Fabaceae). Economic Botany. 1995;**49**(1):78-95

[17] Zimmerer KS. Biological diversity and local development: "Popping beans" in the Central Andes. Mountain Research and Development. 1992;**12**(1):47-61

[18] National Research Council. Nuña (Popping Beans). In: Vietmeyer ND, editor. Lost Crops of the Incas. Washington, DC: National Academy Press; 1989. pp. 173-179

[19] Vorwald J, Nienhuis J. Effects of seed moisture content, cooking time, and chamber temperature on nuña bean (*Phaseolus vulgaris* L.) popping. HortScience. 2009;**44**(1):135-137

[20] Suszkiw P. The nuna bean: 'Power popper' has funny name, serious nutritional benefits. USDA [Internet]. 2015. Available from: https://www.usda.gov/media/blog/2015/08/25/nuna-bean-power-popper-has-funny-name-serious-nutritional-benefits

[21] Ogg JB, Brick MA, Pearson C. Introgression of popping ability into common beans adapted to temperate regions. Annual Report of the Bean Improvement Cooperative. 1998;**51**:114-115

[22] Pietrysiak E, Zhu Y, Gu BJ, Ganjyal GM. Whole nuña bean (Phaseolus vulgaris L.) flour showed higher direct expansion during extrusion processing at relatively lower temperatures. Journal of Food Science. 2020;**85**(7):2134-2142, 2

[23] Pallottini L, Garcia E, Barcaccia G. The Enola and Nuna bean patents in the context of intellectual property rights for plant cultivar. In: Report of the Bean Improvement Cooperative. 2004:155-162. Available from: https://www.bic.uprm.edu/?page_id=77

[24] Kisha TJ, Naratto G, Kodin K. Analysis of genetic and nutritional diversity among selected accessions of dry beans and nuña beans (*Phaseolus vulgaris* L.) from the USDA-ARS National Plant Germplasm System. Abstracts of Annual Meeting Horticultural Society. 2015

[25] Debouck DG. Phaseolus germplasm exploration. In: Gepts P, editors. Genetic Resources of Phaseolus Beans. Current Plant Science and Biotechnology in Agriculture. Dordrecht: Springer; 1988;**6**:3-29. DOI: 10.1007/978-94-009-2786-5_1

[26] Yuste-Lisbona FJ, Santalla M, Capel C, García-Alcázar M, De La Fuente M, Capel J, et al. Marker-based linkage map of Andean common bean (*Phaseolus vulgaris* L.) and mapping of QTLs underlying popping ability traits. BMC Plant Biology. 2012;**12**:136

[27] Sanzi Calvo M, Attienza Del Rey J. Sensory analysis of beans (*Phaseolus vulgaris*). Biotechnology, Agronomy, Society and Environment. 1999;**3**:201-204

[28] Keith J, Heitholt J, Bolak J, Samet-Brown A. Impact of maturation stage and pod color at harvest on

popping bean lines of *Phaseolus vulgaris*. In: Wyoming Agricultural Experiment Station Field Days Bulletin. 2020. pp. 35-37. Available from: https://www.uwyo.edu/uwexpstn/publications/field-days-bulletin/2020-field-day-bulletins-web.pdf

[29] Keith J, Heitholt J. Potential of seed production of photoperiod-sensitive and photoperiod insensitive popping bean lines of *Phaseolus vulgaris* under greenhouse conditions during the winter months. In: Wyoming Agricultural Experiment Station Field Days Bulletin. 2019. pp. 11-12. Available from: https://www.uwyo.edu/uwexpstn/publications/field-days-bulletin/2019-field-day-bulletin.pdf

[30] Heitholt J, Eberle C, Magnuson B, Keith J. Cooperative Dry Bean Nursery Report—SAREC Lingle2020. pp. 61-63. Available from: https://www.uwyo.edu/uwexpstn/publications/field-days-bulletin/2020-field-day-bulletins-web.pdf

[31] Keith J, Heitholt J. The effect of two nitrogen sources (and rates) on seed yield of six greenhouse-grown common bean genotypes that express the 'popping' trait. In: Wyoming Agricultural Experiment Station Field Days Bulletin. 2019. pp. 13-14. Available from: https://www.uwyo.edu/uwexpstn/publications/field-days-bulletin/2019-field-day-bulletin.pdf

[32] Bolak J, Gifford C, Rule D, Heitholt J, Keith J. Nutrient and Popping Characteristics of Wyoming-Grown Peruvian Popping Beans. Food Science & Nutrition. 2022;**12**:4209-4215. DOI: 10.1002/fsn3.3014

[33] Simons CW, Hall CH, Tulbek M, Mendis M, Heck T, Ogunyemi S. Acceptability and characterization of extruded pinto, navy and black beans. Journal of the Science of Food and Agriculture. 2015;**95**:2287-2291

[34] Mkanda AV, Minnaar A, de Kock HL. Relating consumer preferences to sensory and physicochemical properties of dry beans (*Phaseolus vulgaris*). Journal of the Science of Food and Agriculture. 2007;**87**:2868-2879

[35] Mavromatis AG, Arvanitoyannis IS, Chatzitheodorou V, Kaltsa A, Patsiaoura I, Nakas CT. A comparative study among landraces of *Phaseolus vulgaris* L. and P. Coccineus L. based on molecular, physicochemical and sensory analysis for authenticity purposes. Science Horticultural. 2012;**2012**(144):10-18

[36] Balogh P, Békési D, Gorton M, Popp J, Lengyel P. Consumer willingness to pay for traditional food products. Food Policy. 2016;**2016**(61):176-184

[37] Bejaei M, Wiseman K, Cheng KM. Developing logistic regression models using purchase attributes and demographics to predict the probability of purchases of regular and specialty eggs. British Poultry Science. 2015;**56**(4):425-435

[38] Krystallis A, Chryssohoidis G. Consumers' willingness to pay for organic food: Factors that affect it and variation per organic product type. British Food Journal. 2005;**107**(4/5):320-343

[39] Keith J, Heitholt J, Griebel-Thompson A. Wyoming-grown Peruvian popping beans: Sensory analysis and consumer acceptance. In: Bean Improvement Cooperative (BIC) Annual Report. Vol. 63. 2020. p. 99. Available from: http://www.bic.uprm.edu/wp-content/uploads/2020/11/BICv63_2020.pdf

[40] Weston TR, Derner JD, Murrieta CM, Rule DC, Hess BW. Comparison of

catalysts for direct transesterificiation of fatty acids in freeze-dried forage samples. Crop Science. 2008;**48**:1636-1641. DOI: 10.2135/cropsci2007.07.0376sc

[41] Khan MU, Majeed M, Tayyab M, Shariati MA, Rashidzadeh S. Chemical and nutritional properties of some commercial available corn and wheat products. Journal of Microbiology, Biotechnology, and Food Sciences. 2016;**6**:863-866. DOI: 10.15414/jmbfs.2016.6.2.863-866

[42] Nguyen V, Cooper L, Lowndes J, Melanson K, Angelopoulos TJ, Rippe JM, et al. Popcorn is more satiating than potato chips in normal-weight adults. Nutrition Journal. 2012;**11**:71. DOI: 10.1186/1475-2891-11-71

Section 4

Legume Crops for Sustainable Agriculture

Chapter 10

Perspective Chapter: Technology of Growing the Main Vegetables in Ecological Agriculture

Ioana Stanciu

Abstract

Ecological agriculture promotes sustainable, diversified and balanced production systems, in order to prevent crop and environmental pollution. Ecological production in plant culture, without the use of harmful traditional products, has been a special concern for several decades in economically developed countries. Interest in ecological products and production is continuously growing in our country as well. Leguminous plants play a particularly important role in ecological agriculture. The favorable effect of legumes for successor plants has been known since ancient times. This article includes the importance and biological characteristics of legumes.

Keywords: technology, legumes, agriculture, ecological, cultivation

1. Introduction

Ecological agriculture promotes sustainable, diversified and balanced production systems, in order to prevent crop and environmental pollution. Ecological production in plant culture, without the use of harmful traditional products, has been a special concern for several decades in economically developed countries. Interest in ecological products and production is continuously growing in our country as well. It is regrettable that the areas cultivated under ecological conditions in our country are still very small. In order to expand the organic farming system in Romania, national legislative regulations have been established regarding the production, processing and valorization of organic agri-food products, in accordance with international norms in this sense. Among these, we mention: Government Emergency Ordinance no. 34/2000, respectively Law no. 38/2001; H. G. no. 917/2001 for the approval of the methodological rules for the application of these regulations (including the annexes to the methodological rules), etc.

For the implementation of the provisions of these regulations, the National Organic Products Authority (ANPE) functions as a specialized service within the Ministry of Agriculture, Food, Forests and Environment, which ensures compliance with all specific legal provisions and ensures control over the ecological production methods of agro-food products.

Field plant technologies are developed in the spirit of these regulations, economically efficient and non-polluting, with the prevention of environmental damage and the maintenance of the fundamental resources of agriculture. It is recommended to place crops in rotation on plots converted to organic agriculture, following the best practices, some that also improve soil fertility (such as leguminous plants and green manures), the use of only fertilizers allowed in organic agriculture and the exclusion of all pesticides that pollute production and the environment.

Diseases, pests and weeds in this culture system are combated by cultivating the most resistant varieties/hybrids, by suitable crop rotations, mechanical and physical methods of combating, protecting useful entomofauna, etc. The expected technological solutions are based on the knowledge of the biological elements of plants, orienting the specialist in their application in different climate and soil conditions. The optimal vegetation conditions are highlighted, the whole complex of phytotechnical measures, in order to increase the photosynthetic yield of useful biomass production and to value the genetic potential of cultivated varieties and hybrids.

Plants provide a large part of the products necessary for human nutrition and animal feed, also representing an important source of raw materials for various industries producing consumer goods.

2. Biological importance of legumes

The plants included in this phytotechnical group have as their main product the seeds (grains) rich in protein. This includes: peas, beans, soybeans, lentils, chickpeas, beans, lupins, sorghum, peanuts and broad beans. All are part of the order Leguminosales (Fabales), family Leguminosae (fam. Fabaceae or Papilionaceae). Being from the same botanical family, these plants have certain common morphological, biological, chemical, ecological and technological properties. From the Leguminosae (Fabaceae) family, other plants with various uses (forage, vegetables, medicinal, etc.) are cultivated.

Legumes for grains were cultivated with the beginnings of agriculture, as various testimonies attest. Over 5000 years B.C.E. the inhabitants of the lakeside settlements on the current territory of Switzerland cultivated peas, chickpeas and other legumes for grains. In China, more than 3000 years B.C.E. soybeans are cultivated. The ancient Greeks and Romans appreciated the nutritional value of legumes and their favorable effect on the soil [1]. On the territory of our country, carbonized leguminous seeds were discovered in various Neolithic settlements (polished stone age, 5th–2nd millennium BC) [2]. Over the years, gradually, the number of cultivated species and biotypes increased, due to their great importance in human and animal nutrition.

The importance of legumes for grains lies, first of all, in the high protein content of the seeds, giving them a high nutritional value. Some of them (soybeans and peanuts) also have a high oil content, making it possible to extract them through ordinary technical procedures, being valuable oleaginous plants. Soy is the largest producer of vegetable oil on the globe, and peanuts are in third place (after soybeans and sunflowers).

The protein content of leguminous grains is 2–4 times higher than that of cereals. In some of them (soy, lupine) the protein content exceeds the carbohydrate content. The ratio between crude protein and non-protein components is: in soy and lupine of 1/1.7; to peas 1/2.8; to beans 1/2.4 etc. So leguminous grains represent food and fodder concentrated in protein. It is also worth noting the high protein value of the grains, equivalent in some species to proteins of animal origin, containing essential amino acids. The protein from leguminous grains has a high digestibility (about 90%) and does not form uric acids (like some animal proteins) whose accumulation in the body is harmful.

To achieve, quantitatively, 1 kg of animal protein, 5–7 kg of vegetable proteins are consumed. The use of (processed) vegetable protein in higher proportions in the human diet would greatly increase the protein ration of people and including the yield of agricultural production. The protein production obtained per hectare from the main legumes for grains (peas, soybeans, etc.) exceeds that of wheat, corn, potatoes and other plants. The problem of proteins occupies a central place in the concerns regarding the present and the future of human nutrition. They are basic, essential components of food rations. As shown by I. Cerbari [3], although it seems curious, only in the middle of our century did nutritionists establish that "no human nutritional deficiency syndrome is as harmful as protein malnutrition, although since 1838 Mulder – based on finding that these complex substances containing carbon, oxygen, hydrogen, nitrogen and sulfur are found in all forms of life – he concluded that they must occupy the first place in the unfolding of vital phenomena and called them proteins" (from the Greek "proteias" which means primary, first). In 2013, Cerbari [3], in relation to the body's needs, divided amino acids into essential (indispensable for the body) and non-essential (which can be synthesized in the body). Later it was established that (in addition to their presence) the ratio between the essential amino acids also determines the biological value of the food protein. The amount of protein in a balanced ration of an adult should represent about 11–13% of the energy value, i.e. 1–1.2 g/kg/day [3]. Man, for a rational diet, needs all 10 essential amino acids (lysine, methionine, threonine, histidine, valine, isoleucine, leucine, phenylalanine, tryptophan and arginine) found in animal protein, but also plant products (especially legumes for grains) can provide a good part of this requirement.

Ensuring the global protein requirement is increasingly dependent on the contribution of plants rich in these substances. The "Protein Consultative Group" within the F.A.O. mentions "a new green revolution – that of legumes", and as the "main hope" for solving the protein deficit in the world, legumes for grains are considered.

In addition to proteins, leguminous seeds contain carbohydrates, fats, vitamins, mineral salts, etc., which complete their nutritional value.

In human nutrition, the dry seeds (grains) of these plants are used, as well as the pods and green grains, to prepare various dishes or preserves. The dried grains can be used in human nutrition directly or after a prior industrialization. Products similar to: milk, cheese, meat, etc. are obtained from soybeans. The coffee substitute is obtained from chickpeas and soybeans, and the flour of some legumes (soybean) can be mixed, in a certain proportion, with that of wheat, in the manufacture of bread and flour pasta.

In animal feed, the uruiala from leguminous grains is used in a mixture, in certain proportions, with cereals, which improves their nutritional value. Soy cakes are also a concentrated feed. Legumes for grains are used in the feed of animals in the green or ensiled state, alone or in a mixture. The secondary products of legumes (stems, leaves, sheaths) resulting from threshing have a high protein content (8–14%), exceeding 10 times the content of cereal straw (0.7–1.3%). For fodder purposes, pea, lentil, latir, beans and fava beans straws are used in particular. In some legumes, the stems become lignified, being consumed by animals only after possible processing (chopping, ensiling).

Leguminous plants play a particularly important role in ecological agriculture. In crop rotation, the favorable effect of legumes for succeeding plants has been known since ancient times. This effect derives from the symbiosis of leguminous plants with bacteria of the genus Rhizobium, which fix atmospheric nitrogen, enriching the soil with 100–300 kg of nitrogen per ha and realizing an important saving of conventional energy. The nitrogen left after legumes in the soil is in organic form, easily accessible, having a slow, prolonged action. Being spread at different depths in the soil, it favors the development of the root system of the successor plant. Legumes with high solubilizing power for phosphates (lupin, peas, etc.) make this element available to the successor plants in a more accessible form. The lupine manages to use phosphorus from very difficult to dissolve minerals, such as apatite, enriching the soil not only in nitrogen, but also in easily assimilable phosphorus.

Since all parts of leguminous plants are richer in nitrogen than other cultivated plants, they can be used as green manure for soil fertilization. result good results were obtained with lupine on sandy and acid soils. For this purpose, other legumes can be used, on various types of soil.

Leguminous grains, used in human nutrition, also present some shortcomings, among which are mentioned [4, 5]: the thick covering of the grain, with a high cellulose content, which significantly complicates digestibility; long boiling time; deficiencies in the protein composition of non-sulphured amino acids (cystine and methionine) and tryptophan (these being below the requirements indicated by the O.M.S.); the presence of some non-protein antimetabolites, which reduce the degree of digestibility, causing some states of indigestion, etc.

3. Biological peculiarities of legumes

The requirements for temperature and humidity in the process of germination—emergence differ depending on the species. The minimum germination temperature is 1–2°C for peas, 3–4°C for lupine and beans, 6–7°C for soybeans, 10°C for beans, 12°C for peanuts, etc. Below the minimum temperature, the duration of germination and emergence is prolonged, some of the seeds mold, emergence is difficult, uneven and with gaps. The amount of water absorbed by seeds to germinate, relative to their mass, is about 75% for chickpeas, 92–100% for beans, peas and lentils, 106–110% for soybeans and beans, 116–120% for species of lupine etc. This amount of water is absorbed in 24–48 h, depending on the temperature. From the determinations made for white lupine [6], it was found that legumes absorb water more intensively in the hilum area than in the rest of the seed surface, influencing the germination rate.

The emergence of legumes for grains can be epigeic, when the hypocotyl elongates a lot, raising the cotyledons to the surface of the soil (beans, soybeans, broad beans, lupins, peanuts) or hypogeic, when the hypocotyl grows a little, the cotyledons remaining in the soil (peas, beans, lentils, side) (**Figure 1**).

In general, legumes with trifoliolate and palmate leaves have epigeic emergence (with the exception of *Phaseolus multiflorus*), and those with pinnate leaves have hypogeic emergence (with the exception of *Arachis hypogaea*).

At emergence, after the exhaustion of the reserves in the cotyledons, the plants enter a "critical period" (7–12 days after emergence), until the symbiotic bacteria penetrate the root and the symbiotic system begins to function, after which the plant grows normally. In the soils low in nitrogen, small amounts of nitrogen applied at sowing avoid these stagnations in the growth and development of legumes. Higher doses of nitrogen are not recommended, because they inhibit the symbiotic system.

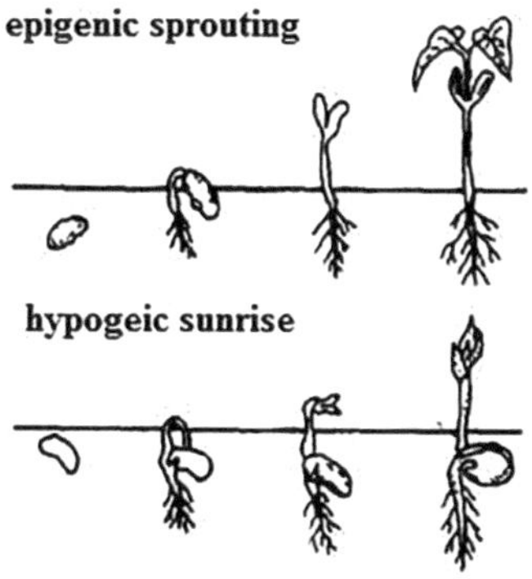

Figure 1.
Epigeic and hypogeic emergence in legumes.

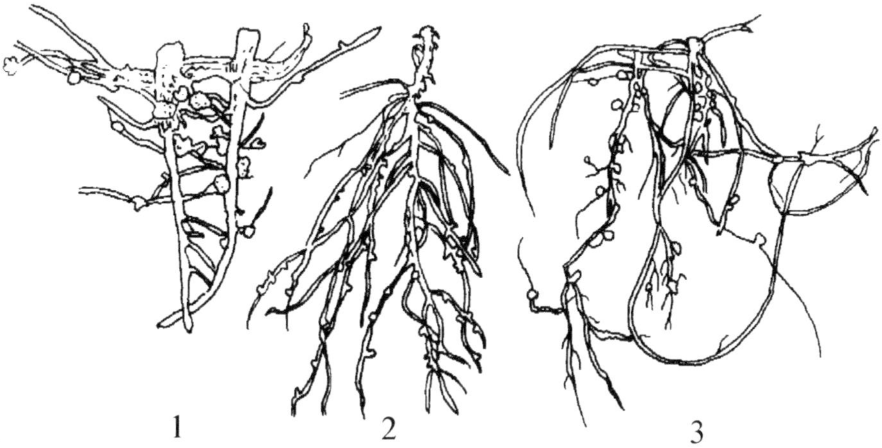

Figure 2.
Legume roots with nodules: 1—lupine; 2—peas; 3—beans.

The root of legumes is taproot. In the first phase of vegetation, a further increase vigorous has the main pivot, which continues its development until almost maturity. Lateral branches develop more slowly at first, then in some species they grow beyond the main pivot. According to the ratio between the main pivot and the secondary branches, Leah and Cerbari [7] distinguishes three types of roots in legumes (**Figure 2**, after [1]).

Type I with thick, strong main pivot that penetrates deep into the soil. The lateral roots are few in number, short and develop later (in the second month of vegetation). Lupine species belong to this type.

Type II with thinner main pivot, having less penetration power than the previous type; instead, the secondary branches are more numerous and develop somewhat earlier, and the first-order branches approach the length of the main pivot. Among the plants that have this type of root, the grain and the chickpea have the pivot better developed than the ramifications, while in peas, lentils and peanuts the first-order ramifications approach the thickness of the main pivot.

Type III has the main root similar to type II, but branches of the 1st and 2nd order they are very numerous. Sometimes the side branches starting from near the base of the stem exceed the main axis in length. Overall, the root appears as fasciculata, being spread more laterally than in depth. Bean and soybean species have this type of root.

Nodulation and symbiotic nitrogen fixation. Some nodules (nodosities) are formed on the roots of legumes, as a result of symbiosis with bacteria of the genus Rhizobium. These bacteria fix atmospheric nitrogen, offering it to the host plant, which in turn provides the bacterium with the carbon dioxide it needs. The value of legumes as soil improvers has been known since ancient times, but only [2] highlighted the role of microorganisms in this phenomenon. Isolates bacteria from nodules, calling them Rhizobium. Klapp (1941, cited by [2]) shows that 15 races of Rhizobium are known, which cannot be replaced and that for each one cultures must be used for inoculation. Later they were considered as different species, as follows [3].

"lupins"—for lupine
"phaseoli"—for beans
"japonicum"—for soybeans, beans, peanuts, etc.

According to some authors, it would be Rh specific for chickpeas. chickpeas, and for peanuts Rh. peanuts [3]. Rh is also known meliloti for lucerne, sulphina and guisdia; Rh. shamrocks for shamrocks; Rh. ornithopes for the seradela, etc.

Bacteria are rod-shaped 1–7 μm long and 0.2–1 μm thick, but their shape and size vary greatly during their life cycle. In the soil, the ciliated forms of the bacteria move, and when they meet the roots of the legumes, they attach to them (specific chemotactism). The bacteria penetrate through the absorbent brushes, forming a cord that crosses the root bark (through the secretion of enzymes, it dissolves the cell walls). Until it begins to fix atmospheric nitrogen, the bacterium lives as a parasite on the host plant, which enters a "crisis" (7–12 days after emergence) of development, which it passes after the symbiotic relationships are established.

The process of biological nitrogen fixation is determined by the existence of a complex enzyme system called nitrogenesis, which mediates the central reaction of breaking the triple bond between molecular nitrogen atoms and their coupling with hydrogen atoms [4]. The entire nitrogen-fixing system is conditioned by a series of endogenous factors. Ammonium produced in nitrogenesis is used in protein synthesis, via fixation by organic acids and following the Krebs cycle, giving birth to amino acids [5].

Rhizobium species are aerobic, adapted to different pHs and are sensitive to soil dryness. They use carbohydrates (glucose, sucrose, maltose, etc.) as energy sources. In the vicinity of the cord (filament) of infection, the cells begin to divide, forming nodules. The shape, size and location of the nodules on the root is different, depending on the plant. In peas, lentils, and milkweed, the nodules are elongated, in grains and beans they are almost spherical, and in lupine they take different shapes. The nodules are arranged especially on the main pivot in lupine, and in beans and soybeans on the branches.

The number of nodules, the size and the depth of formation reach a maximum during the flowering phase of the plants (when the maximum accumulation of dry matter in the roots also occurs). In this phase, the activity of symbiotic bacteria is also at its maximum.

It was found that, when the bacteria become active, a red pigment of hemoprotein nature appears in the nodule [3]. The discovered this porphyrin with iron in the nodosis of legumes, giving it the name leghemoglobin. It is similar to the hemoglobin of the blood of invertebrates, belonging to the group of erythrocrurins [6], being found in nodules in the amount of 0.25–1.0 mg/1 g, fresh substance [1–7]; by oxidation, leghemoglobin is transformed into methemoglobin. Nitrogen fixation is closely correlated with the concentration in leghemoglobin.

Maintaining the biological balance and soil fertility is of great importance in the intensity of nitrogen fixation by bacteria. The irrational use of phytopharmaceutical products, herbicides and other physical-chemical treatments disrupts the activity of nitrogen-fixing microflora. When the conditions of their application are respected, the danger is minimal.

Inoculation (infection) with specific bacteria can also be done artificially. For this, the preparation "Nitragin" (produced in our country), which contains bacterial cultures, is used. It is distributed in colored vials, to protect the bacteria from light, on the label mentioning the species for which the treatment technique can be used. For inoculation, the culture of bacteria from 3 to 4 vials is mixed with 0.5–2 l of water, forming a suspension of bacteria with which the seed used for one ha is sprinkled. The treatment is done in the shade and around the sowing area. Artificial inoculation is necessary especially for leguminous plants newly introduced to culture in certain areas, where specific bacteria are not found in the soil, as is the case of soybeans, on some soils. Good results were obtained, however, with other leguminous plants treated with Nitragin.

Symbiosis, both with bacteria from the soil or artificially inoculated, gives results in favorable conditions of fertilization with N.P.K., Ca and microelements (boron, molybdenum, etc.). In soils with a large amount of assimilable nitrogen, the penetration of bacteria into the roots, the formation of nodules and the nitrogen fixation process are inhibited.

At I.C.C.P.T. A special device was made at the bottom, equipped with nozzles attached to the sowing coulters (S.P.C., S.U., S.U.P.) with which the bacterial suspension can be applied by spraying (pressure 0.5 atmospheres) directly in the furrow (Ştefanic, 1979). Thus, the labor consumption used to treat the seeds is reduced, the treatment does not have to be repeated in case of sowing interruption, the seeds can be treated with fungicides (which are toxic to the bacteria with which the seeds are inoculated). For 1 ha, 18–20 l of liquid are used, when sowing with S.P.C, and 60–80 l when sowing in dense rows with S.U. or S.U.P.

It sometimes happens that the bacterial strains with which the inoculation was done cannot penetrate the roots, due to the invasion of ineffective, more vigorous strains existing in the soil (antagonism between the strains). The existence of bacteriophage viruses, which destroy bacteria of the genus Rhizobium, is also reported, which partly explains the failure of some legumes on certain soils.

The stem differs depending on the genus (species), being: erect in broad beans, soybeans, chickpeas, lupins, broad beans; fickle take climbing or recumbent beans to peas and broad beans. As for the branching, it is more pronounced in soybeans, beans, chickpeas, lupins, latir, groundnuts and less branched in peas and beans. In the cross-section, the stem can be round (lupin and peas), prismatic (lard and beans), or transitional (beans, soybeans and lentils).

The possibility of mechanizing the harvest is conditioned by the erect position of the stem and the place of insertion of the ramifications on the stem.

Legume leaves are compound, having a pair of more or less developed stipules. The leaves are different in type (conformation) and size, shape, color, hairiness of leaflets and stipules. They have even-pinnate leaves: the pea, the bean, the pea, the groundnut, the lentil, and the odd-pinnate leaves: the chickpea. Trifoliate leaves are in beans, soybeans and broad beans, and in lupine they are palmate (**Figure 3**, after [1]).

Stipes in some species are larger than the leaflets (peas), smaller (chickpeas, beans, lentils) or absent (beans). Some legumes (lupin, beans, soybeans and

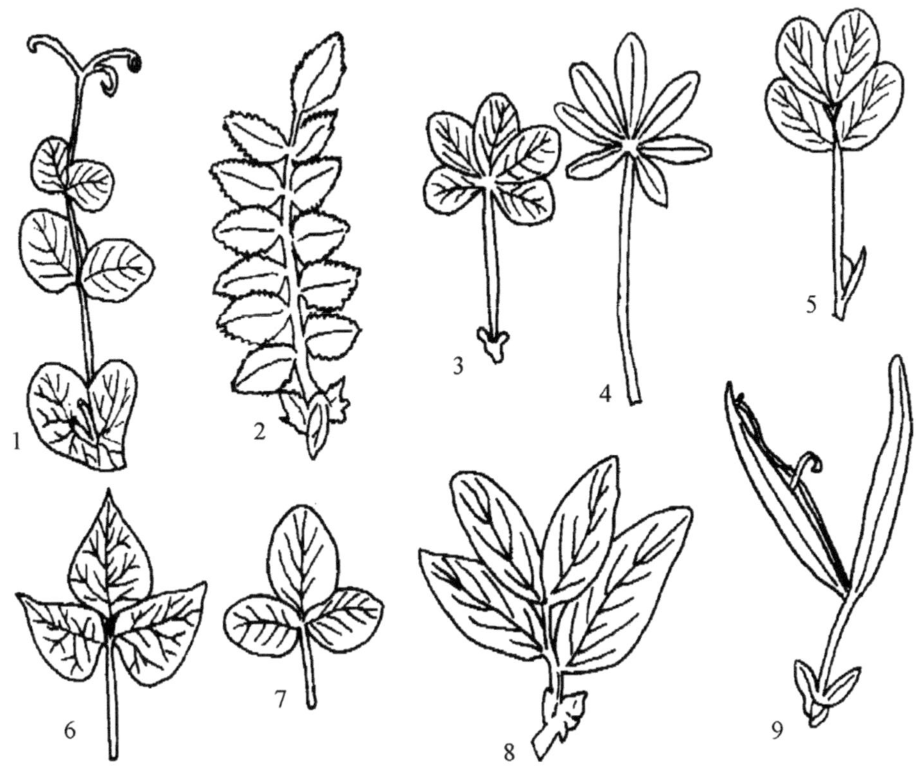

Figure 3.
Legume leaves: 1—peas; 2—chickpeas; 3, 4—lupine; 5—peanuts; 6—beans; 7—soy; 8—grain; 9—lateral.

chickpeas) show heliotropic movements; the leaflets in strong light are arranged obliquely, to avoid it, and in weak light they are oriented perpendicularly, to retain it.

The flowers are grouped in axillary inflorescences (racemes), except for lupine species that have a terminal raceme. The calyx consists of 5 concretized sepals (gamosepal); corolla of 5 free petals (banner, wings and inner part), differently colored; the androecium is composed of 10 stamens, diadelph (9 united + 1 free) or monadelph (all united); gynoecium with superior ovary, monocarpellar, with different number of ovules according to gender, and stigma crushed. The flowers are hermaphrodite, having autogamous pollination (peas, soybeans, lentils, white lupine and blue lupine), with varying degrees of allogamy (peas, soybeans, beans, etc.) or strongly allogamy (yellow and perennial lupine, beans, beans), pollination making it through insects. A common characteristic of legumes is staggered flowering (from the base to the top of the inflorescence) and a low percentage of binding (sometimes it reaches 15–20%). Through a sprinkling irrigation during flowering, fertilization is favored. One of the causes of low fruiting of legumes is the low relative humidity of the air during binding.

The fruit is a pod of different shapes, sizes and colors, dehiscent on the welding line of the valves (sheaths) of the pericarp (beans, peas, broad beans) or indehiscent (chickpeas, lentils, beans) (**Figures 4** and **5**, after [1]).

The seeds (grains) are caught in the fruit on the ventral side, having different shapes, colors and sizes, depending on the species. The hilum—the place of fixation of the ovary ovule, respectively of the pericarp seed—is a characteristic of the species, variety or even variety, being different in placement, shape, size, color. In the hilum

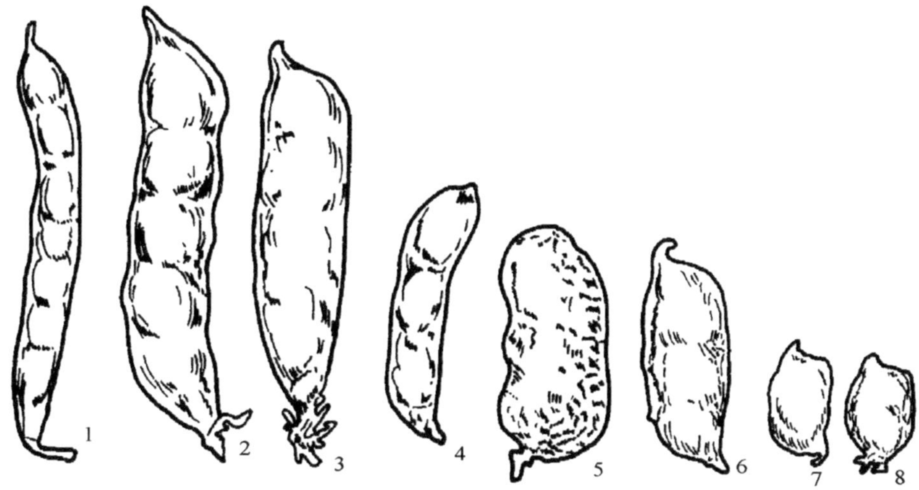

Figure 4.
Legumes: 1—beans; 2—white lupine; 3—peas; 4—soy; 5—peanuts; 6—lateral; 7—chickpeas; 8—lentils.

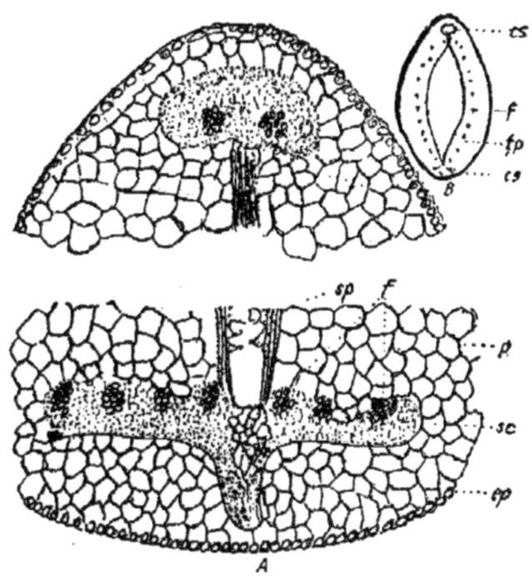

Figure 5.
Transverse section through the pericarp of the pea fruit: A—microscope view; ep—epidermis; sc—sclerenchyma; f—free-woody bundles; sp—parchment layer; p—parenchyma; B—overview; cs—sclerenchymatic cord; tp—parchment tissue; f—free-woody bundles.

area, the two cotyledons are joined by the crucible. Legume seeds do not have endosperm, this being consumed by the embryo during growth; a "rest" of endosperm is found only in chickpeas.

Characteristic for legumes are the so-called "hard seeds", which germinate with difficulty, because they have a palisadic, dense cellular layer in the shell, hardly permeable to water and air (**Figure 6**, after [1]).

In the cotyledons, there are large starch granules and very fine aleurone granules. The formation of the fruit and the seed begins after fertilization, and after a while it enters the three ripening phases:

- green ripening (milk), when the plant, the pods and the seeds are green, continuing to grow, and the content of the seeds is soft, milky;

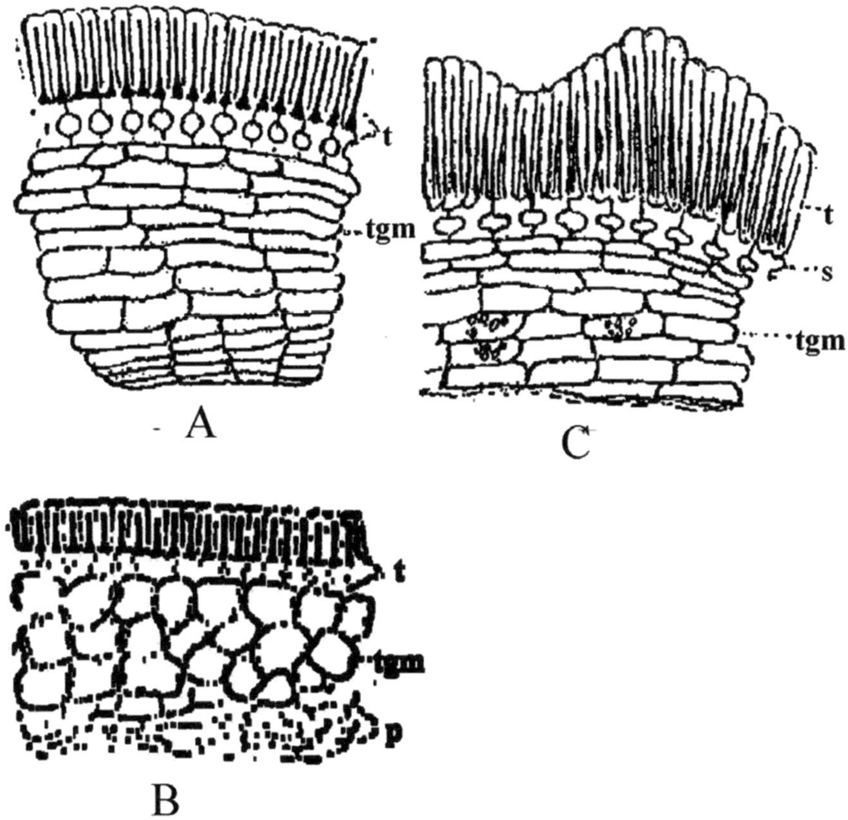

Figure 6.
Transverse section through the integument of the leguminous seed: A—lupine; t—test composed of palisade layer and mossor cell layer; tgm—tegmen; B—beans; t—test composed of palisade layer and mossor cell layer; tgm—tegmen; p—tissue formed by cells with thin walls; C—chickpea; t and s—testa consisting of palisade layer and mossor cell layer; tgm—tegmen.

- yellow ripening (or ripening), when the whole grain becomes yellow, the pods are yellowed, the seeds become consistent like wax, their color being characteristic of the species, the variety and variety to which they belong;
- full ripening, when the vine is completely dry, the fruits and seeds are hard, in dehiscent ones the fruits burst, shaking the seeds and causing losses.

Organogenesis—the differences at the level of the growth cone are barely visible and happen very quickly. At the appearance of the first leaves, the growth tip takes a hemispherical shape, and later secondary cones appear at its base, which gradually give rise to the fruiting organs, the primordia of the flower buds, and then to the flower organs. Details on organogenesis will be presented for each individual plant.

4. Conclusions

Grain-producing leguminous crops: soybeans, beans, peas, chickpeas, lentils, broad beans, broad beans and peanuts are sources of cheap, valuable and absolutely necessary vegetable protein for the national economy.

Romania's sustainable agriculture system can be directed through the implementation, use and modeling of scientific achievements in plant engineering, plant breeding, animal husbandry, economy, biomechanization, environmental protection, fair remuneration of specialists, agricultural producers, etc., with the achievement of quantitative production efficiency and quality of agricultural production.

Author details

Ioana Stanciu
Faculty of Chemistry, Department of Physical Chemistry, University of Bucharest, Bucharest, Romania

*Address all correspondence to: istanciu75@yahoo.com

References

[1] Celac V. Plantele leguminoase–Actualitate şi viitor. Revista de Ştiinţă, Inovare, Cultură şi Artă "Akademos". 2009;**13**(2):77-79

[2] Vozian V, Avadănii L, Iacobuţă M. Soiuri noi de culturi leguminoase elaborate la ICCC "Selecţia". Revista de Ştiinţă, Inovare, Cultură şi Artă "Akademos". 2016;**40**(1):72-76

[3] Cerbari V, Scorpan V, Ţăranu M. Restabilirea stării de calitate a stratului arabil al cernoziomurilor cambice sub influienţa amestecului de ierburi perene graminee şi leguminoase. In: Cernoziomurile Moldovei–evoluţia, protecţia şi restabilirea fertilităţii lo. 2013. pp. 123-126

[4] Mihaela B, Irina T. Studiul comportării unor amestecuri simple de graminee şi leguminoase perene utilizate la înfiinţarea pajiştilor temporare. Revista Lucrări ştiinţifice USAMV Iaşi, seria Agronomie. 2008;**50**(2):236-241

[5] Vozian V, Iacobuţă M, Avadănii L, Ungureanu A. Rolul soiurilor noi în extinderea culturilor leguminoase pentru boabe în sectorul agrar al Republicii Moldova. In: Cercetări la culturile plantelor de câmp în Republica Moldova. 2018. pp. 203-208

[6] Chintea P, Bobeică V, Şveţ S. Saponine steroidice bioactive din deşeuri agroindustriale ale plantelor leguminoase. In: Ecological Chemistry. Vol. 3. 2005. pp. 312-313

[7] Leah T, Cerbari V. Evaluarea agriculturii conservative asupra proprietăţilor solului în rotaţia culturilor cu leguminoase în agricultura Republicii Moldova. In: Agriculture and Food Engineering. 2020. pp. 33-34

Chapter 11

Role of Leguminous Crops by Enhancing Soil Fertility and Plant Nutrition

Ambrin Rajput, Qurban Ali Panhwar and Hafeezullah Babar

Abstract

Food legumes, such as lentils, chickpeas, mung bean, soybean, peas, and beans, have been cultivated worldwide. They are providing essential nutrients and contributing to overall food security. Legumes are rich in protein, ranging from 20 to 45%, and contain essential amino acids, fiber, vitamins, and minerals, making them a nutritionally balanced food source. Biological nitrogen is fixed (90%) by legumes for sustainable agriculture. Meanwhile, the production of various leguminous crops is reduced due to low yield potential, the blend of biotic and abiotic stresses, and environmental changes. The continued cultivation of food legumes in existing cropping patterns is supported by their substantial nutritive values, advantageous cost-benefit ratios, and positive influence on soil health. The strategic integration of legumes into agricultural practices to boost productivity has gained significant attention in response to global food demand. Moreover, legumes play a pivotal role in rejuvenating soil organic matter and addressing problems when included in crop rotations with non-leguminous crops. Research findings that underscore the procedure of N_2 fixation stand out as the most environmentally friendly for meeting the substantial nitrogen requirements. This will increase food production in an eco-friendly manner by reducing reliance on agrochemicals and preserving nutrient balances in the soil.

Keywords: soil nutrient, improving fertility, macro and micro element, chickpea, lentil and mung bean

1. Introduction

In the year 2050, the worldwide population is anticipated to grasp 9.6 billion [1], presenting a myriad of challenges. Foremost among these challenges are the imperatives of achieving food security and addressing the imminent threat of climate change, with a specific focus on mitigating the net release of greenhouse gases into the atmosphere. The consequences of climate change, coupled with the increasing vulnerability of crop systems to both biotic and abiotic stresses, carry significant ramifications for global food production. These challenges underscore the pressing need for the acceptance of sustainable and resilient agricultural practices to ensure a secure and stable food supply for the expanding global population in the approaching decades [2]. In the face of various challenges, mineral fertilizers, particularly those containing nitrogen, have played a crucial role in significantly boosting agricultural yields. However, the continuous processes of agriculture have directed to the depletion and exhaustion of soils, marked by issues such as erosion and the reduction of organic matter and essential nutrients due to the heightened nutrient requirements of cultivated plants. These practices have contributed to environmental problems, including water pollution and the release of trace gases into the atmosphere [3]. The decline in soil organic matter (SOM), a pivotal factor influencing physco-chemical and biological soil properties, has resulted in a considerable decline in overall soil quality. Rasmussen [4] stresses the importance of maintaining and improving soil quality to ensure the sustainability of agricultural activities for future generations. Consequently, there has been a growing interest in adopting environmentally friendly agricultural practices in recent years. This trend is evident in initiatives such as the United Nations (UN) Sustainable Development Goals and the plans outlined in the European Commission (EC) Green Deal [5, 6]. These approaches aim to address the environmental challenges associated with traditional agricultural practices and promote sustainable methods that not only safeguard soil quality but also ensure the long-term viability of agricultural systems. The restoration and enhancement of soil fertility are now widely acknowledged as essential elements in poverty alleviation efforts. Moreover, developing countries grapple with rising costs linked to the use of synthetic fertilizers despite the associated adverse and unpredictable environmental consequences, notably in terms of soil, water, and natural area contamination. In addressing these challenges, legumes emerge as pivotal contributors to the improvement of soil fertility by promoting the activity of microorganisms. These microorganisms play a fundamental role in shaping diverse soil properties, spanning biological, chemical, and physical aspects [7–9]. Certainly, the increasing emphasis on cultivating crops with minimal input and promoting sustainable agricultural systems, when considered from environmental, economic, and managerial standpoints, has led to the ongoing recognition and integration of legumes into agricultural methods [7, 10, 11]. This trend is credited to the beneficial attributes of legumes, such as their capability to fix nitrogen (N_2) and recover soil fertility, while simultaneously disrupting the cycles of diseases and pests affecting crops [12]. Legumes are acknowledged as economical crops, providing advantages in environmental and socioeconomic aspects. Their potential inclusion in modern agricultural systems is particularly relevant, given the common characteristics of reduced crop diversity and heavy reliance on fertilizers and agrochemical inputs in such systems [7]. Legumes are characterized by plants that bear their seeds in pods. While there are over 16,000 known species of legumes, encompassing herbs, shrubs, and trees, nearby 200 are cultivated on a global scale. Over time, the depletion of soil organic reserves occurred due to cereal cropping and frequent fallowing practices. This led to an increased emphasis on green manuring, consequently elevating the importance of legumes in agriculture. However, post-World War II, there was a swift mechanization of farming, and traditional “legume-cereal” rotations fell out of favor. This shift was prompted by the widespread availability of relatively inexpensive nitrogen fertilizers, which were perceived as more efficient than legumes in enhancing grain yields. Despite this, food legumes remain a crucial component of the agriculture sector [13]. Legume Phylogeny Working Group [14, 15]. The latest available data indicates that food legumes are extensively cultivated on a global scale, covering approximately 93.18 million hectares. These crops contribute significantly to agriculture, generating an annual production of around 89.82 million tons. On average, food legumes exhibit a productivity of approximately 963.9 kilograms per hectare. These statistics underscore the widespread cultivation and substantial impact of food legumes to global agricultural output [16]. The leguminosae family is divided into three subfamilies: Papilionoideae, Caesalpinioideae, and Mimosoideae. Notably, the sub-family Papilionoideae is home to various edible legume crops. This includes well-known examples such as chickpeas, mung beans, lentils, peas, and several others. The diverse array of edible legumes within the Papilionoideae sub-family plays a substantial role in global nutrition and agricultural practices, contributing to the food security and dietary diversity of populations worldwide.

2. Leguminous crops

Food legumes are pivotal for food security, contributing to humanoid health, climate change mitigation, and biodiversity enhancement. They serve as significant nutrition source, animal feed, and raw materials. With high nutritional value, legumes are essential plant-based protein sources in humanoid nourishments. Cultivating and consuming legumes not only fosters nutritional well-being but also promotes sustainable agriculture and ecological balance, making them indispensable for a resilient and secure food system. Leguminous crops represent a valuable protein source, notably rich in the essential amino acid lysine, constituting 20–45 percent of the total-protein content. In contrast, cereals are characterized by a high content of sulfur-containing amino acids. Combining legumes and cereals in a diet creates a complementary nutrient profile [17]. This synergistic combination significantly enhances protein uptake in the people. Legumes holds the potential to positively impact the production of dietary food, making nutrition more affordable for low-income groups and addressing issues of malnourishment, especially among lactating women and children, prevalent in many developing countries [17]. Malnourishment remains a pressing issue among lower-income groups in developing countries, primarily due to the inaccessibility of regular animal protein sources like eggs, meat, and milk. Recognizing this challenge, the FAO underscores the importance of incorporating legumes into human diets to address nutritional issues such as malnutrition, micronutrient deficiencies, obesity, and food-related diseases prevalent in many countries. Research efforts have extensively delved into the impact of legumes on improving rice yields. While the use of legumes, especially groundnut stover, as a nitrogen source to enhance rice yields is a focus, a notable challenge is that green manure legumes do not provide immediate economic returns to farmers. Obstacles such as the availability of legume seeds, the need for phosphorus fertilizer, and the integration of green manure legumes into rice production systems impede widespread adoption, particularly among small farmers with limited resources. In contrast, grain legumes like groundnuts, which yield edible grains, are more appealing to farmers. Studies, including the work by McDonagh [18] and Whitmore [19], suggest that groundnuts can seamlessly integrate into production systems, offering economic benefits. Furthermore, Toomsan [20] has explored the positive effect of legumes on soil fertility, concluding that they contribute to increased yields in subsequent crops such as rice, maize, and sugarcane.

2.1 Chickpea

The provided information highlights the significance of pulses, particularly chickpeas (*Cicer arietinum* L.), as essential leguminous food crops. Chickpeas are rich in nutrients and vitamins A, B, and C. Further, it contains dietary protein, carbohydrates, fat, and crude fiber (23, 64, 5, and 6%) [21, 22]. It contributes significantly to the global human food supply, accounting for 20% of the world's food. This makes them a vital source of nutrition for a large population. Chickpeas are a winter-season crop and can thrive in various climatic conditions, including tropics, arid and semi-arid regions, temperate zones, and rainfed areas worldwide. Chickpeas have the ability to fix atmospheric N_2, which contributes to improving soil fertility. This is beneficial for sustainable agricultural practices [23, 24]. Pulses are cultivated in Pakistan on approximately 931 thousand hectares. The national production is reported to be 359,000 tons. The grain productivity is specified as 0.386 ton per ha [25]. In the Sindh province, pulses, particularly chickpeas, are cultured about 22,000 hectares. The total production of the province is reported to be 19,000 tons. The grain yield in Sindh is given as 0.864 ton per hectare [26]. The mean chickpea yield in Pakistan is noted as 0.386 ton per hectare. A comparison is made with other countries, indicating that Pakistan's yield is relatively lower than countries such as China (3.3 ton per hectare), Canada (1.90 ton per hectare), USA (1.70 ton per hectare), Lebanon (2.30 ton per hectare), and Egypt (1.70 ton per hectare) [27, 28]. Chickpea is identified as one of the eight "founder crops." It was shown in the Fertile Around 12,000–10,000 years ago, as indicated by research conducted by Kislev and Bar-Yosef [29], chickpeas became prominent. The earliest indications of chickpeas being utilized as a food source trace back to the eighth millennium BC. Noteworthy archaeological evidence from sites like Tell el-Kerkh and Tell Abu Hureyra in Syria confirms the early consumption of chickpeas. Domesticated chickpeas have been identified in archaeological sites dating back to the Pre-Pottery Neolithic period, initially concentrated in the Fertile Crescent. However, during the late Neolithic era, chickpea cultivation extended beyond the Fertile Crescent and reached modern Greece [30]. Chickpea seeds have even been discovered in the Nile Valley, with indications of their presence dating as far back as the New Kingdom (1580–1100 BC). Additionally, the nutrients nitrogen (N), phosphorus (P), and potassium (K) exhibit a synergistic effect when utilized in combination. The balanced and combined use of these nutrients is highlighted for better crop enhancement and improved productivity. The application of macronutrients (NPK) is suggested to assist plants in a more effective mode, leading to enhancements in plant growth parameters. The overall goal is to contribute to higher crop production. The uptake of these nutrients (NPK) by plants showed a positive response, particularly with the inclusion of potassium. The positive effects on chickpea cultivar growth are mentioned. The combined use of NPK is reported to have a significantly positive impact on chickpea cultivars. This suggests that a balanced application of these nutrients is beneficial for the overall performance of chickpea crops. The information is supported by previous studies conducted by Mondal [31] and Goud [32]. Specific examples include K rates (25 and 30 K_2O kg ha^{-1}) given with NP, which showed positive responses in various leguminous crops. Potassium is described as a third key element that is required in large amounts for various essential plant functions. Further, K serves as an enzymes, contributing to respiration, aiding in photosynthesis, regulating water within plants, and exhibiting synergistic effects with nitrogen and phosphorus [33]. Despite the importance of potassium, the information suggests that research on the use of K has not acknowledged sufficient attention in pulse crops in Sindh, Pakistan [34]. K fertilizer stands out as a financially burdensome contribution (agricultural) in Pakistan, with farmers reportedly refraining from its utilization in crop production due to its exorbitant cost [35]. Specifically in the case of chickpeas, the application of K is deemed negligible. Despite numerous studies highlighting its positive impact on crop's output [36, 37], farmers tend to overlook the utilization of K in chickpea cultivation due to its high price. Various researchers have delved into the effects of different inorganic mineral rates and their collective applications on various crops including chickpea [38–40]. Notably, application of P_2O_5 and K in the chickpea variety (CM-98) have been reported to yield positive effects on various growth parameters, encompassing plant height, pods bearing branches per plant, pods per plant, seed pods per plant, seed per plant, grain, biomass, straw, productivity and 1000-grain weight [37, 41].

The addition of K showed a 7% rise in plant height and a 100% increase in the number of pods per plant compared to the control, marking a 13% improvement over treatments without potassium applications. In the presence of potassium, the plants exhibited a remarkable boost, producing 22% more grain weight (1000 grains), a 23% increase in biomass yield, and a 16% rise in straw yield compared to treatments without potassium. The grain yield experienced a substantial surge from 975 to 1896 kg ha^{-1} compared to without K application (**Table 1**).

The introduction of potassium (K) resulted in a notable 10.4% rise in protein content compared to the treatment without potassium in the study (**Table 2**). There were variations in nutrient uptake among different fertilizer treatments, with the combined fertilizer application (NPK) demonstrating the highest nutrient uptake in the shoots of chickpea plants. Specifically, the NPK treatment exhibited higher concentrations of macronutrients, including N at 3.6 ppm, P at 2.56 ppm, and potassium (K) at 0.68 ppm, in comparison to treatments without K [43]. The utilization of NPK had a significant effect on the (N), (P), and (K) in the shoots, leading to peak values for reproductive parameters, including a seed index of 241.5 g and a biological, straw, and economic yield. Similar positive effects of NPK treatments on growth and yield were observed in other leguminous crops, such as gram, during the winter season (2007–2008). The outcomes indicated the gram cultivar Paidar 91, subjected to NPK

Treatment	Plant height (cm)	Pods $plant^{-1}$	1000-grain weight (g)	Biomass yield (kg ha^{-1})	Straw yield (kg ha^{-1})	Harvest index	Grain productivity (kg ha^{-1})
Control	62.2C	21.7C	375C	4060C	3085B	24.0C	975C
NP	79.2B	38.7B	421B	5005B	3420A	31.5B	1584B
NPK	85.0A	43.8A	455A	5205A	3309A	36.5A	1896A

Rajput [42].

Table 1.
Effect of K application on growth and productivity parameters of chickpea.

Treatment	Protein	Shoot N	Shoot P	Shoot K	N uptake	P uptake	K uptake
Control	8.75C	1.2C	0.2C	0.3C	44.3C	8.7C	19.8C
NP	14.28B	1.8B	0.3A	0.3C	80.1B	14.3B	35.5B
NPK	15.77A	2.1A	0.3A	1.3A	95.3A	16.4A	49.9A

Modified from Rajput [42].

Table 2.
Effect of K application on protein and NPK uptake (%) of chickpea.

treatments, recorded uppermost plant tallness (104.7 cm), seed numbers per pod (1.76), and 1000-grain mass (280.08) compared to other cultivars [37, 38]. It is worth noting that in certain regions of Pakistan, like Sindh, the soil is generally alkaline and calcareous, with low organic matter and imbalanced fertilization. This condition may lead to macro and micronutrient deficiencies. The study underscores the global concern about micronutrient deficiencies, affecting over 40% of the world's population and causing various health problems, economic costs, and learning disabilities, particularly in children. Micronutrient deficiencies in plants have become a growing global concern due to intensive farming practices.

2.2 Mung bean

Mung bean (*Vigna radiata* L.) is a versatile herbaceous belonging to the family Phaseoleae [44]. In Pakistan, it serves multiple purposes, being utilized as a cereal-based food grain, an ingredient in cuisine, a vegetable, green manure, livestock feed, and even for medicinal purposes [45]. The mung bean is recognized as an energetically ironic source of easily edible protein and lysine [46]. The nutritional composition of mung bean seeds is notable, comprising proteins (24.70), fat (0.65), fiber (0.95), and ash (3.75) % [47]. Moreover, mung bean seeds boast a rich array of vitamins and minerals. The vitamin content encompasses vitamin A (74.370), vitamin C (4.80), and folate 625.0 mg. Regarding minerals, mung bean seeds are a source of calcium (132), phosphorus (367), and zinc (2.7) mg/100 g [47]. This nutrient-dense profile underscores the significance of incorporating mung bean into diets for potential health benefits [48]. Mung bean holds the second position after chickpea among pulse family [49, 50]. In Pakistan, the cultivation area for mung bean was reported at 0.163 million hectares, yielding 118 thousand tons with an mean value of 723 kg ha^{-1} in a specific year [46]. In a subsequent reporting period, the area, production, and yield of mung bean in Pakistan increased to 186 thousand hectares, 133 thousand tonnes, and 711 kg ha^{-1}, respectively, showing a 12.6% increase in production over the previous year [51]. Despite the rise in production, the average yield of mung bean in Pakistan remains significantly below the potential yield level of 2650 kg/ha compared to other countries. Notably, approximately 98% of cultivators in Sindh, Pakistan abstain from using phosphorus mineral for mung bean crops, because due to the common thought among growers, as a leguminous crops, it does not need mineral nutrition. In Pakistan, about 90% of soils suffer from P deficiency. Even when phosphorus is applied, its efficacy is limited, ranging from 15 to 20% at best, particularly in alkaline and calcareous soils [52]. These factors contribute to the suboptimal yield of mung bean in the country.

The issue of P inadequacy in Pakistani soils is widespread due to P sorption with $CaCO_3$ and Ca ions in soils with higher pH [53]. Addressing production gaps necessitates careful consideration of the balance and efficient utilization of phosphorus minerals alongside nitrogen and potassium applications [33]. Despite the prevailing trend among farmers to overlook fertilizer application due to high costs, the application of phosphorus (P) has demonstrated positive outcomes in enhancing crop production. Local studies have illustrated the combined effects of P with N and K mineral application on the productivity of mung bean and other pulse and oil seed crops [40, 42]. Research conducted by Fageria [54] emphasized the increase in dry bean grain productivity with the application of P mineral nutrients. Likewise, Higgs [55] observed a global grain production increase of 30–50% since the 1950s with the application of fertilizers, including P.

Numerous studies have highlighted the crucial role of the combined application of nitrogen, phosphorus, and potassium (NPK) in enhancing plant height, nodulation, agronomic progression, productivity, and value of mung bean, which have been extensively studied by researchers [56, 57]. Moreover, the combination of P with N has consistently been shown to produce the highest productivity, fruit-bearing branches per plant, and protein content [58–60]. Additionally, various methods of applying different P levels, such as broadcast, banding, and fertigation, have been investigated in mung bean crops has been studied which has shown that fertigation methods with increasing P levels provide the highest yield, uptake, and agronomic efficiency [61]. This emphasizes the importance of optimizing P application methods for improved mung bean production in Pakistan. Similarly, the combined application of nitrogen, phosphorus, and potassium (NPK) with rhizobacteria demonstrated the highest yield in mung bean crops [62]. Furthermore, Malik [63] concluded that the highest productivity was achieved with a phosphorus (P) rate of 50 kg ha^{-1}. Sadeghipour and Tajali [59] concluded that the maximum productivity (224.2 kg ha^{-1}) was obtained with 120 kg P_2O_5 ha^{-1} along with nitrogen (N) at 90 kg ha^{-1}. In contrast, Kumar [64] showed that the maximum yield (10.78 kg ha^{-1}) and straw productivity (26.63 kg ha^{-1}) were obtained with a P_2O_5 application of 45 kg ha^{-1}. Moreover, studies by Meena and Varma [65] and Kaysha [66] consistently indicated that higher yields were observed with increased rates of phosphorus application. The information on the importance of the nitrogen-phosphorus (N:P) ratio, water use efficiency, and the necessity of simultaneous application of nitrogen and phosphorus for maximum crop production is insightful. Additionally, the role of nitrogen in enhancing grain quality and crop production, as highlighted by various studies, adds depth to the understanding of nutrient management practices. Nitrogen and phosphorus stand out as the most deficient nutrients in the soil. Hence, their concurrent application is essential to achieve optimal crop production. Nitrogen boosts the quality of grain with higher crop production. Before using fertilizer, it is critical to identify the soil's nutritional values including nutrient uptake in plants. Additionally, the role of nitrogen in improving grain quality and crop production, as highlighted by various studies, adds depth to the understanding of nutrient management practices. Moreover, the emphasis on the importance of assessing the soil's nutritional condition and plant nutrient uptake before using fertilizers reinforces the need for a targeted and informed approach to nutrient management in agriculture. This information contributes to a well-rounded understanding of the factors influencing mung bean productivity and the significance of nutrient optimization strategies.

The plant height exhibited a notable increase, rising from 67.05 cm under control conditions to 75.0 cm with the application of 150 kg of P_2O_5 per hectare. Various levels of phosphorus (P) application demonstrated an increasing trend compared to the control plot [42]. The addition of 50 kg ha^{-1} of P_2O_5 showed in the maximum number of pods per plant (88), representing a 49% increase compared to the control plot (**Table 3**). This trend continued to escalate with higher P levels. The maximum count of fruit-bearing branches (32) was observed at the P level of 50 kg P_2O_5 ha^{-1}, marking a 20% increase over the control plot where the minimum count (17) was documented. The lowest productivity (859 kg ha^{-1}) was noted in the plot wherever P was not used, and only N and K were used. The results revealed that the maximum yield was calculated with the application of 75 kg ha^{-1} of P_2O_5 in combination with N and K, resulting in a maximum grain productivity of 2089 kg ha^{-1} at the rate of 50 kg P_2O_5 ha^{-1}. The biomass and straw productivity increased from 1260 and 401 kg ha^{-1} in the control plot to 2986 and 508 kg ha^{-1}, respectively, with the application of

Treatment $NP_2O_5K_2O$ Kg ha^{-1}	Plant height (cm)	No. of plants $treatment^{-1}$	No. of fruit-bearing branches	No. of pods $plants^{-1}$	1000 grain Weight (g)	Grain yield (Kg ha^{-1})	Biomass yield (kg ha^{-1})	Straw yield (kg ha^{-1})
Control (25–0-25)	67.47A	74.0A	17.1A	59B	52.9D	993.3E	6000F	368.3C
1/2 P (25–25-25)	69.0A	70.0A	18.3A	74AB	52.4CD	2025.3D	7367E	717.1AB
3/4P (25–38-25)	70.5A	78.0A	18.5A	80A	55.7C	2478.7C	7667D	871.6AB
Full P (25–50–25)	70.2A	79.3A	22.9A	88A	57.5B	2922.7B	8467C	1022.9 BC
1.5P (25–75-25)	69.5A	74.3A	22.4A	85AB	55.4A	2911.3A	8133B	1017.3 BC
2P (25–100–25)	69.9A	63.3A	22.0A	85AB	51.7A	3036.3A	9733B	1057.8ABC
2.5P (25–125–25)	75.4A	76.7A	22.7A	82AB	58.3A	3190.0A	9700A	1110.1A
3P (25–150–25)	66.1A	90.3A	22.3A	83AB	59.6A	3190.0A	9800A	1110.9A

Modified from Rajput and Memon [42].

Table 3.
Phosphorus application on yield contributing parameters of mung bean.

30–75-30 kg ha^{-1} of NP_2O_5K2O. Hence, biomass and straw productivity enhanced by >100 and 26%at 75 P_2O_5 kg ha^{-1} over control plot [42].

2.3 Lentil

Lentil, derived from the Latin word "lens," accurately describes the character of the seed. The name Lens culinaris to lentil was given by the German botanist Medikus in 1787, as mentioned by Cubero [67]. Lentils hold the position of the second significant cool-season food legume, following chickpeas. They have been cultivated for over 8000 years, signifying one of the early domesticated species, as indicated by Dhuppar [68] and Cokkizgin and Munqez [69]. Lentils are thriving in semi-arid regions of the world, notably in the Indian subcontinent and the dry areas of the Middle East, as highlighted by Malik [70]. Asia dominates lentil cultivation, representing 80 percent of the global area and contributing to 75 percent of world production [71]. Lentils, known locally as Masoor, is a nutritionally rich food source, offering 59% carbohydrates, 25% proteins, 0.7% fat, 2.1% minerals, 0.7% fiber, vitamins, and minerals such as K, P, Ca, Fe, and Zn. They are also a good source of dietary fiber and provide a high energy value, as noted by DeAlmeida Costa [72]. Lentil seeds contain significant amounts of oleic, linoleic, and palmitic acids, as reported by Roy [73]. Additionally, the nutritional profile makes lentils a substantial and well-rounded food source [73]. Globally, lentils are grown as a rainfed pulse in about 52 countries, holding approximately 3.85 million hectares with a production of 3.59 million tonnes, as reported by Erskine et al. in 2011. Major contributors to lentil cultivation include Canada, India, Turkey, Australia, the United States, Nepal, China, and Ethiopia, with diverse climates and agricultural practices. Lentils are particularly significant in regions with marginal lands, where resource-poor farmers cultivate them. They thrive in areas with low fertility and water-holding capacity, making them adaptable to subtropical foothill soils. Their suitability for cultivation in Ladakh, known for its frost and severe winters, underscores their hardy nature. They have become an integral part of farming practices in the Ladakh region, fitting well into the local climate conditions. Lentils show potential as an inter-crop in young orchards, demonstrating their adaptability to different farming systems. Their ability to thrive with minimal irrigation further enhances their value for local farmers. Considering the specific conditions of Ladakh, the integrated use of chemical fertilizers with organic compost is recommended. This approach aims to enhance lentil production and maintain soil health, particularly in cold arid conditions. Investigating the optimal proportion levels of these fertilizers becomes crucial for achieving the highest productivity in Ladakh.

2.4 Soybean

Soybean (*Glycine max* L.) stands out as a widely cultivated legume crop on a global scale. In 2019, its production reached a substantial 334 m metric tons, covering an extensive harvested area of 121 m ha [74]. The U.S. Midwest, known for its agricultural productivity, significantly contributes to over 34% of the world's soybean output. Belonging to the Fabaceae family, soybeans play a crucial role in supplying approximately 50% of the world's edible oil [75]. Their applications extend beyond human consumption to include uses in animal feed and various non-food sectors, highlighting their versatility. Despite its prominence, a notable point of concern is that commercial fertilizers are given (less than 40% of soybean acreage). The limited use of commercial fertilizers emphasizes the potential impact of improved nutrient management practices on soybean cultivation. Efficient fertilizer application and soil management practices are crucial for addressing challenges related to soil health and nutrient deficiencies. This practice contributes to challenges such as shortage of micronutrients, poor soil fertility, and a decline in soybean productivity over time [76]. Addressing these challenges through enhanced fertilizer application and sustainable soil management practices is crucial for ensuring the long-term viability of soybean cultivation worldwide. This not only contributes to improving soybean yields but also aligns with broader goals of agricultural sustainability. In conclusion, soybeans play an active part in global agriculture, and nutrient management tasks are essential for sustaining and enhancing their productivity. As the demand for soybeans continues to rise, adopting effective and sustainable agricultural practices becomes paramount for the future of soybean cultivation.

Legume-based cropping systems play a crucial role in enhancing soil fertility and health, with soybeans being significant contributors to these benefits. Soyabean provides availability of soil organic matter (SOM). SOM is rich in N and P nutrients, which are important nutrients for plant growth [77]. Soybeans actively contribute to the enhancement of SOM by providing biomass, organic carbon, and nitrogen. These contributions foster the proliferation of nodule-forming bacteria, specifically Rhizobia. The presence of Rhizobia is crucial for N fixation, a process that converts atmospheric nitrogen into a form usable by plants [78, 79]. The benefits of legume-based cropping systems, including soybeans, are diverse and include, increased SOM levels, which improve overall soil fertility, soil structure, water infiltration and root growth, helping to maintain a stable pH level, diversification of microscopic soil flora and fauna, contributing to a more robust and balanced soil ecosystem and disruption of pest and disease cycles, which can occur due to the interactions within the soil environment [80, 81]. Soil organic matter, which is boosted by legume-based systems like those involving soybeans, performs a pivotal function in micronutrient accessibility and plant uptake. Micronutrients are essential for various biochemical processes in plants, and their availability is crucial for plant health and growth. In essence, the incorporation of legume-based cropping systems, particularly those involving soybeans, has multifaceted benefits for soil health. These systems contribute to the enrichment of soil organic matter, improve overall soil structure, and deliver nutrient cycling and availability. The resulting enhancement of soil fertility and health has positive implications for sustainable and resilient agricultural practices. To enhance soybean production, factors beyond fertilization, such as the adoption of high-yielding varieties and consideration of optimal growing conditions, play a crucial role. Soybeans thrive in specific climatic conditions, and optimal cultivation involves: ideally, a temperature range of 25°-27°C is favorable for soybean cultivation; average air humidity of around 65% supports optimal growth; soybeans require at least 10–12 hours of sunlight per day for effective photosynthesis. The ideal monthly rainfall falls between 100 and 200 mm, ensuring an adequate water supply for the plants. Soybeans prefer loose, moist, textured soil with a slightly acidic to neutral pH range of 6–6.8. While they can tolerate slightly more acidic conditions (down to pH 5.5), growth may be significantly hampered due to aluminum poisoning. In such cases, lime application is recommended to mitigate the adverse effects of low pH on soybean growth [76]. The historical variability in soybean yield response to nutrient applications is influenced by factors such as genetic diversity, management practices, and environmental circumstances, including soil properties, pH, organic matter, moisture, temperature, and aeration. To optimize soybean production, it is crucial to identify and understand the genetic and environmental factors influencing micronutrient uptake and crop removal. This knowledge serves as the foundation for effective nutrient management strategies. Research on micronutrients like zinc (Zn) and Iron (Fe) has demonstrated positive effects on yield and yield components in various crops [82]. Micronutrient applications, including foliar spray of Fe and Mo (Molybdenum) on soybeans, have been shown to reduce damages caused by water deficit conditions, resulting in improved yields compared to control treatments without these micronutrients [83, 84]. Studies by Ross [85] emphasize the importance of boron (B) application. A notable increase in soybean yield, ranging from 4 to 130%, was observed when B was applied to plants showing visual deficiency symptoms and grown under low B levels. The findings underscore the importance of considering micronutrients in soybean nutrient management strategies. Tailoring micronutrient applications to address deficiencies identified in the local environment can significantly contribute to enhancing soybean yields and overall crop health.

3. Soil fertility and plant nutrition

3.1 Soil fertility

Sustainable crop production requires the application of N and P fertilization in a definite ratio. Ensuring a balanced N:P ratio is crucial not only for current agricultural practices but also for investment in soil fertility and an environmentally friendly approach for the benefit of future generations. Continuous cropping of various varieties in regions like Pakistan has led to the depletion of necessary elements from the soil, resulting in poor soil fertility. Without proper nutrient replenishment, continuous cultivation can adversely affect soil productivity over time. Maintaining a balanced N:P ratio in fertilization practices is essential for improving agricultural production and contributing to sustainable agriculture. This balanced approach helps ensure the long-term health and fertility of the soil, providing a foundation for consistent and resilient crop yields. The significance of this balanced nutrient management approach becomes even more critical globally, especially as the availability of agricultural land is decreasing due to urbanization and industrialization. The pressure on limited land resources emphasizes the need for sustainable and eco-friendly agricultural practices [86]. In summary, the imperative for balanced nutrient management intensifies in the face of global challenges related to shrinking agricultural land. By prioritizing sustainability and eco-friendliness, agricultural practices can contribute to both food security and the responsible stewardship of limited land resources [87]. Despite numerous attempts, the task of increasing crop production in developing countries, especially for smallholder farmers, remains a complex and unresolved issue. Various practices intended to boost production often prove unsustainable, carrying elevated economic, ecological, and environmental costs. Smallholder farmers, who constitute a significant portion of agricultural activities in developing nations, face particular challenges due to the economic, ecological, and environmental implications of certain agricultural practices. These challenges can act as barriers, hindering their ability to adopt and sustain these approaches. With the global population projected to reach 10 billion in the next 50 years and Sub-Saharan Africa (SSA) expected to host around 2.12 billion people by 2065, there is an urgent need to address these challenges. In response to these challenges, there is a pressing need for innovative and sustainable agricultural practices. These practices should not only address the economic and ecological aspects but also be socially equitable. The aim is to come across the future food supply of a growing global population without compromising the well-being of farmers or the health of ecosystems. New agricultural approaches, including innovative fertilization techniques, alternative soil amendments, and harnessing beneficial bacteria and fungi, offer compelling and sustainable solutions for enhancing soil health, optimizing fertility, and promoting robust crop growth. These methods are key to ensuring adequate and high-quality agricultural yields, thereby advancing the cause of sustainable farming practices.

3.2 Plant nutrition

For their growth and production, plants necessitate 17 essential nutrients, with hydrogen, oxygen, and carbon obtained from water and air. Macronutrients, comprising N, P, K, S, Ca, and Mg, are essential in higher amounts, whereas micronutrients like iron (Fe), boron (B), manganese (Mn), zinc (Zn), copper (Cu), molybdenum (Mo), chlorine (Cl), and nickel (Ni) are required in smaller amounts. Leguminous crops like soybeans typically experience fewer micronutrient deficiencies compared to macronutrient deficiencies. However, micronutrients are vital for critical cell functions in plants [88–90]. Despite their lower quantity requirement, micronutrient deficiencies can result in reduced plant development, productivity, and quality, impacting the health and efficiency of both animals and humans dependent on these crops [91–93]. Micronutrient deficiency in arable soil poses a pervasive worldwide challenge [10, 11]. In recent years, there has been a heightened focus on micronutrients, spurred by elevated nutrient removal rates resulting from the cultivation of newly developed high-yielding cultivars [94]. The soil's micronutrient availability profoundly influences plant uptake [95]. Positive yield responses, observed across diverse crops such as soybeans, underscore the significance of incorporating micronutrients alongside macronutrients [96]. Crop production contends with an array of environmental anxieties, encompassing disease and pest infestations, low soil fertility, and insufficient water supply [97, 98]. Addressing micronutrient deficiencies emerges as a crucial aspect of optimizing crop yields, particularly considering evolving agricultural practices and the development of high-yielding crop varieties. The intricate relationship between soil health, nutrient availability, and environmental stresses underscores the necessity for a holistic approach to crop management and nutrient supplementation. In the past few years, the scientific community has increasingly emphasized the importance of soil quality and soil health, recognizing their crucial roles in producing food, conserving water, nutrient cycling, climate change mitigation, and biodiversity conservation [99]. The thought of soil fertility and quality gained prominence in the 1980s as a holistic approach that goes beyond just managing fertility to tackle various soil degradation issues. The FAO has been a key player in promoting awareness and understanding in this regard [100]. defines soil health as the "capacity of soil to function as a living system, within the ecosystem and land-use boundaries, to sustain plant and animal productivity, maintain or enhance water and air quality, and promote plant and animal health." The shift toward soil health management signifies a holistic consideration of the crucial nutrients and interconnected soil physical, chemical, and biological practices needed to produce crops and provide low-cost overall farm care (**Figure 1**). This approach acknowledges the dynamic and multifaceted nature of soils, emphasizing their critical role in maintaining ecological balance and supporting sustainable agricultural practices. Legume-based cropping systems enhance soil fertility by increasing the accessibility of soil organic matter

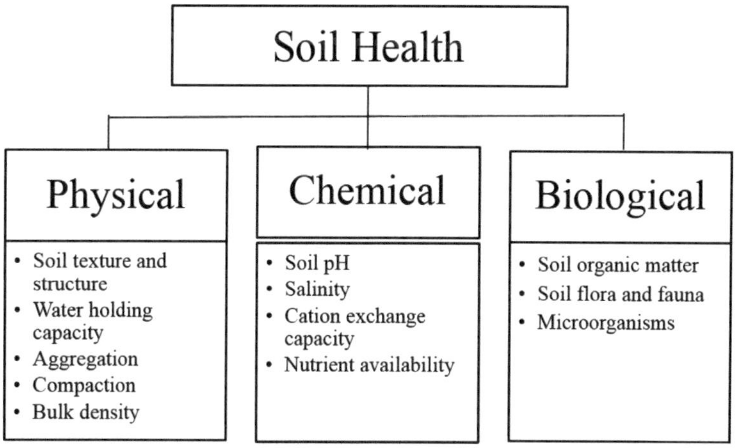

Figure 1.
Soil health management processes. Source: modified from Moebius-Clune [101] and Hills et al. [102].

(SOM) with enriched N and P nutrients [103]. Legumes, such as soybeans, are key contributors to this improvement, supplying biomass, organic carbon, and nitrogen [104]. They also foster the growth of nodule-forming bacteria, particularly Rhizobia, further enhancing soil health [78]. The advantages of legume cultivation extend to various aspects, including increased SOM, improved soil porosity and structure, nutrient recycling, pH buffering, and disruption of pest and disease cycles [79]. Soil organic matter (SOM) shows a pivotal function in facilitating the availability and uptake of micronutrients by floras [81]. The incidence of chelating organic compounds in soils plays a significant role in enhancing the accessibility and solubility of micronutrients. Chelation of metal elements, including zinc and iron, with SOM is crucial for facilitating their transport to the plant's root system [105]. Legume-based cropping systems emerge as valuable contributors to soil health, fostering a dynamic interplay of soil properties, nutrient dynamics, and overall ecosystem well-being in agriculture.

3.3 Role of legumes in soil health improvement

3.3.1 Biological nitrogen fixation

Legume crops are ironic in protein, a feature directly attributed to their ability to fulfill a significant portion of their nitrogen requirements with the support of symbiotic Rhizobia microorganisms. When legumes are inoculated with the appropriate strain of Rhizobia bacteria, they can independently supply up to 90% of their nitrogen requirements. After germination, Rhizobia bacteria enter the root hairs and infiltrate the root, leading to multiplication and the formation of distinct pale pink nodules on the roots. Within these nodules, the microorganisms bind nitrogen gas from the soil air. They utilize carbohydrates produced by the above-ground plant during photosynthesis as a nutrient source. The symbiotic relationship allows legumes to efficiently use atmospheric nitrogen, converting it into a beneficial form for their growth. This process contributes to the high-protein content in legume tissues [106]. Bacteria in symbiosis with legumes produce ammonia (NH_3), utilizing hydrogen from the plant's carbohydrates and atmospheric nitrogen. The produced ammonia serves as a vital nitrogen source for the plant's growth, contributing to the production of high-protein seeds or forage crops. This mutualistic interaction allows both the bacteria and the legume to thrive, showcasing the efficiency of nitrogen fixation in leguminous plants. While legumes can fix atmospheric nitrogen, they are also capable of absorbing significant amounts of soil nitrogen if available. Soil microorganisms play a crucial role in decomposing nitrogen-rich organic material, releasing nitrogen into the soil upon their death. On average, approximately two-third of N_2 fixed by a legume crop turns out to be accessible in the following growing season when it is incorporated into a crop rotation [107].

3.3.2 Carbon sequestration

The process involves capturing atmospheric CO_2 and incorporating it into the soil through plant activity, providing benefits such as enhanced biodiversity, improved nutritional security, increased renewability, water quality improvement, and strengthened nutrient recycling [108]. Agricultural production processes are major contributors to greenhouse gas emissions, amplifying global warming. Significant factors encompass the manufacturing and utilization of fertilizers, machinery usage, agronomic practices, and diverse soil processes [109]. Improved soil organic carbon (SOC): Leguminous crops, with their ability to shed leaves and robust biomass production underground, play a significant role in sequestering soil organic carbon. This positive impact is further heightened when legume farming practices are integrated into mixed farming and crop rotation. The potential for carbon storage is maximized when agronomic practices of legume cultivation are combined with other techniques such as no-tillage and mixed cropping [110]. Influence on Soil Biology: Legumes actively shape soil biology by fostering the formation and upkeep of soil aggregates. Consequently, this contributes to enhanced soil carbon storage by safeguarding it from decomposition. The overall influence of leguminous crops on soil organic carbon and other structural aspects relies on the intricate interactions between soil components and crop residues [111].

3.3.3 Reduction in greenhouse gas emission

The potential of forage legumes, particularly those rich in condensed tannins, to reduce methane emissions in ruminants is an interesting area of research. While the initial findings are promising, ongoing studies are crucial to fully understand the mechanisms involved, optimize the utilization of tannin-containing forages, and assess their long-term environmental impact and sustainability. Implementing such strategies not only has the potential to benefit livestock management by reducing methane emissions but also aligns with wider efforts to highlight climate change and endorse sustainable agricultural practices. It's important to consider the multifaceted aspects of these interventions, including their feasibility, economic implications, and compatibility with overall farm management practices. Continued research, collaboration between scientists, farmers, and policymakers, and the integration of innovative approaches into agricultural systems will contribute to the development of effective and sustainable strategies for mitigating greenhouse gas emissions from livestock [112]. The relationship between the use of mineral nitrogen fertilizer and nitrous oxide emissions is highlighted, indicating a direct correlation. This underscores the importance of managing nitrogen fertilizer application to mitigate nitrous oxide emissions. The positive impact of legumes, such as alfalfa and soybeans, on reducing nitrous oxide emissions is noted. This aspect is crucial for national inventories of greenhouse gas emissions from agriculture, suggesting that incorporating legumes can be an environmentally beneficial practice. The comparison with perennial grass and the lower nitrous oxide emissions observed in legume systems highlight the potential environmental advantages of growing legumes. This comparison provides insights into the role of specific crops in influencing greenhouse gas emissions. The recognition of legumes as soil-amendment crops with positive effects on soil health is important. The observation that nitrous oxide emissions were not strictly correlated with soil mineral nitrogen pools underlines the complexity of the interactions and factors involved in emissions, extending beyond nitrogen quantity. The concluding emphasis on considering legumes differently in national inventories stresses the need to recognize the unique contributions of legumes to lowering nitrous oxide emissions and improving overall soil health. Incorporating legumes into farming systems is presented as a valuable and beneficial practice with potential environmental advantages [109]. Legumes enhance agricultural productivity, suggesting improved yields when incorporated into cropping systems. Legumes play a role in soil conservation, preventing erosion and maintaining soil structure for long-term fertility. Legumes positively impact soil biology, enhancing microbial activity and diversity essential for nutrient cycling and overall soil ecosystem functioning. Legumes contribute to the accumulation of soil organic carbon and nitrogen stocks, which are crucial for soil fertility and supporting plant growth. Legumes positively influence soil chemical and physical properties, potentially improving nutrient availability, soil structure, and water retention. Legumes fix atmospheric N_2 by symbiotic relationships, improving N_2 availability in the soil. Legumes help mitigate nitrous oxide emissions and reduce nitrate leaching, contributing to environmental sustainability and water quality. Legumes reduce the dependency on chemical fertilizers, offering economic benefits and promoting sustainable and environmentally friendly agricultural practices.

3.3.4 Nutrient availability in soil

Increased nitrogen addition to the soil may result in elevated nitrogen losses through leaching, potentially limiting the strengthening of the soil nitrogen pool. Legume farming is suggested as a strategy to reinforce the soil nitrogen pool. The effectiveness of this depends on legumes fixing more nitrogen than is removed during harvesting. Phosphorus is highlighted as essential for plant growth, and its deficiency can significantly reduce photosynthesis rates, impacting leaf growth and overall photosynthetic capability. Phosphorus deficiency is expected to have severe implications for the symbiotic associations of plant roots with rhizobacteria, affecting the development and functioning of nodules involved in biological N_2 fixation, an energy-intensive process [113]. The passage highlights the strong connection between aerial metabolic and developmental events in plants and the symbiotic associations of plant roots with rhizobacteria. Phosphorus deficiency is indicated to negatively impact the development and proper functioning of nodules, crucial components in the symbiotic relationship between plants and rhizobacteria. The passage emphasizes that the process of biological nitrogen fixation (BNF) is described as highly energy-intensive [114]. Nitrogen is contrasted with phosphorus, emphasizing the nonrenewable nature and depletion of inorganic phosphorus reserves over time. Developing legume cultivars with efficient N_2-fixing capability, particularly in P-deficient conditions, is highlighted as crucial for enhancing soil health and overall sustainability.

Plants have evolved adaptive morphological and physiological approaches to enhance phosphorus acquisition, especially when facing phosphorus constraints in symbiotic relationships. Adaptive strategies operate at multiple levels, including maintaining higher phosphorus levels in nodules, increasing root surface area, enhancing root exudate excretion, and regulating the expression of transporters and aquaporin to improve phosphorus uptake. Compensatory mechanisms are mentioned, such as an increase in fixed nitrogen per unit of nodule to compensate for a decrease in nodule number. Improved phosphorus acquisition actively contributes to the efficiency of biological nitrogen fixation (BNF) in nodules. Efficient BNF in nodules significantly contributes to maintaining the nitrogen balance in the soil. Approximately all legume cultivars adopted by farmers incorporate various methodologies to enhance phosphorus acquisition. The adoption of these approaches by legume cultivars is presented as a superior way to retain soil health, highlighting the importance of sustainable agriculture. The overall goal is to improve both nitrogen and phosphorus availability in the soil. Legumes play a vital role in effectively managing dependence on inorganic fertilizers. The conclusion emphasizes that legumes contribute to improving overall soil fertility.

3.3.5 Role of legumes in improving soil properties

3.3.5.1 Improving soil physical properties

The root systems of leguminous plants help to fix the soil particles composed, plummeting the risk of erosion. This is particularly important in areas prone to soil erosion due to wind or water. Legumes contribute to increased soil organic matter that boosts the soil's water retention capacity. The organic matter acts like a sponge, holding water for plant use during dry periods and reducing the risk of water runoff. Legumes, through their symbiotic relationship with nitrogen-fixing bacteria, improve soil fertility. This promotes a healthier and more diverse soil ecosystem, fostering the growth of various microorganisms and supporting plant growth. Leguminous plants release organic compounds into the soil through root exudates. These compounds act as binding agents, enhancing soil structure by improving aggregation and stability. This, in turn, facilitates better root penetration and nutrient availability. The incorporation of organic matter from leguminous plants can reduce soil bulk density, improving porosity. This is important for aeration and root development. Increased porosity allows water and air to move more freely through the soil [115]. The organic matter added by leguminous plants contributes to soil aggregation, enhancing combination stability. Moreover, the improvement in soil texture occurs as a result of the organic materials breaking down over time, influencing the composition of soil particles. Legumes positively impact various water-related processes, such as reducing runoff and erosion. The enhanced soil structure and organic matter content improve water infiltration rates and water-holding capacity, making water more available to plants. The vegetation cover provided by leguminous plants can act as a natural insulator, buffering temperature fluctuations in the soil. This helps maintain more stable soil temperatures, creating a promising atmosphere for plant development [116]. Leguminous cover crops contribute significantly to soil health by producing large amounts of biomass. This biomass serves as a substrate for soil organic processes, facilitating the development and accumulation of soil organic matter. Increased organic matter content enhances soil structure, water retention, and nutrient availability [117]. Moreover, leguminous cover crops play a dual role in protecting the soil.

They help prevent the loss of plant nutrients by covering the soil surface and reducing leaching and nutrient runoff. Additionally, the dense cover provided by these crops acts as a protective layer against erosion, safeguarding the topsoil from being washed away. Green manure plants, particularly leguminous ones, are intentionally grown to improve soil physical properties. When these plants are incorporated into the soil as green manure, they contribute organic matter, enriching the soil and enhancing its fertility. This practice is a proactive approach to soil management and improvement. Some leguminous plants have the ability to physically modify the soil profile. This could involve changes in soil structure, texture, or composition influenced by the specific characteristics of the legume species. Leguminous cover crops influence soil structure by promoting aggregation. The root systems and residues of these crops contribute to the formation of soil aggregates, which enhance soil porosity, aeration, and water movement. This, in turn, improves the overall physical condition of the soil. Leguminous cover crops play a role in maintaining an appropriate carbon-to-nitrogen (C/N) ratio in the soil. This balance is crucial for the efficient decomposition of organic matter and the overall health of the soil microbial community. The presence of leguminous cover crop residues in the soil facilitates increased water infiltration. Direct effects of crop residues on soil structure and aggregation, as well as their role in preventing surface crusting, which can impede water movement [118–121].

3.3.5.2 Improving soil chemical properties

Soil chemical properties play an important part in sustainable agriculture by providing essential nutrients to crops and managing harmful substances. A balanced and fertile soil is fundamental for the long-term health and productivity of agroecosystems. Cation exchange capacity (CEC) is an indicator of the soil's capacity to retain and exchange cations, which are positively charged ions influencing nutrient availability to plants. Soil pH affects nutrient availability and microbial activity. Different crops thrive in different pH ranges, making pH management crucial for successful agriculture. Monitoring levels of essential nutrients like nitrogen, phosphorus, and potassium is critical for understanding soil fertility and making informed nutrient management decisions. Organic carbon is a key component of SOM, influencing soil structure, water retention, and nutrient availability. Leguminous crops and soil chemical properties: leguminous crops have a significant impact on soil chemical properties due to their special capability to fix atmospheric nitrogen (N_2) through symbiotic relationships with nitrogen-fixing bacteria. This nitrogen fixation process contributes to increased nutrient availability in the soil, particularly nitrogen. The nitrogen-fixing ability of leguminous crops allows for a rapid improvement in soil chemical properties. By converting atmospheric nitrogen into a form that plants can use, legumes contribute to enhanced nutrient levels in the soil. Legume-based rotations induce changes in the pH of the rhizosphere—the soil zone influenced by the roots. This alteration in pH can impact nutrient availability and microbial activity, influencing overall soil health. Legumes release organic compounds through root exudation, and the alteration or release of organic acids on the root surfaces can enhance the availability of P in the soil. This is crucial for the nutrient uptake by plants. The capability of legumes to fix nitrogen, influence pH, and enhance nutrient availability makes them valuable components in strategies aimed at improving soil fertility and ensuring the long-term productivity of agricultural systems [122]. Legume-added organic matter can both increase and decrease soil pH. Organic acids and CO_2 from decomposition decrease pH, while the reduction of H^+ into H_2O and CO_2 raises pH [123].

The pH changes significantly affect microorganisms, which are crucial for nutrient cycling. Leguminous green manure enriches soil with organic matter. It increases nitrogen and releases phosphorus, renews organic carbon, and enhances soil chemistry. Incorporating legume residues boosts soil organic carbon, aiding in carbon sequestration. Leguminous cover plants as green manure enhance the availability of N, P, K, and trace elements to subsequent plants. This is facilitated by the pH decrease resulting from CO_2 during decomposition [124]. When pulses and oil seed crops are utilized as green manure mixed with soil, their remains enhance the availability of N, K, and minor nutrients, which enhance plant growth. This increased availability is attributed to the dropping of soil pH caused by the CO_2 formed during the decomposition method. Overall, the use of leguminous green manure appears to have multifaceted effects on soil properties, nutrient cycling, and microbial activity, with potential benefits for both agricultural productivity and environmental sustainability.

3.3.5.3 Enhancing soil biological properties

Soil microorganisms play a significant part in bridging the connection between plant yield and nutrient accessibility. They actively participate, both directly and indirectly, in nutrient cycling by converting both inorganic and organic forms of nutrients [125]. Legumes are essential components for increasing soil microbial biomass. They actively contribute to key strategies such as nutrient cycling and breakdown of soil organic matter, leading to improved crop productivity and soil sustainability. Microorganisms in the rhizospheric zone, physically interacting with leguminous plants, can positively enhance crop productivity by promoting plant development and improvement. The profusion and variety of microbes show a direct correlation with plant production. The occurrence and copiousness of microbial life are unswervingly linked to soil element availability. Microorganisms play a role in facilitating the transfer of nutrients from the soil to plants, thereby improving soil fertility and the overall health of the plant-soil system [126]. Intercropping with legumes contributes to the formation of diverse root types. This practice results in a change in the complete root distribution and architecture within the soil. Intercropping with legumes modifies the process of exudation in the rhizosphere. Rhizospheric exudates play a vital part in influencing soil microbial communities and their interactions with plants. The changes in root development, distribution, and exudation strongly influence the microbial communal in the soil. This, in turn, affects the interactions between microbes and plants. Intercropping with legumes is stated to promote beneficial interactions between the microbial community and plants. The practice of intercropping cereals with legumes is noted to promote complementarity and facilitation in agroecosystems. This suggests that the combined growth of cereals and legumes has positive effects on the overall agricultural system [126].

A long-term study was indicated to assess the rotation of cereals and leguminous plants and cover crops in a specific area. The study revealed that although there were slight improvements in physical properties, significant soil improvements required a cropping pattern spanning five years. Future research should focus on better understanding the impacts related to species selection, seeding rates, termination methods, and other factors. This research will be essential in guiding farmers to choose the most suitable cropping options to maximize benefits [127]. Furthermore, subsequent studies have revealed that incorporating leguminous crop residues into the soil leads to increased levels of phospholipid-derived fatty acids, total bacterial abundance, fungal root colonization, and spore density compared to residues placed on the surface or removed. This effect is likely attributed to the enhanced contact between residues and soil microbes, which improves soil nutrient status, particularly by increasing the availability of carbon (C) and nitrogen (N) substrates [128].

For the improving soil directly incorporating fresh red clover into the soil was found to be more effective in enhancing soil microbial biomass and enzyme activity compared to using biogas slurry from fermented red clover or composted red clover [129]. Moreover, biological interventions involve the use of leguminous cover crops and other plant materials to enhance the availability of nutrients for the benefit of the environment and ultimately improve crop yields. The success of a successful biological intervention management program is determined by the objectives and purpose of a chosen leguminous crop or shrub. Site specifications, timing, and cropping history are all factors to consider when selecting the most appropriate leguminous crop species to introduce into a system. Specific reasons to incorporate leguminous crops into a crop management regime include slowing erosion, improving soil health, enhancing water infiltration, suppressing weeds, controlling pests and diseases, and increasing biodiversity [130]. The loss of soil organic matter (SOM) in agroecosystems, in particular, which is a key factor in affecting the physical, chemical, and biological soil properties, has determined a massive decrease in soil quality [131].

4. Conclusion and future strategies

Sustainable agriculture aims to meet the requirements of the existent population without compromising the productive potential for future generations. The role and importance of legumes in sustainable agriculture can be enhanced through emerging research opportunities. Careful selection and introduction of legume species and cultivars across diverse cropping systems is crucial. Striking a balance between achieving economic returns through yield and considering environmental and agronomic benefits is essential when contemplating legume cultivation. Assessing the N_2 fixation ability of grain legumes in relation to soil, climate, plant characteristics, and management conditions is vital for optimal enhancements. In future cropping systems, legumes with the ability to recover unavailable forms of soil phosphorus could prove to be valuable assets. Therefore, it is imperative to explore and integrate such legumes to improve overall agricultural sustainability. Proper management of leguminous crops is essential to raise fertility levels and sustain crop production. Appropriate tillage practices for leguminous crops can contribute to soil improvement. Clear identification and selection of practices suitable for the agroecological zone are crucial. In particular, growing land pressure has led to shorter fallow periods. Enhanced fallow systems necessitate the integration of rapidly growing leguminous crops that can generate substantial biomass or efficiently fix nitrogen in brief periods. Encouraging seed production of important leguminous crops ensures their availability and affordability.

Conflict of interest

The authors declare no conflict of interest.

Author details

Ambrin Rajput[1*], Qurban Ali Panhwar[2] and Hafeezullah Babar[1]

1 Soil Fertility Research Institute, Agriculture Research Center, Tandojam, Sindh, Pakistan

2 Soil and Environmental Sciences Division, Nuclear Institute of Agriculture (NIA), Tandojam, Sindh, Pakistan

*Address all correspondence to: ambrin2004@gmail.com

References

[1] United Nations (UN), Department of Economic and Social Affairs, Population Division. New York World Population Prospects: The 2012 Revision, Key Findings and Advance Tables. Working paper no. ESA/P/WP.227. 2013

[2] Van Zwieten L, Rose T, Herridge D, Kimber S, Rust J, Cowie A, et al. Enhanced biological N2 fixation and yield of faba bean (*Vicia faba* L.) in an acid soil following biochar addition: Dissection of causal mechanisms. Plant and Soil. 2015;**395**:7-20

[3] Ludwig B, Geisseler D, Michel K, Joergensen RG, Schulz E, Merbach I, et al. Effects of fertilization and soil management on crop yields and carbon stabilization in soils. A review. Agronomy for Sustainable Development. 2011;**31**:361-372. DOI: 10.1051/agro/2010030

[4] Rasmussen PE, Goulding KWT, Brown JR, Grace PR, Janzen HH, Körschens M. Long-term agroecosystem experiments: Assessing agricultural sustainability. Global Change Science. 1998;**282**:893-896. DOI: 10.1126/science.282.5390.893

[5] United Nation (UN). Transforming our World: The 2030 Agenda for Sustainable Development. New York: United Nations; 2015. Available from: https://www.un.org/sustainabledevelopment/development.agenda/ [Accessed: November 5, 2021]

[6] European Commission (EC). Communication from the Commission. The European Green Deal. 2019. Available from: https://ec.europa.eu/info/strategy/priorities-2019-2024/european-green-deal_en [Accessed: November 5, 2021]

[7] Stagnari F, Maggio A, Galieni A, Pisante M. Multiple benefits of legumes for agriculture sustainability: An overview. Chemical and Biological Technologies in Agriculture. 2017;**4**:1-13. DOI: 10.1186/s40538-016-0085-1

[8] Nanganoa LT, Njukeng JN, Ngosong C, Atache SKE, Yinda GS, Ebonlo JN. Short-term benefits of grain legume fallow systems on soil fertility and farmers livelihood in the humid forest zone of Cameroon. International Journal of Sustainability Agricultural Research. 2019;**6**:213-223. DOI: 10.18488/journal.70.2019.64.213.223

[9] Vasconcelos MW, Grusak MA, Pinto E, Gomes A, Ferreira H, Balázs B. The biology of legumes and their agronomic, economic, and social impact. In: The Plant Family Fabaceae. Singapore: Springer; 2020. pp. 3-25. DOI: 10.1007/978-981-15-4752-2_1

[10] Ghosh P, Bandyopadhyay K, Wanjari KK, Manna RH, Misra MC, Mohanty AK. Legume effect for enhancing productivity and nutrient use-efficiency in major cropping systems–an Indian perspective: A review. Journal of Sustainable Agriculture. 2007;**30**:59-86. DOI: 10.1300/J064v30n01-07

[11] Kebede E. Grain legumes production in Ethiopia: A review of adoption, opportunities, constraints and emphases for future interventions. Turkish Journal of Agriculture Food Science and Technology. 2020b;**8**:977-989. DOI: 10.24925/turjaf.v8i4.977-989.3254

[12] Kebede E. Grain legumes production and productivity in Ethiopian smallholder agricultural system, contribution to livelihoods and the way forward. Cogent Food

& Agriculture. 2020a;**1722353**:6. DOI: 10.1080/23311932.2020.17.22353

[13] Lewis GP. Legumes of the World. London, UK: Royal Botanic Gardens Kew; 2005. 22p

[14] LPWG. A new subfamily classification of the leguminosae based on a taxonomically comprehensive phylogeny. Taxon. 2017;**66**:44-77

[15] Zhao Y, Zhang R, Jiang KW, Qi J, Hu Y, Guo J. Nuclear phylotranscriptomics and phylogenomics support numerous polyploidization events and hypotheses for the evolution of rhizobial nitrogen-fixing symbiosis in fabaceae. Molecular Plant. 2021;**14**:748-773. DOI: 10.1016/j.molp.2021.02.006

[16] FAOSTAT. Available from: http://faostat.fao.org/site 2020;May 31

[17] Staniak M, Księżak J, Bojarszczuk J. Mixtures of legumes with cereals as a source of feed for animals. In: Croatia VP, editor. Organic Agriculture towards Sustainability. London, UK: InTech; 2014. pp. 123-145. DOI: 10.5772/58358

[18] McDonagh JF, Toomsan B, Limpinuntana V, Giller KE. Grain legumes and green manures as pre-rice crops in Northeast Thailand. I. Legume N2 -fixation, production and residual nitrogen benefits to rice. Plant and Soil. 1995a;**177**:111-126

[19] Whitmore AP, Cadisch G, Toomsan B, Limpinuntana V, van Noordwijk M, Purnomosidhi P. An analysis of the economic values of novel cropping systems in N.E. Thailand and S. Sumatra. Netherlands Journal of Agricultural Science. 2000;**48**:105-114

[20] Toomsan B, Viriya L, Sanun J, Aran P, Prabhakar P. Suhas PW, Sahrawat KL. Role of Legumes in Improving Soil Fertility and Increasing Crop Productivity in Northeast Thailand. See Discussions, Stats, and Author Profiles for this Publication. 2014. Available from: https://www.researchgate.net/publication/267574760

[21] Soltani A, Robertson MJ, Mohammad-Nejad YA, Rahemi-Karizaki. Modeling chickpea growth and development: Leaf production and senescence. Field Crops Research. 2006;**99**:14-23. DOI: 10.1016/j.fcr.2006.02.005

[22] Deppe C. The Resilient Gardener: Food Production and Self-Reliance in Uncertain Times. Chelsea Green Publishing; 2010. p. 241

[23] Siddiqui ZA, Mahmood I. Effects of rhizobacteria and root symbionts on the reproduction of meloidogyne javanica and growth of chickpea. Bioresource Technology. 2001;**79**:41-45. DOI: 10.1016/S0960-8524(01)

[24] Kantar F, Hafeez FY, Shivakumar BG, Sundaram SP, Tejera NA, Aslam A, et al. Chickpea: Rhizobium management and nitrogen fixation. In: Chickpea Breeding and Management. Wallingford, UK: CABI; 2007. pp. 179-192

[25] GoP. Agricultural statistics of Pakistan. Ministry of Food and Agriculture (economic Wing), Finance Division GoP, Islamabad. 2020

[26] GoP. Agricultural statistics of Pakistan. Ministry of Food and Agriculture (economic Wing), Finance Division, GoP, Islamabad. 2021

[27] FAO. FAOSTAT. 2009. Available from: http://faostat.fao.org/site/567/Desktop Default

[28] Anonymous. Economic Survey, Government of Pakistan, economic

advisor's wing, finance division, Islamabad, Pakistan. 2010

[29] Kislev ME, Bar-Yosef O. The legumes: The earliest domesticated plants in the near east? Current Anthropology. 1998;**29**:175-179. DOI: 10.1086/203623

[30] Redden RJ, Berger JD. History and origin of chickpea. Chickpea Breed Management. 2007;**1**:1-13

[31] Mondal SS, Mandal P, Saha M, Bag A, Nayak S, Sounda G. Effect of potassium and Sulphur on the productivity, nutrient uptake and quality improvement of chickpea. Journal of Crop and Weed. 2005;**2**(1):64-66

[32] Goud VV, Konde NM, Mohod PV, Kharche VK. Response of chickpea to potassium fertilization on yield, quality, soil fertility and economic in vertisols. Legume Research. 2014;**37**(3):311-315. DOI: 10.5958/j.0976-0571.37.3.047

[33] Sahai VN. Mineral Nutrients. In: Fundamentals of Soil. 3rd ed. New Delhi, India: Kalyani Publishers; 2004. pp. 151-155

[34] Singh K, Bansal SK, Mouinuddin. Effect of continuous cropping for twenty years on some properties of the intensively cultivated alluvial soils and nutrient indexing of rice. Journal Indian Society of Soil Science. 2007;**55**:265-269

[35] Nawaz I, Zia-ul-hassan RAM, Arshad M. Exploiting genotypic variation among fifteen maize genotypes of Pakistan for potassium uptake and use efficiency in solution culture. Pakistan Journal of Botany. 2006;**38**:1689-1696

[36] Boulbaba I, Bouaziz S, Mainassara ZA, Zourgui L, Mokhtar L. Response of Chick pea (*Cicer arietinum* L.) to potassium fertilization. Journal of Agriculture and Social Science. 2005;**1**(l):7-9

[37] Ganga N, Singh RK, Singh RP, Choudhury SK, Upadhyay PK. Effect of potassium level and foliar application of nutrient on growth and yield of late sown chickpea (*Cicer arietinum* L.). Environment and Ecology. 2014;**32**(1):273-275

[38] Rashid A, Ishaque M, Hameed K, Shabbir M, Ahmad M. Growth and yield response of three chickpea cultivars to varying NPK levels. Asian Journal of Agriculture Biology. 2013;**1**:95-99

[39] Badini SA, Khan M, Baloch SU, Baloch SK, Baloch HN, Bashir W, et al. Effect of phosphorus levels on growth and yield of chickpea (*Cicer aretinum* L.) varieties. Journal of Natural Sciences Research. 2015;**5**:169-176

[40] Memon M, Rajput AN, Rajput A, Memon N, Jamro GM, Kumbhar MI. Response of chickpea cultivars to phosphorus application. Soil & Environment. 2016;**35**(1):22-29

[41] Kumar P, Kumar P, Singh T, Singh AK, Yadav RI. African Journal of Agricultural Research. 2014;**9**(8):728-734. DOI: 10.5897/AJAR2013.7883

[42] Rajput A. Potassium application on chickpea crop under irrigated area. Sarhad Journal of Agriculture. 2018;**34**(4):941-947. DOI: 10.17582/journal.sja/2018/34.4.941.947

[43] Rahman MH, Hara M, Hoque S. Growth and nutrient uptake of grain legumes as affected by induced compaction in andisols. International Journal of Agriculture and Biology. 2005;**7**(5):740-743

[44] Aziz-ur-Rehman M, Kaukab S, Saeed S, Aqeel M, Riasat G, Rafiq CM. Prospects of mungbean as an additional crop in rice wheat system of Punjab Pakistan. Universal Journal of

Agricultural Research. 2019;7:136-141. DOI: 10.13189/ujar.2019.070303

[45] Cheng XZ. Mungbean Production Technology Books. Beijing: China agricultural Publishing House; 2016

[46] GOP. Government of Pakistan. Pakistan Economic Survey. Vol. 2. Islamabad: Ministry of Finance; 2019. pp. 24-25

[47] Monem R, Mirtaheri SM, Ahmadi A. Investigation of row orientation and planting date on yield and yield components of mung bean. Annals of Biological Research. 2012;**3**(4):1764-1767

[48] Mbeyagala KE, Amayo R, Obuo JP, Pandey AK, War AR, Nair RMA. Manual for Mungbean (Green Gram) Production in Uganda. National Agricultural Research Organization (NARO); 2007. 32 p

[49] Raina SK, Govindasamy V, Kumar M, Singh AK, Rane J, Minhas PS. Genetic variation in physiological responses of mungbeans (*Vigna radiata* (L.)) wilczek to drought. Acta Physiologiae Plantarum. 2016;**38**(11):1-12. DOI: 10.1007/s11738-016-2280-x

[50] Amin R, Laskar RA, Khan S, Deb CR, Tomlekova N, Wani MR, et al. Lentil (*Lens culinaris* Medik.) diversity, cytogenetics and breeding. In: Adv. Plant Breed Strategies Legumes. Cham: Springer; 2019. pp. 319-369. DOI: 10.1007/978-3-030-23400-3_9

[51] GOP. Government of Pakistan. Pakistan Economic Survey. Vol. 2. Islamabad: Ministry of Finance; 2020. pp. 24-25

[52] Mehta P, Walia A, Kakkar N, Shirkot CK. Tricalcium phosphate solubilisation by new endophyte bacillus methylotrophicus CKAM isolated from apple root endosphere and its plant growth-promoting activities. Acta Physiology Plantarum. 2014;**36**:2033-2045

[53] Smith VH, Schindler DW. Eutrophication science. Where do we go from here? Trends in Ecology & Evolution. 2009;**24**(4):201-207

[54] Fageria NK, Baligar VC, Jones CA. Growth and mineral nutrition of field crops. 2nd ed. Marcel Dekker: Inc NY USA. Higgs B, Johnston AE, salter JL, Dawson C J. Some aspects of achieving sustainable phosphorus use in agriculture. Journal of Environmental Quality. 1997;**29**:80-87

[55] Dixon M, Simonne E, Obreza T, Liu G. Crop response to low phosphorus bioavailability with a focus on tomato. Agronomy. 2020;**10**(5):617

[56] Meena RS. Response to different nutrient sources on green gram (*Vignaradiata* L) productivity. Indian Journal of Ecology. 2013;**40**(2):353-355

[57] Dubey SN, Singh R, Kumar R, Dubey S. Effect of phosphorus and PSB on growth, nodulation and fertility status in different mungbean (*Vigna radiata* L.) varieties and its residual effect on fodder yield of sorghum in indo-gangetic plain zone of India. 2018:196-201

[58] Malik AM, Saleem MF, Ali A, Mahmood I. Effect of nitrogen and phosphorus application on growth yield and quality of mungbean (*Vigna radiate* L.). Pakistan Journal of Agricultural Sciences. 2003;**40**:3-4

[59] Sadeghip OR, Monem JAA. Production of mungbean (*Vigna radiata* L.) as affected by nitrogen and phosphorus fertilizer application. Journal of Applied Sciences. 2010;**10**(10):843-847. DOI: 10.3923/jas.2010.843.847

[60] Razzaque MA, Haque MM, Karim MA. Effect of nitrogen on different genotypes of mungbean as affected by nitrogen level in low fertile soil. Bangladesh Journal of Agricultural Research. 2017;**42**(1):77-85. DOI: 10.3329/bjar.v42i1.31981

[61] Shah KH, Aslam M, Khan P, Memon MY. Effect of different methods and rates of phosphorus application in mungbean. Soil Environment. 2006;**25**(1):55-58

[62] Figueiredo MVB, Burity HA, Martinez CR, Chanway CP. Alleviation of drought stress in common bean (*Phaseolus vulgaris* L.) by co-inoculation with *Paenibacilluspolymyxa* and *Rhizobiumtropici*. Applied Soil Ecology. 2008;**40**:182-188. DOI: 10.1016/j.apsoil.2008.04.005

[63] Malik MA, Hussain S, Warraich S, Habib EA, Ullah SA. Effect of seed inoculation and phosphorus application on growth, seed yield and quality of mungbean (*Vigna radiata* L.) cv. NM-98. International Journal of Agriculture and Biology. 2002;**4**:1560-8530

[64] Kumar R, Singh YV, Singh S, Latare AM, Mishra PK. Effect of phosphorus and Sulphur nutrition on yield attributes, yield of mungbean (*Vigna radiata* L. Wilczek). Journal of Chemical and Pharmaceutical Research. 2012;**4**(5):2571-2573

[65] Meena RS, Varma D. Mungbean yield and nutrient uptake performance in response of NPK and lime levels under acid soil in Vindhyan region. Journal of Applied Natural Sciences. 2016;**8**(2):860-863. DOI: 10.31018/jans.v8i2.886

[66] Kaysha K, Shanka D, Bibis MO. Performance of mung bean (*Vigna radiata* L.) varieties at different NPS rates and row spacing at Kindo Koysha district, southern Ethiopia. Cogent Food & Agriculture. 2020;**6**(1):1771112. DOI: 10.1080/23311932.2020.1771112

[67] Cubero JI. Orgin, taxonomy and domestication. In: Webb C, Hawtin GC, editors. Lentils. England: Commonwealth Agricultural Bureau, Slough; 1981

[68] Dhuppar P, Biyan S, Chintapalli B, Rao S. Lentil crop production in the context of climate change: An appraisal. Indian Research Journal of Extension Education. 2012;**2**(Special Issue):33-35

[69] Cokkizgin A, Munqez JY. Lentil origin, cultivation techniques, utilization and advances in transformation. Agricultural Science. 2013;**1**(1):55-62

[70] Malik R. Genetic divergence analysis in lentil (*Lens culnaris* Medik) [M.Sc. thesis]. Department of Agricultural Botany Ch: Charan Singh University Meerut (U.P.) India; 2005

[71] Anonymous. Project coordinators report. In: All India Coordinated Research. Project on MULLaRp (Rabi Crops, 2010-11). Indian: Institute of Pulses Research Kanpur; 2011. p. 16

[72] De Almeida Costa GE, da Silva Queiroz Monici K, Pissini Machado Reis SM, de Oliveira AC. Chemical composition, dietary fibre and resistant starch contents of raw and cooked pea, common bean, chickpea and lentil legumes. Food Chemistry. 2006;**94**(32):7-330

[73] Roy A, Aich SS, Bhowmick MK, Biswas PK. Response of lentil varieties to sowing time in plains of West Bengal. Journal of Crop and Weed. 2009;**5**(2):92-94

[74] Adamic S, Leskovsek R. Soybean (*Glycine max* (L.)) growth, yield and nodulation in the early transition period from conventional tillage to conservation and No-tillage systems. Agronomy. *2021*;**11**(12):2477

[75] Akparobi SO. Evaluation of six cultivars of soybean under the soil of rainforest agro-ecological zones of Nigeria. Middle East Journal of Science Research. 2009;**4**:6-9

[76] USDA. Oil Crops Sector at a Glance. 2021. Available from: https://www.ers.usda.gov/topics/crops/soybeans-oil-crops/oilcrops-sector-at-a-glance/ [Accessed: May 5, 2021]

[77] Jensen ES, Ambus P, Bellostas N, Boisen S, Brisson N, Corre-Hellou G, et al. Intercropping of cereals and grain legumes for increased production, weed control, improved product quality and prevention of N-losses in European organic farming systems. In: Proceedings of the International Conferences Joint Organic Congress—Theme 4. Denmark: Crop Systems and Soils Odense; 2006. pp. 30-31

[78] Lemke RL, Zhong Z, Campbell CA, Zentner RP. Can pulse crops play a role in mitigating greenhouse gases from north American agriculture? Agronomy Journal. 2007;**99**:1719-1725

[79] La Favre JS, Focht DD. Conservation in soil of H2 liberated from N2 fixation by H up-nodules. Applied and Environmental Microbiology. 1983;**46**:304-311

[80] USDA. Legumes and Soil Quality; Technical Note No. Soil Quality; Technical Note No.: Washington DC; USA. 1998. 998

[81] Rengel Z, Batten G, Crowley D. Agronomic approaches for improving the micronutrient density in edible portions of field crops. Field Crops Research. 1999;**60**:27-40

[82] Heidarzade A, Esmaeili M, Bahmanyar M, Abbasi R. Response of soybean (Glycine max) to molybdenum and iron spray under well-watered and water deficit conditions. Journal of Experimental Biology and Agricultural Science. 2016;**4**:37-46

[83] Heidarian AR, Kord H, Mostafavi K, Lak AP, Mashhadi FA. Investigating Fe and Zn foliar application on yield and its components of soybean (*Glycine max* L) at different growth stages. Journal of Agricultural Biotechnology and Sustainable Development. 2011;**3**:189-197

[84] Ekhtiari S, Kobraee S, Shamsi K. Soybean yield under water deficit conditions. Journal of Biodiversity and Environmental Sciences. 2013;**3**:46-52

[85] Schon MK, Blevins DG. Foliar boron applications increase the final number of branches and pods on branches of field-grown soybeans. Plant Physiology. 1990;**92**:602-607

[86] Ajwa HA, Tabatabai MA. Decomposition of different organic materials in soils. Biology and Fertility of Soils. 1994;**18**:175-182

[87] Brookes PC, Landman A, Pruden G, Jenkinson DS. Chloroform fumigation and the release of soil nitrogen: A rapid direct extraction method for measuring microbial biomass nitrogen in soil. Soil Biology and Biochemistry. 1985;**17**:837-842

[88] Malakouti MJ. The effect of micronutrients in ensuring efficient use of macronutrients. Turkish Journal of Agriculture and Forestry. 2008;**32**:215-220

[89] Mallarino AP, Kaiser DE, Ruiz DA, Laboski CAM, Camberato JJ, Vyn TJ.

Micronutrients for Soybean Production in the North Central Region; CROP-3135. Ames IA, USA: Iowa State University; 2017

[90] Rajput A, Memon M, Memon KS, Tunio S, Sial TA, Khan MA. Nutrient composition of banana fruit as affected by farm manure, composted pressmud and mineral fertilizers. Pakistan Journal of Botany. 2017;**49**(1):101-108

[91] Cakmak I. Plant nutrition research: Priorities to meet human needs for food in sustainable ways. Plant and Soil. 2002;**247**:3-24

[92] Malakouti MJ. Zinc is a neglected element in the life cycle of plants: A review. Middle Eastern and Russian Journal of Plant Science and Biotechnology. 2007;**1**:1-12

[93] Monreal CM, DeRosa M, Mallubhotla SC, Bindraban PS, Dimkpa C. Nanotechnologies for increasing the crop use efficiency of fertilizer-micronutrients. Biology and Fertility of Soils. 2016;**52**:423-437

[94] Oliver MA, Gregory P. Soil, food security and human health: A review. European Journal of Soil Science. 2015;**66**:257-276

[95] Maharjan B, Shaver TM, Wortmann CS, Shapiro CA, Ferguson RB, Krienke BT, et al. Micronutrient Management in Nebraska, Nebraska Extension. Lincoln NE, USA: University of Nebraska-Lincoln; 2018

[96] Fageria NK, Filho MB, Moreira A, Guimarães CM. Foliar fertilization of crop plants. Journal of Plant Nutrition. 2010;**32**:1044-1064

[97] Rietra RPJJ, Heinen M, Dimkpa CO, Bindraban PS. Effects of Nutrient Antagonism and Synergism on Fertilizer Use; VFRC Report 2015/5. Washington DC, USA: Virtual Fertilizer Research Center; 2015

[98] Dimkpa CO, Bindraban PS, Fugice J, Agyin-Birikorang S, Singh U, Hellums D. Composite micronutrient nanoparticles and salts decrease drought stress in soybean. Agronomy for Sustainable Development. 2017;**37**:5

[99] Adisa IO, Pullagurala VLR, Peralta-Videa JR, Dimkpa CO, Elmer WH, Gardea-Torresdey JL, et al. Recent advances in nano-enabled fertilizers and pesticides: A critical review of mechanisms of action. Environmental Science: Nano. 2019;**6**:2002-2030

[100] McBratney A, Field D, Koch A. The dimensions of soil security. Geoderma. 2014;**213**:203-213

[101] Moebius-Clune BN, Moebius-Clune DJ, Gugino BK, Idowu OJ, Schindelbeck RR, Ristow A, et al. Comprehensive Assessment of Soil Health-the Cornell Framework, Edition 3.2. Ithaca NY, USA: Cornell University; 2016

[102] Hills K, Collins H, Yorgey G, McGuire A, Kruger C. Safeguarding Potato Cropping Systems in the Pacific Northwest through Improved Soil Health. Pullman WA, USA: Center for Sustaining Agriculture and Natural Resources Washington State University; 2018

[103] FAO. An International Technical Workshop. Investing in Sustainable Crop Intensification: The Case for Improving Soil Health. Rome, Italy: Food and Agriculture Organization of the United Nations; 2008

[104] Jensen ES, Ambus P, Bellostas N, Boisen S, Brisson N, Corre-Hellou G, et al. Intercropping of cereals and grain

legumes for increased production, weed control, improved product quality and prevention of N-losses in European organic farming systems. In: Proceedings of the International Conferences: Joint Organic Congress (Theme 4). Odense, Denmark: Crop Systems and Soils; 2006

[105] Schulin R, Khoshgoftarmanesh A, Afyuni M, Nowack B, Frossard E. Effects of soil management on zinc uptake and its bioavailability in plants. In: Banuelos G, Lin Z, editors. Development and Uses of Biofortified Agricultural Products. Boca Raton FL, USA: CRC Press; 2009

[106] Naab JB, Chimphango SMB, Dakora FD. N2 fixation in cowpea plants grown in farmers' fields in the upper west region of Ghana, measured using 15N natural abundance. Symbiosis. 2009;**48**:37-46

[107] Deakin WJ, Broughton WJ. Symbiotic use of pathogenic strategies: Rhizobial protein secretion systems. Applied Soil Ecology. 2009;**7**:312-320

[108] Lal R, Negassa W, Lorenz K. Carbon sequestration in soil. Current Opinion in Environment Sustainability. 2015;**15**:79-86

[109] Dhakal Y, Meena RS, Kumar S. Effect of INM on nodulation, yield, quality and available nutrient status in soil after harvest of green gram. Legume Research. 2016;**39**(4):590-594

[110] Veloso MG, Angers DA, Tiecher T, Giacomini S, Dieckow J, Bayer C. High carbon storage in a previously degraded subtropical soil under no-tillage with legume cover crops. Agriculture, Ecosystems and Environment. 2018;**268**:15-23

[111] Oliveira M, Barre P, Trindade H, Virto I. Different efficiencies of grain legumes in crop rotations to improve soil aggregation and organic carbon in the short-term in a sandy. Cambisol Soil Tillage Research. 2019;**186**:23-35

[112] Dhakal Y, Meena RS, De N, Verma SK, Singh A. Growth, yield and nutrient content of mung bean (*Vigna radiata* L.) in response to INM in eastern Uttar Pradesh, India. Bangladesh Journal of Botany. 2015;**44**(3):479-482

[113] Chaudhary MI, Adu-Gyamfi JJ, Saneoka H, Nguyen NT, Suwa R, Kanai S, et al. The effect of phosphorus deficiency on nutrient uptake, nitrogen fixation and photosynthetic rate in mashbean, mungbean and soybean. Acta Physiologiae Plantarum. 2008;**30**:537-544

[114] Sulieman S, Tran LSP. Phosphorus homeostasis in legume nodules as an adaptive strategy to phosphorus deficiency. Plant Science PSL. 2015;**9221**:1-8

[115] Chen J, Heiling M, Resch C, Mbaye M, Gruber R, Dercon G. Does maize and legume crop residue mulch matter in soil organic carbon sequestration? Agriculture, Ecosystems and Environment. 2018;**265**:123-131

[116] Lal R. Climate change and soil degradation mitigation by sustainable management of soils and other natural resources. Agricultural Research. 2012;**1**:199-212

[117] Lal R. Restoring soil quality to mitigate soil degradation. Sustainability. 2015a;**7**:5875-5895

[118] Liang B, Lehmann J, Sohi SP, Thies JE, O'Neill B, Trujillo L. Black carbon affects the cycling of non-black carbon in soil. Organic Geochemistry. 2010;**41**:206-213

[119] Srinivasarao C, Venkateswarlu B, Lal R. Long-term effects of soil fertility management on carbon sequestration in a rice-lentil cropping system of the indo-Gangetic plains. Soil Science Society of America Journal. 2012;**76**(1):167-178

[120] Meena RS, Meena V, Meena SK, Verma JP. The needs of healthy soils for a healthy world. Journal of Cleaner Production. 2015;**102**:560-561

[121] Meena RS, Meena PD, Yadav GS, Yadav SS. Phosphate solubilizing microorganisms: Principles and application of microphos technology. Journal of Cleaner Production. 2017;**145**:157-158

[122] Varma D, Meena RS, Kumar S, Kumar E. Response of mung bean to NPK and lime under the conditions of Vindhyan region of Uttar Pradesh. Legume Research. 2017;**40**(3):542-545

[123] Buragohain S, Sharma B, Nath JD, Gogaoi N, Meena RS, Lal R. Impact of ten years of bio- fertilizer use on soil quality and rice yield on an Inceptisol in Assam. India Soil Research. 2017;**56**(1):49-58

[124] Ayarza M, Barrios E, Rao IM, Amezquita E, Rondón M. Advances in improving agricultural profitability and overcoming land degradation in savanna and hillside agroecosystems of tropical America. In: Bationo A, Waswa B, Kihara J, Kimetu J, editors. Advances in Integrated Soil Fertility Research in Sub-Saharan Africa: Challenges and Opportunities. The Netherlands: Springer; 2007. pp. 209-229

[125] Bargaz A, Lyamlouli K, Chtouki M, Zeroual Y, Dhiba D. Soil microbial resources for improving fertilizers efficiency in an integrated plant nutrient management system. Frontiers in Microbiology. 2018;**9**:1606

[126] Duchene O, Vian JF, Celette F. Intercropping with legume for agroecological cropping systems: Complementarity and facilitation processes and the importance of soil microorganisms. Agriculture, Ecosystems and Environment. 2017;**240**:148-161

[127] Çerçioğlu M, Anderson SH, Udawatta RP, Alagele S. Effect of cover crop management on soil hydraulic properties. Geoderma. 2019;**343**:247-253. DOI: 10.1016/j.geoderma.2019.02.027

[128] Muhammad I, Wang J, Sainju UM, Zhang S, Zhao F, Khan A. Cover cropping enhances soil microbial biomass and affects microbial community structure: A meta-analysis. Geoderma. 2021;**381**:114-696. DOI: 10.1016/j.geoderma.2020.114696

[129] Elfstrand S, Båth B, Mårtensson A. Influence of various forms of green manure amendment on soil microbial community composition, enzyme activity and nutrient levels in leek. Applied Soil Ecology. 2007;**36**:70-82. DOI: 10.1016/j.apsoil.2006.11.001

[130] Issaka RN, Buri M, Sekyi-Annan E, Musah M, Dugan E. Enhancing soil productivity in Ghana through the use of leguminous cover crops and tillage practices. West African Journal of Applied Ecology. 2021;**29**(2):24-32. Available from: https://www.ajol.info/index.php/wajae/article/view/220049

[131] Scavo A, Fontanazza S, Restuccia A, Pesce GR, Abbate C, Mauromicale G. The role of cover crops in improving soil fertility and plant nutritional status in temperate climates. A review. Agronomy for Sustainable Development. 2022;**42**(5):93. DOI: 10.1007/s13593-022-00825-0

Chapter 12

Increase of Corn Grain Yield by Improving Nitrogen Supply from Green Manure Hairy Vetch

Mohammad Zarif Sharifi

Abstract

A three-year field experiment (2007–2009) at Tokyo University of Agriculture and Technology examined the effects of nitrogen (N) released from hairy vetch (HV) on corn growth and yield. The study evaluated N release synchrony with corn uptake and aimed to establish HV usage guidelines. Four treatments were tested: control (A), chemical fertilizer (D), HV residue incorporated into the soil (B), and HV residue used as mulch (C), with three replications each. In 2007, no significant differences were observed among treatments in above-ground biomass (AGB), N uptake, total soil N, or grain yield, likely due to residual soil fertility. In 2008, AGB, grain yield, and N uptake significantly increased in the B and C plots. Correlations between soil properties, AGB, grain yield, and inorganic N levels were significant. In 2009, residue-recycling effects were inconsistent, but total and inorganic soil N levels remained higher in B and C plots. The study concluded that HV effectively supplies N to corn, reducing reliance on chemical fertilizers and promoting sustainable farming. Incorporating HV residue into the soil enhances soil fertility and reduces environmental impacts associated with synthetic fertilizers.

Keywords: hairy vetch, corn AGB, N fertilizer, green manure, sustainable farming

1. Introduction

A 50% increase in global population will significantly strain existing agricultural systems. Doubling food production is necessary to meet the demands of a larger and wealthier global population [1].

Recent successes in Latin America and Southeast Asia demonstrate the potential of GMCCs in increasing crop yields and improving soil fertility. In many cases, legumes like HV, Mucuna, crotalaria, and sesbania are cultivated as GMCCs to supply organic matter and N fixed by rhizobium to soil by their incorporation [2].

GMCCs, like hairy vetch, contribute high amounts of organic matter to the soil, improving soil structure by promoting aggregation and increasing water infiltration capacity. This, in turn, supports the root growth of subsequent crops and helps control soil erosion [3, 4]. GMCCs play a crucial role in nutrient management by temporarily storing nutrients, preventing leaching, and ensuring their availability for succeeding crops [5]. This is particularly important for nitrogen (N), as cover crops like hairy vetch can capture residual N after the main crop harvest and return it to the soil [6–8]. Proper management of GMCC residues is critical for maintaining soil health and ensuring the sustainability of conservation cropping systems [9, 10]. The way residues are managed, including tillage practices, influences soil properties and the effectiveness of the GMCCs in subsequent cropping cycles [11, 12].

Legumes, such as hairy vetch, have the ability to fix atmospheric N, reducing the need for synthetic N fertilizers. Hairy vetch, in particular, is noted for its high N content and low C-to-N ratio, which allows it to decompose rapidly and provide a significant amount of N to the soil [4, 13]. This can lead to corn yields similar to those achieved with synthetic N fertilizers, making it an economically viable option [1, 14].

The use of hairy vetch as a GMCC can provide most of the N requirements for crops like corn, leading to competitive yields without the need for large amounts of synthetic fertilizers [2, 15]. Research indicates that hairy vetch can produce between 135 and 225 kg N ha^{-1}, which is sufficient to meet the majority of corn's N needs [15–17]. Farmers who carefully manage their GMCCs, particularly winter annual legumes, can maximize the benefits these cover crops offer [18]. By selecting the appropriate method of N cover crops with main crops and managing the residues effectively, farmers can improve soil health, reduce reliance on synthetic fertilizers, and increase the yield of their primary crops [19]. This integrated approach is part of sustainable agricultural practices, helping to balance productivity with long-term soil and environmental health.

Of many ways of GMCCs management, it is recently focused on the incorporation or the placement a soil surface, respecting to the synchrony of N release from the residues with the N uptake of corn [20, 21]. Therefore, our objectives were (i) to evaluate the amounts of released N from HV and its effect on the growth and yield of corn, (ii) to examine the synchrony of N release from HV and the N uptake of corn, and (iii) to establish the guideline to use GMCCs (HV) in corn production. We hypothesized that the grain yield of corn would be similar or greater with HV in comparison with that applied with recommended fertilizer N.

2. History, benefits, and botanical properties of hairy vetch

Hairy vetch, *Vicia villosa,* is a winter annual legume noted for its winter hardness and productivity as forage and winter green manure [10]. HV has climbing, prostrate, or trailing properties (**Figure 1**). HV belongs to bean or pea family (*Fabaceae*), and it grows 30–90 cm feet length, branching occasionally and tending to sprawl. There are about 150 species of vetch, several of which have agricultural importance [22, 23]. Some 25 species are native in the United States, and commercial species (*Vicia villosa* Roth) are originated to Europe or Western Asia [16]. The round stems with spreading white hairs and prominent ridges. The alternate, compound leaves that can reach up to 10 inches long and 2 inches across consist of 8–12 pairs of leaflets. The leaflets are accompanied by a terminal tendril, which helps the plant climb and support itself on nearby vegetation [24]. Blooms from early to mid-summer, with a blooming period lasting 1–2 months, although some plants may bloom later in the year. The flowers can produce flat-sided seedpods up to 2 inches long, each containing several round seeds [17].

Figure 1.
Hairy vetch plant in the experimental field 2007.

Hairy vetch thrives best in environments with full or partial sun, favoring moist to mesic conditions and fertile, loamy soil. These conditions support its growth and development, allowing it to effectively fix N and improve soil health [25]. However, the plant is quite adaptable and can also be found in drier locations, such as slopes or areas with less ideal soil conditions. This adaptability makes HV a versatile choice for various agricultural and ecological settings, particularly in areas where soil improvement is needed or where erosion control is a concern [26]. Like many other introduced vetches, HV can spread aggressively (hairy vetch). Vetch can tolerate a range of adverse soil conditions, from poor to fertile soils, and is capable of thriving under both low and high rainfall. Additionally, it can grow across a wide pH range, from 4.9 to 8.2, which further enhances its suitability for diverse agricultural settings [26]. The cover grows slowly in fall, but root development continues over winter and rapidly in the spring (Guire, Cover crop for the Columbian basin).

Hairy vetch is typically planted in summer or early fall after summer crops are harvested [16] and adapted to plant at this time of a year because the optimum germination temperature ranges from 15 to 23°C and the optimum temperature for early root growth is 20 to 25°C. It is adapted to grow in early spring because of the soil temperature for optimum growth, well water use efficiency [10]. The timing for killing HV in the spring is crucial, especially during April and May when its growth rate significantly accelerates. Research indicates that delaying the kill date by approximately 2 weeks between late April and early May can increase biomass production by 61% [26]. This additional biomass can enhance soil fertility and structure, making it beneficial for the subsequent corn crop [22, 27]. Terminating HV in late April or early May is often preferable for corn planting. This timing allows for substantial biomass accumulation while ensuring that the vetch does not interfere with corn growth [28]. HV is relatively easy to kill through mechanical methods like mowing or rolling, particularly when it is flowering. However, these methods may be less effective if applied to vegetative stands, as the vetch might continue growing if not fully terminated [28]. Overall, managing the timing of HV termination is critical to maximizing its benefits as a cover crop while ensuring a smooth transition to corn planting.

Among the GMCCs, HV had shown to provide several benefits to corn production systems. Some of these include that 50 to 120 kg N per hectare [16] is provided *via* the surface-residue mulch, reducing runoff and conserving soil moisture [27]. It is anticipated that the combined effects of the extra moisture and N will result in a higher yield potential than fertilized corn without HV residue [10].

3. Materials and methods

A three-year, 2007–2009, field experiments were conducted at an upland field of 432 m^2 in the Field Science Centre of TUAT, Fuchu, Tokyo (139° 28′W, 35°41′N). Soil of the experimental site was a volcanic ash soil, *Slic Andisols*, and the total annual rainfall and temperatures during the experiment period are presented in **Figure 2**. Prior to the experiments, the land was under cultivations of upland crops for a long time.

3.1 Experimental design

The treatments were lied out by a randomized block design with three replications. Based on the experimental design, the treatments consisted of hairy vetch residue removed (control, A), hairy vetch residue incorporated (B), hairy vetch placed on inter-rows (C), and chemical fertilizer (D) (**Figure 3**). The treatments were subsequently planted with corn (*Zea mays*). Each plot size was 6 × 6 m.

4. Field operation

Before planting HV, the experimental plots were tilled to a depth of 20 cm. In this study, HV was sawn in rows with 15 cm spacing in all the experimental plots using a "Gon-bei" (seeding machine) in October 2006, but in 2007 and 2008, it was seeded just on the previous residue plots (B and C).

On May 22, 2007, HV samples were taken out randomly from 1 x 1 m quadrates in triplicate from one plot, and on May 12, 2008, and 2009, HV was sampled similarly. To measure biomass yield, fresh samples of HV were first weighed. A sub-sample, approximately 100 g, was taken from the total sample for further analysis. This sub-sample was used to determine the dry matter yield and N concentration, providing

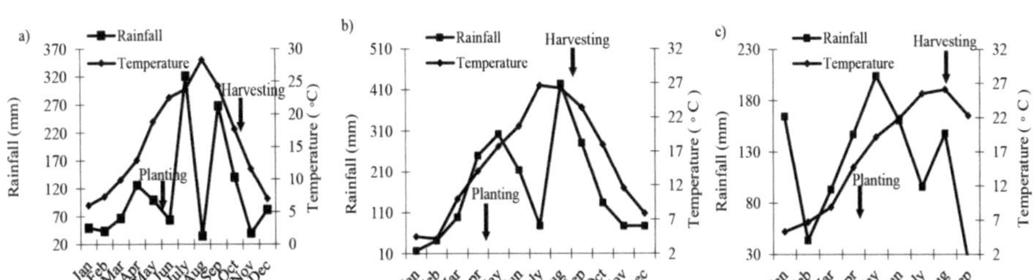

Figure 2.
Temperatures and rainfall recorded in (a) 2007, (b) 2008, and (c) until September of 2009 in field areas.

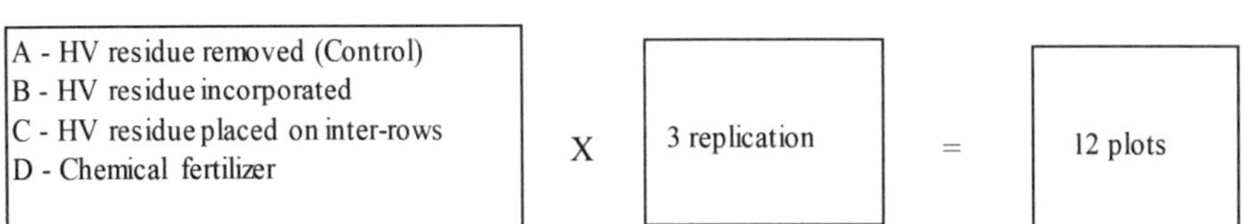

Figure 3.
Experimental design and treatments.

key data on the nutrient content and overall biomass produced. The remainder of the biomass was then returned to the harvested area, where it was evenly spread by hand.

All the sample pieces were transferred to the laboratory and dried in an oven at 70°C for 2 days. The dry weights were measured immediately and then milled to <1-mm powder, and the N concentration was determined. Nitrogen and carbon were analyzed by using the NC analyzer machine (NC-80 AUTO; Sumika Chemical Analysis Service, Ltd.).

After sampling, HV of B plot was mowed with a rotary mower (HAMMER knife HR55OB) to break the plants into small pieces to prevent from dragging, which makes tillage and seeding work hard. Hairy vetch of A, C, and D plots was slashed by a hand mower in 2007. Based on the experimental design, the residues were completely removed from A and D treatments, but, for C plot, their residue was removed temporary to till the plots and then replaced on the inter-rows. For B treatment, the residue was incorporated by a rotary till into the soil two times. Residues were allowed to decompose in the soil for 2 weeks before corn planting.

Corn seeds (var. Golddent KD77 SP, Kaneko co. Ltd) were planted by "Gon-bei" machine about 18–20 cm distance at a depth of 2–3 cm in all the plots at a density of 62,500 seeds ha^{-1} in June 2007 and 71,428 seeds ha^{-1} at the end of May in 2008 and 2009. The size of each plot was 6 by 6 meters. All the plots were covered with a net to avoid the corn seed from pigeons and other birds. Weeds were adequately controlled by a hand-hoe and hand weeding two times until 8-leaf stage.

For D treatment, nitrogen fertilizer was applied as $(NH_4)_2SO_4$ for 2007 and 2009 and the compound fertilizer (N, P_2O_5, K_2O = 14-14-14%, respectively) for 2008 at a rate of 100 kg/ha^{-1} in a band 5-cm below and 5-cm laterally from the corn plant. To control corn borer and other corn insects, Parathion was sprayed in three different times during the corn growth in 2007, Diazinon powder was applied two times during the 8–12 leaf stage, and Parathion was sprayed at the silky stage in 2008 and in 2009.

5. Measurements

5.1 Plant height

Regarding plant height measurement, two plants were randomly chosen from each row of entire plots, and their heights were measured every week. The plant height is considered one of the most representative indicators of corn growth. Measurements were performed on July 13, 16, and 8 in 2007, 2008, and 2009, respectively, and continued until the end of the silky stage, when the plant height was assumed to reach its final value. The height measurements were conducted using a regular ruler from the soil surface up to the end of plant growth point.

5.2 Leaf color

Leaf color of the top leaf, which is fully opened, is widely used as an indicator of chlorophyll content [29]. Two plants were chosen randomly from each row of whole plots to measure the leaf color. The leaf color was measured for the first time on July 13, 16, 8, in 2007, 2008, and 2009, respectively, by the SPAD chlorophyll meter. The biomass yield measurement continued throughout the growing period until the time of harvest. During this period, in addition to measuring biomass, the number of leaves on the corn plants was also counted. This was done to monitor and assess the growth stage of the corn, providing valuable information about the plant's development in relation to the timing of the biomass measurements and the influence of the hairy vetch cover crop. Tracking leaf count alongside biomass yield allowed for a more comprehensive understanding of how the presence of vetch residue and its associated nutrients impacted corn growth throughout the season.

5.3 Above ground biomass (dry weight)

Corn biomass yields were determined by harvesting the above-ground portion of the six randomly selected plants in 2007 and 7 plants in 2008 and 2009 from the different places of each plot of each treatment on August 18, 2007, July 30, 2008, and July 24, 2009, during at 12-leaf stage for first sampling. The whole plants weighted separately, and their fresh weights were measured, then among the six weighted plants, just one of them was selected to represent the approximate average of six plants. The subsamples were chapped and oven-dried at 70°C for 3 days. The dry weights were measured immediately after removing the samples from the oven. The second time sampling for biomass determination was conducted on September 9, 2007, and August 13, 8 in 2008 and 2009, respectively, during the silky stage as the prior time procedure. The third sampling was established at the harvesting time on October 3, 2007, and at ripening stage on September 17, 2008, and September 3, 2009, for the purpose of biomass estimation. In 2009, corn biomass was measured at the harvesting time on September 18, 2009, as well. For 2007 experiment, the last sampling procedure was different. Firstly, 18 randomly selected plants from each plot of whole treatments were hand-harvested, and their fresh weights were measured separately for biomass yield determination. Secondly, the mean of each three plant of 18 randomly plants was taken, then among the six weighted plants, just one of them was selected to represent the approximate average of six plants, and the rest of the procedure was the same as prior steps. After oven-dried the dry weights were measured and ground to pass 1 mm for determination of N amount. The amount of N and carbon was analyzed by using the NC analyzer machine (NC-80 AUTO; Sumika Chemical Analysis Service, Ltd.).

5.4 Grain yield

Corn grain yields were determined by hand harvesting 18 plants on October 4, 2007, 21 plants on September 25, 2008, and 6 plants on September 18, 2009, from the middle of each row in each treatment. After harvesting, the corn ears were separated from the biomass parts and their husks removed, and then, the fresh weights were measured. In 2007, for the purpose of grain yield determination, the three ears were randomly selected from the 18 harvested ears of each treatment, and their fresh weights were measured. For 2008 experiment, from 21 harvested ears the average of three ears from each row of each plot (one plot included seven average) which was seven ears taken out, and their fresh weights were measured. After that, all corn ears samples were transferred to the laboratory and put in the oven at 80°C for 2 days. After oven drying the dry weights were measured, all grains were detached from corn cobs by hand, and then, the dry weight of seeds and cobs were measured. Additionally, in 2008, the grain number was counted for calculation of corn grain yield. Then detached corn seeds and their cobs were separately ground for determination of N amount throughout the years. The amount of N and carbon of corn seeds and cobs were analyzed by using the NC analyzer machine (NC-80 AUTO; Sumika Chemical Analysis Service, Ltd.). Then, the N uptake was calculated.

5.5 Soil samples

After harvesting the corn plants, the soil samples were taken in two increments (soil depths of 0 to 15 and 15 to 25 cm) from the two midways of each plot in October 2007. The five different places of each plot in one increment (soil depth of 25 cm) on October 3, 2008, and from four different places in two increments (soil depth of 0 to 15 and 15 to 25 cm) in September 2009. The soil samples were air-dried and crushed to pass through a 2-mm sieve. Total soil N was analyzed by using the NC analyzer machine (NC-80 AUTO; Sumika Chemical Analysis Service, Ltd.). Inorganic N in soil was extracted using a 20 mM L^{-1} calcium chloride solution and determined using a total carbon analyzer with an N unit (TOC-VCSH+TNM; Shimadzu Corp.). Soil pH value was determined using (Personal pH meter, model pH 81, Yokogawa Electric Cooperation).

Statistical analysis was carried out by standard ANOVA procedures using Excel Statistics ver. 1.1 (SRI, Tokyo).

6. Results and discussion

6.1 Hairy vetch biomass and nitrogen production

The mean of dry matter of hairy vetch biomass, which was collected approximately 7 months after planting, was 8.6 Mg ha^{-1} $year^{-1}$. This was greater than the minimum rate for a high cover crop residue in corn production, 5.29 Mg ha^{-1}, reported by some researches [3]. The HV residue in this experiment contained nitrogen 278 kg N ha^{-1} (**Table 1**). This N content is comparable and, in some cases, higher than the values for some another covers crops.

6.2 Plant height

Corn plant height was not influenced by the incorporation of HV in 2007 (**Figure 4a**). Similar increases in plant height in most treatments indicate that the soil fertility in all plots was uniform in 2007. In 2008, the plant height in B plots was increased from 53 DAP by the incorporation of hairy vetch and followed by C and D plots up to end of growth period, while A plot showed less plant height from 67 DAP (**Figure 4b**). In 2009, the plant height was resulted in almost the same as that in 2008, but there was a very clear difference between the residue-applied plots (B and C) and no residue plots (A and D) (**Figure 4c**).

Year	Dry matter	N concentration	N content
	Mg ha^{-1}	mg g^{-1}	kg ha^{-1}
2007	8.1 ± 2.8	33.0 ± 7.1	267
2008	7.9 ± 0.4	37.9 ± 3.4	299
2009	9.9 ± 0.4	27.0 ± 4.5	267
Mean	8.6	32.6	278

Table 1.
Hairy vetch dry matter production, nitrogen concentration, and nitrogen content measured at the blossom stage during the 2007–2009 growing season.

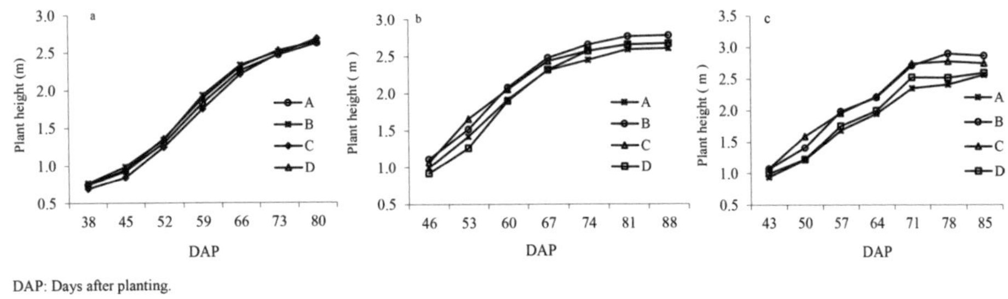

Figure 4.
Effect of hairy vetch on plant height (a) in 2007, (b) in 2008, and (c) in 2009.

We observed that the plant height was significantly affected by both the incorporation and the surface placement of HV residue, suggesting that HV residue had a readily available characteristic in rather short period.

6.3 Above ground biomass (AGB)

AGB of corn was not affected by the treatments significantly in the silky stage and the harvesting time, but there was a significant deference at the ripening stage in 2007 (**Figure 5**). There was a significant difference between A and B, B and C as well as between B and D treatments at ripening stage. The slight decrease of AGB in C plot, which received HV residue by the surface placement, might be due to the less supply of N from the residue. It was considered that the incorporation of HV had a larger N supplying effect than the surface placement, so the incorporation of HV is a preferable method than the other. It has been reported that the release of HV residue is faster when mixed into soil through tillage compared to when it is left on the soil surface without tillage [30]. AGB of A and D plot had reached the level of the other plots at harvesting time, suggesting that the N supply which was an available fraction of total N was run out at the last stage (**Figure 6**).

In 2008, AGB was not affected by the treatments at 12-leaf and silky stages, but a significant difference occurred between the treatments at ripening stage (Figure 5.6 and **Table 2**). AGB of B plot was slightly higher than the other plots in silky stage, and there were significant differences at ripening stage between A, B and A, C treatments

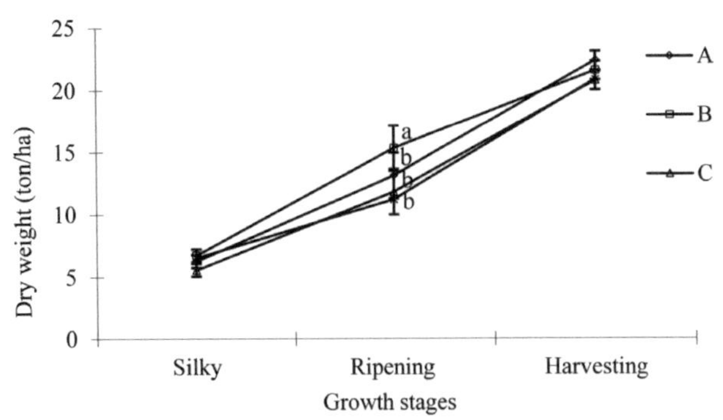

Figure 5.
Overall weight of above ground corn biomass throughout the growth stages in 2007. Unique characters represent important distinctions at a significance level of P < 0.05 (Fischer's PLSD).

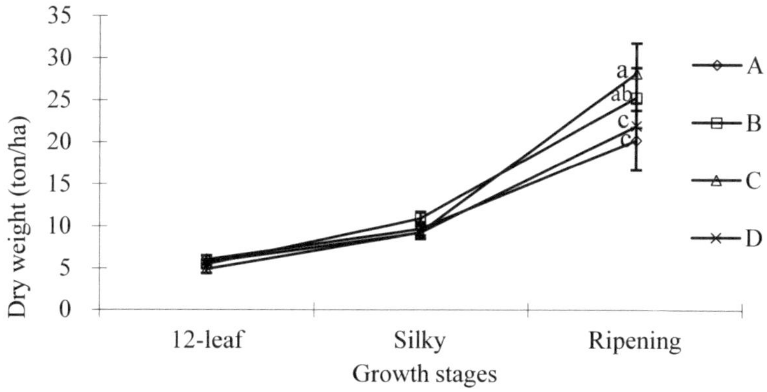

Figure 6.
Overall weight of above ground corn biomass throughout the growth stages in 2008. Unique characters represent important distinctions at a significance level of P < 0.05 (Fischer's PLSD).

as well as between C and D treatments; AGB of C plot was higher compared with other plots and followed by B plot. At 12-leaf stage, the N starvation by the addition of the residue might be a cause for the slightly less AGB in B and C plots. It seemed that HV residue decomposed synchronizing with corn growth in all stages either in B and C plot, and plants absorbed the N derived from HV residues. It was reported that HV residue could decompose rapidly in the soil because of its high N concentration, a low C/N ratio [5, 31].

In 2009, AGB was affected by the treatments significantly from the early stage, but the differences disappeared at harvesting time (Figure 5.7 and **Table 2**). In 12-leaf stage, there has already been a significant difference between residue plots (B, C) and control (A). It was considered that since no fertilizers were applied to A, B, and C plots, N fertility in the soil had gradually decreased in the third year, 2009, and the effect of N supply from the residue became much clearer.

At silky stage, also significant differences were existed between A and B, and between A and C, as well as between B and D, C and D, respectively (**Figure 7**). In this stage, B and C plots tended to increase AGB than A and D plots because of N

	2007			2008			2009		
Treatments	**AGB ton ha^{-1}**	**Grain yield ton ha^{-1}**	**N uptake kg ha^{-1}**	**AGB ton ha^{-1}**	**Grain yield ton ha^{-1}**	**N uptake kg ha^{-1}**	**AGB ton ha^{-1}**	**Grain yield ton ha^{-1}**	**N uptake kg ha^{-1}**
A	22.2	14.4	182.9	20.2 c	13.6 b	235.7 c	23.8	11.8	341.2
B	22.0	11.6	208.3	25.3 ab	16.2 a	333.6 ab	25.2	12.9	390.1
C	20.5	11.2	205.4	28.2 a	15.2 ab	373.6 a	25.3	12.7	379.7
D	20.8	10.4	203.2	21.9 c	15.4 ab	280.4 c	25.7	12.7	409.2

(AGB) above ground biomass; (N): nitrogen; (HV): hairy vetch.
In 2008, the AGB and N uptake were measured during at the ripening stage.
Different litters indicate significant differences at P < 0.05 (Fischer's PLSD).

Table 2.
Total AGB, grain yield, and N uptake of corn plants at harvesting time in 2007, 2008, and 2009.

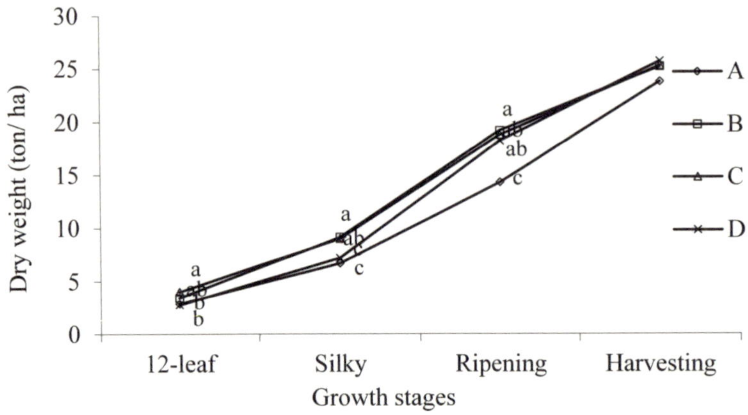

Figure 7.
Overall weight of above ground corn biomass throughout the growth stages in 2009. Unique characters represent important distinctions at a significance level of $P < 0.05$ *(Fischer's PLSD).*

Figure 8.
Condition of corn plant during (a) at 8-leaf stage and (b) at 12-leaf stage in 2009.

supply from HV residues, but for D plot the same condition as 12-leaf stage continued. The researcher found that the immobilization of additional N was consistently higher from high-value residues compared to fertilizer. This was because the close connection between C and N in residues led to a slower decomposition rate than inorganic fertilizer N [8].

At the ripening stages, AGB was also affected by the treatments significantly (**Figure 7**). There was a clear difference of AGB between A and D plots too, though no difference of N application was in both plots in this year. At harvesting time, the differences of AGB disappeared, but B, C, and D plots showed almost the same results to each other, and they had higher amount of AGB than A plot (**Figure 8**).

7. Nitrogen uptake

There was no significant difference among treatments in N absorption in 2007 (**Table 2**). Nitrogen uptake from A plot was less than others at the harvesting time. This inferiority may be due to the lack of N supply from HV residue or fertilizer. B and C plots showed a slightly higher amount of N uptake (**Figure 9**).

In 2008, there was a significant difference in N uptake between B and C treatments as well as between B and D treatments, while no significant difference between A and B plots at 12-leaf stage (**Table 3**). In general, HV residue incorporated into soil tends to decompose and release nutrient (N) much faster than the residue placed on soil surface, because the incorporation increases microbes' access to a small piece of plant residue [32]. Researchers have found that HV residue decomposes faster when mixed into the soil through tillage than when left on the soil surface without tillage [30, 33].

There was no significant difference among treatments in N uptake at silky stage (**Table 4**). However, N uptake in B and C plots is larger than A and D plots, in which received N was less than B and C.

At the ripening stage, there was a significant difference in N uptake between A and B, and between A and C plot, as well as between C and D plot, while no significant difference between B and C plots (Table. 5.5). The total N uptake was within the range of 236 (A: minimum) and 374 (C: maximum) kg N ha^{-1}, and the difference 138 kg N ha^{-1} was estimated as the amount N released from the incorporated HV residue, if the released N was completely absorbed by corn plant.

This value is about 50% of N applied as HV residue in 2008, meaning that the rest 50% of N remained in the soil. The N availability of 50% is in correspondence with that of fertilizers though the mean N availability of fertilizer reaches more than 60% in upland agriculture.

In 2009, there was a significant difference between A and B, B and D plots also between C and D plots in N uptake at 12-leaf stage (**Table 5**). However, in 2007 and 2008, the AGB of corn was not affected by the treatment significantly in 12-leaf stage but, in 2009, it might be caused by continuous HV sowed in residues plots. The decrease observed in the D plot, which received N at this stage, depends on the N supply procedure in 2009, as mentioned earlier. However, since there is no difference between the A and D plots, this decrease might be due to similar N supply conditions in both plots.

(c) (d)

Figure 9.
Condition of corn plant during (c) at silky stage and (d) at the ripening stage in 2009.

	Treatments			
	A	B	C	D
	kg ha^{-1}			
	131.2	124.8	79.7	93.6
	121.4	118.5	109.3	77.0
	135.2	174.5	90.1	111.9
Mean	129.3 ab	139.3 a	93.0 b	94.2 b
STD	7.1	30.6	15.0	17.4

Different letters mean significant differences at P < 0.05 (Fischer's PLSD).

Table 3.
Amount of nitrogen absorbed by corn biomass at the 12-leaf stage in 2008.

	Treatments			
	A	B	C	D
	kg ha^{-1}			
	167.1	141.9	172.7	147.2
	170.8	163.9	166.5	154.4
	128.7	241.6	187.2	151.1
Mean	155.5	182.5	175.5	150.9
STD	23.3	52.4	10.6	3.6

Table 4.
Amount of nitrogen absorbed by corn biomass at the silky stage in 2008.

	Treatments			
	A	B	C	D
	kg ha^{-1}			
	77.0	97.8	91.4	57.9
	59.5	106.9	105.2	68.3
	80.0	97.3	79.5	85.3
Mean	72.2 b	100.7 a	92.0 ab	70.5 b
STD	11.1	5.4	12.9	13.8

Table 5.
Amount of nitrogen absorbed by corn biomass at the 12-leaf stage in 2009.

During at the silky stage, the rate of N uptake increased in residue plots, there was a significant difference between B and C plots followed by D plot; however, A plot was also received N, but its N was less than other plots (**Table 6**).

The trend of N uptake at the ripening stage was almost the same as the silky stage (Table. 5.7), but there was no difference between the residue-added plots (B and C) and the other plots (A and D). At harvesting time, there was no significant difference between treatments in N uptake (**Table 2**).

	Treatments			
	A	B	C	D
	kg ha^{-1}			
	113.2	152.1	194.5	183.3
	131.6	165.6	221.6	167.8
	102.5	174.6	208.6	210.3
Mean	115.8 c	164.1 b	208.2 a	187.1 ab
STD	14.7	11.3	13.5	21.5

Table 6.
Amount of nitrogen absorbed by corn biomass at the silky stage in 2009.

8. Grain yield

In 2007, there was no significant difference in grain yield among treatments, although the yield of A plot (control) was the highest. The tendencies among the treatments were almost the same as those of the dry matter yield (**Table 2**). In 2008, there was a significant difference between A and B treatments in grain yield, while no significant difference among other plots. The greatest corn grain yield was resulted in B plot, in which HV residue was incorporated. It can probably be explained by the improved soil quality due to the HV residue addition. This has induced a faster establishment/development of corn plant, the greater number of leaves, and the highest SPAD value of corn (data was not shown) throughout the growth period. This result shows that HV is a potential source of N for corn grain yield. It was reported that HV residue breaks down quickly in the soil due to its high N content, a low C to N ratio, and improves corn yield similar to N fertilizer rates [30].

In 2009, there was no significant difference between treatments in the corn grain yield, but the highest value was recorded in B plot, followed by C and D plots (**Table 2**). The residue treatments tended to increase the grain yields.

8.1 Leaf color

Leaf color (LC) began to increase at 8–10 leaf stages (45–52 DAP) in 2007. In 12-leaf stage (55–66 DAP), however, the LC of top leaves increased gradually in every treatment, especially but B and D plots (**Figure 10a**). Leaf color of all treatments was reached the height level at silky stage (80 DAP), but there was no significant change except for the slight increase of LC in B and C plots. Those highest levels of LC were maintained for about 20 days up to the ripening stage and then rapidly decreased. There was no difference among treatment.

In 2008, LC started to increase at 8–10 leaf stages (46–55 DAP). In 12-leaf stage (60–70 DAP), LC of top leaf in B plot received HV residue increased rapidly, but in other treatments it gradually decreased until 74 DAP (**Figure 10b**). This increase of LC in B plot must be due to the N derived from HV residue. It was also reported that more than 50% of the N in HV residue was released during 4 weeks after HV desiccation [32]. Leaf color of all treatments reached the highest level at silky stage (88 DAP) with a slight change between treatments. Those highest levels of LC were maintained for about 16 days up to the ripening stage and then rapidly decreased. There was no difference among treatment.

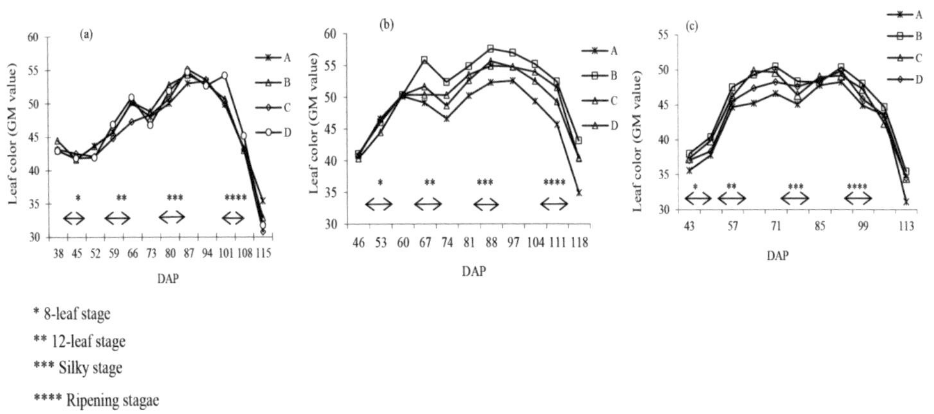

Figure 10.
Effect of hairy vetch on leaf color during the growing period of corn (a) in 2007, (b) in 2008, and (c) in 2009.

In 2009, LC stared to increase from 50 to 67 DAP with the highest value of B plots accompanied by C plot and reached the highest value until the end of 12-leaf stage (71 DAP) (**Figure 10c**). The differences between the treatments were almost the same as 2008, but there was a clearer observable tendency of LC increase in residue plots. LC of top leaves in B and C plots were maintained their high level to the later stages, while that of A and D decreased rapidly (**Table 7**).

8.2 Soil total nitrogen

There was no significant difference in soil total N among the treatments, as well as between the upper and the lower soil layer through 3 years (**Table 8**). It may be reasonable because when HV was shown in 2007, the soil fertility of every plot was almost the same.

In 2008, the residues plot (B and C) showed higher level of soil N (8.1 and 7.8 mg g^{-1}) than A and D plots (6.5 and 6.9 mg g^{-1}), suggesting that soil N was enriched by the addition of HV residue, notwithstanding the different ways.

In 2009, soil total N of the residue plots has the same tendency as that in 2008, although there was no significant difference among all treatments.

	Treatments			
	A	B	C	D
		kg ha^{-1}		
	211.0	340.2	400.0	326.2
	218.2	289.8	328.8	296.1
	275.7	354.8	412.6	271.3
Mean	235.0 b	328.3 ab	380.5 a	297.9 b
STD	35.46	25.64	34.47	18.89

Table 7.
Amount of nitrogen absorbed by corn biomass at the ripening stage in 2009.

		Treatments			
		A	B	C	D
		mg g^{-1}			
2007	Upper	5.4	5.5	5.7	5.6
	Lower	4.8	5.0	5.0	5.0
2008		6.5	8.1	7.8	6.9
2009	Upper	5.4	5.7	5.7	5.4
	Lower	5.3	5.2	5.5	5.3

Table 8.
Total nitrogen reaming in soil after harvesting in 2007, 2008, and 2009.

9. Inorganic nitrogen remaining in soil

Addition of HV residues affected the inorganic N level in soil. The residues plot (B and C) showed the highest level, while it decreased in control plot (C), and there were significant differences among the treatments in 2008 (**Figure 11**). Apparently, the residues of HV had a high share of the easily decomposable fractions of organic N, and the characteristic made high level of inorganic N in B and C plots in comparison with that in A and D plots. Reports indicated that HV residue decomposes faster when tilled into the soil compared to when left on the surface without tillage. This distinction is important for managing nitrogen availability in cropping systems [34].

In 2009, the characteristics of HV residues clearly affected the inorganic N level in soil, especially in B plots that followed by C plots, and there was a significant difference among all treatments except of A and D plots (**Figure 12**). The difference between residue plots seems to be due to the incorporation effect of HV in B plots. Reports indicated that the immobilization of additional N was consistently higher in HV residues compared to fertilizer. This was attributed to the close connection between C and N in HV residues, unlike inorganic fertilizer N, which lacks C. Thus, it is recognized that N originating from HV residues is readily recovered [35].

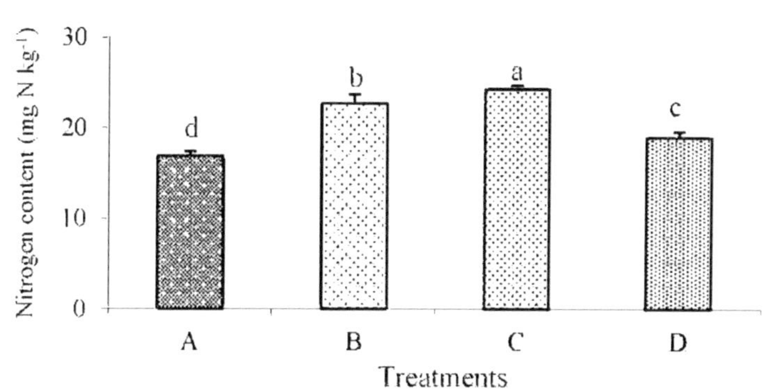

Figure 11.
Changes of inorganic nitrogen in soils applied with fertilizer and hairy vetch residues in 2008.

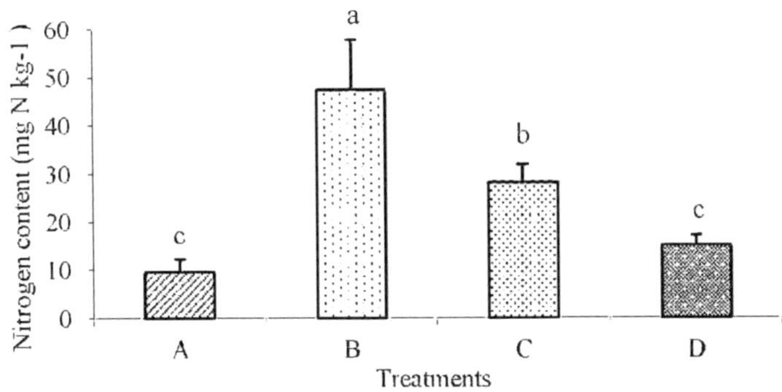

Figure 12.
Changes of inorganic nitrogen in soils applied with fertilizer and hairy vetch residues in 2009.

10. Soil pH value

Soil pH is an important factor in determining crop performance, and it influences nutrient solubility, microbial activity, and root growth. Low pH level below 7.0 affects the nutrient availabilities by converting them into forms that are not readily available to crop [36]. Also, low soil pH stunts the growth and development of corn, and sometimes kills corn plants in an extreme situation [34]. The best pH range for the greatest corn yield in this study was 6.05–6.22 in 2008 and 6.1–6.3 in 2009 (**Table 9**). Researchers were reported that the best soil pH range for corn growth is between 6.0 and 7.0. Nutritional and growth problems start to emerge when soil pH is lower than 5.5 [31].

The pH value of residue plots (between 6.0 and 6.4) becomes reasonable when HV residue applied. We observed that incorporation of GMCC residue like HV maintained the soil pH as a normal range, 6.0 to 6.4, that was the major pH in this research. If the land is used to continue the crop production for a long time without using GMCCs, the soil condition will become acidic in the rain-rich climate like Japan. Therefore, it is preferable for farmers to use GMCCs in crop rotations on their fields.

10.1 Relationship between soil chemical properties and corn AGB and grain yield

There was a close relationship between AGB of corn and the inorganic N remaining in soil as well as satisfactory relationship between the grain yield and total N and inorganic N in 2008. There was a significant negative relationship between the soil pH and the grain yield as well as between AGB and total N. Because the pH is affected by the removal of alkaline nutrients by plant, the lower pH would be caused by the greater crop growth. GMCCs could mitigate the soil

	Treatments			
Years	**A**	**B**	**C**	**D**
2008	6.22	6.05	6.12	6.08
2009	6.32	6.12	6.14	6.10

Table 9.
Soil pH level after harvesting corn plant when hairy vetch residues applied in 2008 and 2009.

	AGB	Grain Yield	Total N	Inorganic N	Soil pH
AGB	—	− 0.6563	− 0.9642*	0.9842*	− 0.7156
Grain yield	—	—	0.8057	0.6829	− 0.9890**
Total N	—	—	—	0.9378	− 0.7156
Inorganic N	—	—	—	—	− 0.6035
Soil pH	—	—	—	—	—

*Significant differences at $P < 0.05$ (Fischer's PLSD).
**Significant differences at $P < 0.1$ (Fischer's PLSD).

Table 10.
Relationship between total N, inorganic N, and soil pH and AGB and grain yield when hairy vetch residues applied as a source of N in field experiment 2008.

	AGB	Grain yield	Total N	Inorganic N	Soil pH
AGB	—	0.9623*	0.7057	0.7857	− 0.8696
Grain Yield	—	—	0.5002	0.7121	− 0.9660*
Total N	—	—	—	0.5039	− 0.3368
Inorganic N	—	—	—	—	− 0.5259
Soil pH	—	—	—	—	—

*Significant differences at $P < 0.05$ (Fischer's PLSD).

Table 11.
Relationship between total N, inorganic N, and soil pH and AGB and grain yield when hairy vetch residue applied as a source of N in field experiment 2009.

acidification by preventing soil nutrients from leaching, but the pH will decrease when the output is great (**Table 10**).

In 2009, the condition of soil pH with other aspects was same as 2008, but there was a close relationship between AGB and grain yield and satisfactory relation with total N and inorganic N with AGB. However, grain yield and total N and inorganic N also had a relationship with each other (**Table 11**). The reason of significant relationship between AGB and grain yield is absorbed nutrients from N sources added.

11. Conclusion

Both treatments of the incorporation and the surface placement of HV residue increased the corn AGB, grain yield, and N uptake in the experiments in 2008 and 2009, although in 2007, no response of corn growth was detected. The grain yield increased significantly in plots where HV residue was incorporated into the soil (referred to as B plots). This increase in yield indicates that the N derived from HV residue was effectively utilized by the corn crop, alongside the N provided by fertilizer. Total N and inorganic N in soil were determined at the high level in the residue plots (B, C) even after harvesting. In 2009, the growth-promoting effects of the residue-recycling were not always clear, but total N and inorganic N in soil were detected in higher level in these plots (B and C).

It is concluded that adding HV residue to the soil either through incorporation or surface placement has a high potential for providing the necessary N for future corn crops, but the supplement of minerals is needed at the time to make the most of the HV's possibility in corn production.

Acknowledgements

I am deeply indebted to Dr. Shoji Matsumura, Faculty of Agriculture, Tokyo University of Agriculture and Technology (TUAT). His generous advice and recommendations in getting the Monbukagakusho scholarship which gave me the opportunity to grow professionally are highly appreciated. I would like to express my thanks to former professors, students, and colleagues of Field Science Centre (FSC), TUAT, who contributed in a way during this time and for their distinctive kindness and unforgettable helps.

Author details

Mohammad Zarif Sharifi
Department of Agronomy, Faculty of Agriculture, Kabul University, Afghanistan

*Address all correspondence to: sharifimz@yahoo.com; mohammad.sharifi@ncl.ac.uk

References

[1] Miler FP. After 10,000 years of agriculture, whither agronomy? Agronomy Journal. 2008;**100**:22-34. DOI: 10.2134/agronj2007.0013

[2] Ruth M, Sjoerd W, Duiker WS, Curran DW. Hairy vetch management for no-till organic corn production. Agronomy Journal. 2010;**102**(1):355-362. DOI: 10.2134/agronj2009.0183

[3] Singer JW, Kohler KA. Rye cover crop management affects grain yield in a soybean-corn rotation [Online]. Crop Management. 2005;**4**:1-6. DOI: 10.1094/CM-2005-0224-02-RS

[4] Sogbedji JM, van Es HM, Agbeko KL. Cover cropping and nutrient management strategies for maize production in Western Africa. Agronomy Journal. 2006;**98**:883-889. DOI: 10.2134/agronj2005.0025

[5] Sharifi MZ, Shoji M, Tadashi H, Masakazu K. Apparent nitrogen mineralization rate of several green manures incorporated in soil and the application effects on growth of Komatsuna plants. Japanese Journal of Farm Work Research. 2009;**44**:163-172. DOI: 10.4035/jsfwr.44.163

[6] Bollero GD, Bullock M, Cender R, Fransworth T, Holsapple TH. Evaluation of nitrogen fertilizer rate, planting date, tillage, and winter cover crops in a summer feed grain/soybean system. In: Illinois Fertilizer Conference Proceedings January 25-27. Bloomington, Illinois: Illinois Fertilizer & Chemical Association (IFCA); 1993

[7] Ruffo ML, Bollero GA. Modeling Rye and hairy vetch residue Decomposition as a function of degree-days and decomposition-days. Agronomy Journal. 2003;**95**:900-907. DOI: 10.2134/agronj2003.9000

[8] Snapp SS, Swinton SM, Labarta R, Mutch D, Black JR, Leep NRJ, et al. Evaluating cover crops for benefits, costs and performance within cropping system niches. Agronomy Journal. 2005;**97**:322-332. DOI: 10.2134/agronj2005.0322a

[9] Quemada M. Predicting crop residue decomposition using moisture adjusted time scales. Nutrient Cycling Agroecosystems. 2004;**70**:283-291. DOI: 10.1007/s10705-005-0533-y

[10] Teasdale JR, Devine TE, Mosjidis JA, Bellinder RR, Beste CE. Growth and development of hairy vetch cultivars in the Northeastern United States as influenced by planting and harvesting date. Agronomy Journal. 2004;**96**:1266-1271. DOI: 10.2134/agronj2004.1266

[11] Morris NL, Miller PCH, Orson JH, Willliams RJF. The effect of wheat straw residue on the emergence and early growth of sugar beet (*Beta vulgaris*) and oilseed rape (*Brassica napus*). European Journal of Agronomy. 2009;**30**:151-162. DOI: 10.1016/j.eja.2008.09.002

[12] Nielsen B. Symptoms Associated with Low Soil pH in Corn. West Lafayette, IN, USA: Purdue Plant and Pest Diagnostic Laboratory; 2005. p. 1. Available from: http://www.ppdl.purdue.edu/PPDL/html [Accessed: May 26, 2009]

[13] Komatsuzaki M, Kotaro S. Influence of cover crop seeding date and their amounts on mitigation of wind erosion. Japanese Journal of Farm Work Research. 2009;**44**(4):189-199. DOI: 10.4035/jsfwr.44.189

[14] Coppens F, Patricia G, Antoine F, Roel M, Sylvie R. Decomposition of mulch versus incorporated crop residues: Modelling with PASTIS clarifies interaction between residue quality and location. Soil Biology and Biochemistry. 2007;**39**:2339-2350. DOI: 10.1016/J.SOILBIO.2007.04.005

[15] Abdul-Baki AA, Teasdale JR. Sustainable Production of Fresh-Market Tomatoes and Other Vegetables with Cover Crop Mulches. (Farmers' Bulletin No. 2280). U.S. Department of Agriculture, Agricultural Research Service; 2007. p. 31

[16] Renzi JP, Garayalde AF, Brus J, Pohanková T, Smýkal P, Cantamutto MA. Environmental and agronomic determinants of hairy vetch (*Vicia villosa* Roth) seed yield in rainfed temperate agroecosystems. European Journal of Agronomy. 2023;**147**:1-9. DOI: 10.1016/j.eja.2023.126822

[17] Verhallen A, Hayes A, Taylor T. Cover Crops: Hairy Vetch. Ontario: Ministry of Agriculture Food and Rural Affairs; 2008. Available from: http://www.omafra.gov.ca/english/crops/facts/cover/hairyvetch.htm [Accessed: December 16, 2008]

[18] Brandsæter LO, Hugh H, Riley H, Stubhaug E, Henrilsen TM. Winter survival, biomass accumulation and nitrogen mineralization of winter annual and biannual legumes sown at various times of year in northern temperate regions. European Journal of Agronomy. 2008;**28**(3):437-448. DOI: 10.1016/j.eja.2007.11.013

[19] Odhiambo JJO, Bomke AA. Grass and legume cover crop effects on dry matter and nitrogen accumulation. Agronomy Journal. 2001;**93**:299-307. DOI: 10.2134/agronj2001.932299x

[20] Balkcom KS, Reeves DW. Sunn-hemp utilized as a legume cover crop for corn production. Agronomy Journal. 2005;**97**:26-31. DOI: 10.2134/agronj2005.0026

[21] Decker AM, Andrew JC, John J, Meisinger F, Ronald M, Marla SM. Legume cover crop contribution to no-tillage corn production. Agronomy Journal. 1994;**86**:126-135. DOI: 10.2134/agronj1994.00021962008600010024x

[22] Clark A. Managing cover crops profitability. In: Sustainable Agriculture Network. 3rd ed. Beltsville, MD: U.S.D.A; 2007. pp. 9-11

[23] Kavdir Y, Smucker AJM. Soil aggregate sequestration of cover crop root and shoot-derived nitrogen. Plant and Soil. 2005;**272**:263-276. DOI: 10.1007/s11104-004-5294-x

[24] Teasdale JR, Abdual-Baqi AA, Park YB. Sweet corn production and efficiency of nitrogen use in high cover crop residue. Agronomy for Sustainable Development. 2008;**28**:559-565. DOI: 10.1051/agro:2008029

[25] Zotarelli L, Avila L, Scholberg JMS, Alves BJR. Benefits of vetch and rye cover crops to sweet corn under no-tillage. Agronomy Journal. 2009;**101**(2):252-260. DOI: 10.2134/agronj2008.0033x

[26] Sattell R, Dick R, Luna J, McGrath D, Peachey E. Common Vetch (*Vicia sativa* L.). Corvallis, Oregon, USA: Oregon State University Extension Service; 1998. Available from: https://ir.library.oregonstate.edu/concern/open_educational_resources/n

[27] Clark AJ, Decker AM, Meisinger JJ, Mulford FR, McIntosh MS. Hairy vetch kills date effects on soil water and corn production. Agronomy Journal. 1995;**87**:579-585. DOI: 10.2134/agronj1995.00021962008700030031x

[28] Teasdale JR, Richard CR. Mechanical versus herbicidal strategies for killing a hairy vetch cover crop and controlling weeds in minimum-tillage corn production. American Journal of Alternative Agriculture. 2003;**18**(2):95-102. DOI: 10.1079/AJAA200240

[29] Argenta G, Silva PRF, Bortolini CG. Leaf relative chlorophyll content as an indicator parameter to predict nitrogen fertilization in maize. Revista Brasileira de Fisiologia Vegetal. 2004;**16**:271-282. DOI: 10.1590/S0103-84782004000500009

[30] Sainju UM, Singh BP, Whitehead WF, Wang S. Accumulation and crop uptake of soil mineral nitrogen as influenced by tillage cover crops, and nitrogen fertilization. Agronomy Journal. 2007;**99**:682-691. DOI: 10.2134/agronj2006.0177

[31] Sanchez GC, Gatan MG. How to Grow Corn? Quezon City, Philippines: Open Academy for Philippine Agriculture, Department of Agriculture of the Philippines; 2007. Available from: http://www.openacademy.ph/ [Accessed: March 10, 2008]

[32] Horimoto S, Hajime A, Mitsuaki I, Michiaki I. Growth and yield of tomatoes in hairy vetch-incorporated and-mulched field. Japanese Journal of Farm Work Research. 2002;**37**(4):231-240

[33] Vaughan JD, Evanylo GK. Corn response to cover crop species, spring desiccation time, and residue management. Agronomy Journal. 1998;**90**:536-544. DOI: 10.2134/agronj1998.00021962009000040016x

[34] Sainju UM, Whitehead WF, Singh BP. Biculture legume-cereal cover crops for enhanced biomass yield and carbon and nitrogen. Agronomy Journal. 2005;**97**:1403-1412. DOI: 10.2134/agronj2004.0274

[35] Seo JH, Meisinger JJ, Lee HJ. Recovery of Nitrogen-15-labeled hairy vetch and fertilizer applied to corn. Agronomy Journal. 2006;**98**:245-294. DOI: 10.2134/AGRONJ2005.0013

[36] Heiniger RW, Robert GM, David EC. Using soil electrical conductivity to improve nutrient management. Agronomy Journal. 2003;**95**(3):508-519. DOI: 10.2134/agronj2003.5080

Chapter 13

Weed Control in Peanut (*Arachis hypogaea* L.) Using Carfentrazone Plus Pyroxasulfone Herbicide Systems

William James Grichar, Peter A. Dotray and Todd A. Baughman

Abstract

Field studies were conducted during the 2017 through 2019 growing seasons in Texas and Oklahoma to determine weed control when using herbicide systems containing the pre-mixture of carfentrazone plus pyroxasulfone (C + P) applied preemergence (PRE), early postemergence-peanut cracking (EPOST), or postemergence (POST). When pendimethalin was not used as a base PRE herbicide treatment, C + P applied PRE controlled Texas millet [*Urochloa texana* (Buckl.)] ≤75%; however, the addition of pendimethalin to C + P applied PRE increased control to >85%. Palmer amaranth (*Amaranthus palmeri S. Wats.*) control was >70% with most C + P systems while smellmelon (*Cucumis melo L.* var. *Dudaim Naud.*) control was never <97% with any C + P system. Pitted morningglory (*Ipomoea lacunose L.*) control in systems with C + P applied either PRE or POST was never <79% while ivyleaf morningglory (*Ipomoea hederacea Jacq.*) control with C + P systems applied PRE varied from 73 to 90%, applied EPOST from 53 to 95%, and POST from 84 to 98%. The use of the premix of C + P provided excellent season—long residual control of several broadleaf weeds including Palmer amaranth, smellmelon, and morningglory spp. This herbicide mixture offers peanut growers another option to control ALS- and glyphosate-resistant Palmer amaranth, which is becoming a major problem in many areas of Texas and Oklahoma.

Keywords: annual grasses, broadleaf weeds, glyphosate resistance, groundnut, preemergence applications, early-postemergence applications, herbicide systems, palmer amaranth (*Amaranthus palmeri* S. Wats.), postemergence applications, smellmelon (*Cucumis melo* L.), Texas millet [*Urochloa texana* (Buckl.) R. Webster]

1. Introduction

Weeds associated with peanut production are predominantly comprised of annual herbaceous species that are propagated by seeds. Because of the disturbance of soil in land preparation and the use of conventional and reduced tillage before planting, not only small-seeded weed species {e.g., pigweed (*Amaranthus* spp.), eclipta (*Eclipta prostrata* L.), goosegrass [*Eleusine indica* (L.) Gaertn.], crabgrass (*Digitaria* spp.)}, but also large seeded species {e.g., sicklepod [*Senna obtusifolia* (L.) H.S. Irwin & Barneby], Benghal dayflower (*Commelina benghalensis* L.), morningglories (*Ipomoea* spp.) and common ragweed (*Ambrosia artemisiifiolia* L.)} can successfully emerge from deeper in the soil. Most of these weed species have temperature and soil moisture germination requirements similar to peanut and this favors their emergence and establishment before peanut canopy closure [1, 2]. Perennial weed species are fairly common in peanut fields and those that are successful frequently produce vegetative propagules such as tubers and rhizomes {e.g., bermudagrass (*Cynodon dactylon* L.), nutsedge (*Cyperus* spp.) and johnsongrass [*Sorghum halepense* (L.) Pers.]}, which can be greatly reduced with the use of tillage and cultivation [3, 4].

Weed management in peanut is challenging because of the prostrate growth habit of the peanut plant, which allows weeds to become established if weed control practices are not properly implemented [5, 6]. Weed control can be influenced by the ability of peanut to compete with weeds, cultural practices that minimize the soil seed bank and weed infestation, mechanical practices such as primary tillage prior to planting, cultivation during the growing season, and also by efficacy of herbicides. Weeds interfere with peanut through direct competition for light, water, nutrients, essential gases, and space [5, 6]. Weeds also can interfere with peanut growth through allelopathy [7–9].

The process of digging and inverting the vines and pods prior to mechanical harvest makes the need for effective season-long weed control a necessity. The tight fibrous root system of annual grasses such as Texas millet [*Urochloa texana* (Buckl.) R. Webster] and southern cragbrass (*Digitaria ciliaris* L.) and broadleaf weeds such as morningglory spp. (*Ipomoea* spp.) and pigweed spp. (*Amaranthus* spp.) becomes intertwined with the peanut plant, causing peanut pods to be stripped from the vine during digging and can also slow down the drying process. Peanuts that become detached from the plant remain unharvested in or on the soil surface [5, 6].

In most instances, fungicides are applied multiple times to peanut fields to control various diseases including stem rot (caused by *Sclerotium rolfsii* Sacc.), early leaf spot [caused by *Passaloray arachidicola* previously known as *Cercospora arachidicola* (Hori)] and late leaf spot [caused by *Nothopassalora personata* previously known as *Cercosporidium personatum* (Berk. & M.A. Curtis) Deighton]. Herbicides are often applied in combination with fungicides to control both weeds and diseases [10–12]. Weeds can also interfere with uniform deposition of fungicides reducing disease control [13].

There are many weeds found in Texas and Oklahoma peanut production areas (**Table 1**); however, some of the most common include Texas millet, horse purslane (*Trianthema portulacastrum* L.), smellmelon (*Cucumis melo* L.), morninglory spp., and Palmer amaranth (*Amaranthus palmeri* S. Wats.). Texas millet is a large seeded, vigorous, fast growing annual grass commonly found in peanut fields in parts of Florida, South Carolina, Oklahoma, and Texas [14]. It is listed as one of the most troublesome weeds in all peanut growing states except Alabama and Georgia [15].

Horse purslane occurs in tropical and subtropical areas throughout the world [16] and has cylindrical green leaves. The seeds germinate at 20–45 C [17] and have essentially no dormancy and can germinate soon after they mature [16]. Although common purslane (*Portulaca oleracea* L.) was rated as one of the ten most common weeds found in Texas peanut fields as early as 1989 [18], horse purslane only recently has become a problem in certain peanut growing areas of south Texas (author's personal observation). Horse purslane can be a stronger competitor with peanut early in

Common name	Latin name	Bayer code
Barnyardgrass/Junglerice	*Echinochloa crus-galli* (L.) Beauv.	ECHCG
Broadleaf signalgrass	*Brachiaria platyphylla* (Griseb.) Nash.	BRAPP
Carolina horsenettle	*Solanum carolinense* L.	SOLCA
Carpetweed	*Mollugo verticillata* L.	MOLVE
Coffee senna	*Cassia occidentalis* L.	CASOC
Common bermudagrass	*Cynodon dactylon* (L.) Pers.	CYNDA
Common cocklebur	*Xanthium strumarium* L.	XANST
Common purslane	*Portulaca oleracea* L.	POROL
Common sunflower	*Helianthus annuus* L.	HELAN
Devil's-claw	*Proboscidea louisianica*	PROLO
Eclipta	*Eclipta alba* L.	ECLAL
Fall panicum	*Panicum dichotomiflorum* Mich.	PAQIN
Golden crownbeard	*Verbesina encelioides* (Cav.) Benth. & Hook. f. ex A. Gray	VEREN
Hophornbean copperleaf	*Acalypha ostryifolia* Riddell	ACCOS
Horse purslane	*Trianthema portulacastrum* L	TRIPO
Ivyleaf morningglory	*Ipomoea hederacea* (L.) Jacq.	IPOHE
Johnsongrass	*Sorghum halepense* (L.) Pers.	SORHA
Kochia	*Bassia scoparia* (L.) A. J. Scott	BASSC
Large crabgrass	*Digitaria sanguinalis* (L.) Scop.	DIGSA
Palmer amaranth	*Amaranthus palmeri* S. Wats	AMAPA
Pitted morningglory	*Ipomoea lacunose* L.	IPOLA
Prairie/Maximillian sunflower	*Helianthus maximiliani* Schrad.	HELMA
Prickly sida	*Sida spinose* L.	SIDSP
Prostrate pigweed	*Amaranthus blitoides* S. Watson	AMABL
Purple nutsedge	*Cyperus rotundus* L.	CYPES
Red morninglory	*Ipomoea coccinea* L.	IPOCO
Russian-thistle	*Salsola tragus* L.	SALTR
Sandbur	*Cenchrus* spp.	CENZZ
Sicklepod	*Senna obtusifolia* (L.) Irwin & Barneby	CASOB
Silverleaf nightshade	*Solanum elaeagnifolium* Cav.	SOLEL
Southern crabgrass	*Digitaria ciliaris* (Retz.) Koel.	DIGSP
Texas millet	*Urochloa texana* (Buckl.) R. Webster	UROTE
Tropic croton	*Croton glandulosus* Muell. Arg.	CVNGS
Tumble pigweed	*Amaranthus albus* L.	AMAAL
Wild poinsettia	*Euphorbia heterophyll* L.	EPHHL
Yellow nutsedge	*Cyperus esculentus* L.	CYPRES

Table 1.
Weeds found in Texas and Oklahoma peanut growing areas.

the growing season than common purslane due to a more upright growth habit [19]. For example, in competition studies with mung bean [*Vigna radiate* (L.) R. Wilcdz.], horse purslane reduced yield 50–60% when left untreated [20]. In earlier work on peanut, Grichar [19] reported a yield increase when controlling horse purslane with postemergence (POST) herbicides.

Hand-hoeing is a common practice for horse purslane control in mung bean in India [16]. Fomesafen {5-[2-chloro-4-(trifluoromethyl) phenoxy]-N-(methylsulfonyl)-2-nitrobenzamide} controlled horse purslane 57–87% in mung bean [20]. Grichar [19] reported that acifluorfen, lactofen, and 2,4-DB controlled horse purslane at least 78% in peanut. He also reported that tank-mixing these herbicides with other broadleaf herbicides did not improve horse purslane control over acifluro-fen, lactofen, or 2,4-DB alone.

Smellmelon is becoming more of a problem in south Texas peanut production fields and has become a problem in several crops along the Texas Gulf Coast [21]. The range of smellmelon stretches from Georgia to the southern part of California and as far north as Arkansas [22]. Smellmelon can be a problem at peanut harvest as the melon can become broken apart during mechanical harvest, increasing drying time because of the high moisture content of the melon [23]. Thompson et al. [24] reported that imazapic at 0.07 and 0.14 kg ai ha^{-1} applied preemergence (PRE), early postemergence (EPOST), or late postemergence (LPOST) in corn (*Zea mays* L.) controlled smellmelon greater than 90%. Tingle and Chandler [25] reported that smellmelon control was at least 93% with various herbicide systems in a corn-cotton (*Gossypium hirsutum* L.)-corn rotation. Glyphosate and glufosinate systems have provided effective smellmelon control [26, 27] in cotton. Tingle et al. [28] reported when smellmelon was allowed to compete with cotton for at least 6 wks, yield was reduced 7% compared to the weed free check, but when smellmelon was allowed to compete for 10–12 wks, cotton yield was reduced 22 and 27%.

Palmer amaranth is ranked as a major weed in several crops in the U.S. and is a common weed in many major crops around the world [14, 15, 21, 29]. Pigweed is listed as one of the 10 most common weeds in most all major peanut-growing states in the U.S., with Palmer amaranth among the 10 most common weeds in Alabama, Arkansas, Florida, Georgia, Mississippi, North Carolina, Oklahoma, South Carolina, and Texas [14, 15]. Palmer amaranth is listed as the most troublesome weed in peanut in 8 peanut producing states [14]. Most of the current distribution of Palmer amaranth is in the southern half of the U.S.; however, it has now been found in 28 states and as far north as Minnesota [30]. In Texas, Palmer amaranth can be found in all areas of the state [31] as well as in Oklahoma (author's personal observation). It is a severe problem in most fields when not properly controlled [32].

Morningglory spp. (*Ipomoea* spp.) are found throughout the peanut growing regions of Texas and Oklahoma and are some of the most common and troublesome weeds to control [33]. The two most common species are ivyleaf [*Ipomoea hederacea* (L.) Jacq.] and pitted morningglory (*Ipomoea lacunose* L.). Morningglories are typically a late-season issue and help to reduce crop yield [34]; however, the main issue in peanut is that they make digging and harvesting difficult because the vines can become intertwined in the digging and harvesting equipment [35]. Barbour and Bridges [36] concluded from field studies that shading from weed escapes leads to reduced peanut yield. Crop shading leads to irreversible adaptations in the photosynthetic processes such as altered partitioning of photosynthetic assimilates and this results in fewer pods set [37].

Pyroxasulfone is in the class of herbicides called isoxaxolines [38]. Herbicides in this class are a site of action, Resistance Action Committee (HRAC) Group 15 herbicide, along with the chloroacetamide herbicides including acetochlor (Warrant®), *S*-metolachlor (Dual Magnum®), and dimethenamid (Outlook®) [38–41]. Such herbicides are shoot and root growth inhibitors and control susceptible germinating seedlings before or soon after they emerge from the soil by reducing the biosynthesis of very-long-chain fatty acids, which cause a buildup of fatty acid precursors [38–40]. Pyroxasulfone is registered in the U.S. for either preplant, preplant incorporated (PPI), PRE, or EPOST use in corn, cotton, soybean (*Glycine max* L.), and wheat (*Triticum aestivum* L.). Application timing is crop specific [41]. It controls many broadleaf weeds (*Amaranthus* spp., *Lolium* spp.) and annual grasses (*Urochloa* spp., goosegrass [*Eleusine indica* L.], crowfootgrass [*Dactyloctenium aegyptium* L.], and *Digitaria* spp. [42–47]. Although pyroxasulfone has a similar weed control spectrum as *S*-metolachlor and dimethenamid-P, it has a higher specific activity allowing for use rates approximately eight times lower than dimethenamid-P [40]. In previous research, pyroxasulfone has shown to have good peanut crop tolerance and provides control of problem weeds [23, 48–50]; however, pyroxasulfone applied PRE to peanut has been documented to cause early-season stunting but no yield loss [48].

Carfentrazone is an aryl triazolinone herbicide [51], and the mode of action is the inhibition of protoporphyrinogen oxidase (Protox) [52, 53] in the chlorophyll biosynthesis pathway that results in the accumulation of protoporphyrin IX (PPIX) in the cytosol (site of action Group 14) [54, 55]. PPIX is photoactive and involved in the light-dependent formation of singlet oxygen, which is responsible for plant death via membrane oxidation [56]. It is a rapid-acting contact herbicide with little or no residual activity [57], and susceptible weeds begin to desiccate within hours of treatment followed by necrosis and plant death within days.

Monoculture production systems and the repeated use of the same or similar herbicides have led to herbicide resistance in weeds [58–60]. *Amaranthus* species are very sensitive to ALS-inhibiting herbicides and possess characteristics that predispose them to have herbicide resistant biotypes such as high genetic variability, prolific seed production, and efficient pollen and seed distribution [61]. The use of soil-applied and POST herbicides with alternative sites of action is necessary to reduce the rate of development of herbicide-resistant weed populations [62].

The premix of carfentrazone + pyroxasulfone (C + P) was labeled for use on peanut in the U.S. as Anthem Flex® by the FMC Corporation [63] in time for the 2020 growing season. It is labeled in peanut for EPOST or POST use only. Proxasulfone systems have provided excellent season-long control of Palmer amaranth and smellmelon, which are broadleaf weeds that can cause peanut growers considerable problems and are hard-to-control with current herbicides [23, 49, 50]. This premix will control ALS- (HRAC Group 2 herbicide) and glyphosate- (HRAC Group 9 herbicide) resistant Palmer amaranth, which is becoming more widespread across southwestern peanut producing areas [64]. Control of annual grasses such as Texas millet is limited, and full-season control of this annual grass in peanut typically requires the postemergence use of clethodim® (WSSA Group 1 herbicide) or other graminicides [64]. Due to questions on weed efficacy of C + P systems on various weeds found in the southwestern U.S. peanut production area, research was undertaken to determine weed efficacy with C + P herbicide systems.

2. Materials and methods

2.1 Field studies

Studies were conducted in the 2017 through 2019 growing seasons to determine weed efficacy of C + P herbicide systems in the southwestern U.S. peanut growing areas. Locations in 2017 through 2019 included the Texas A&M AgriLife Research Site near Yoakum (29.1642°N, −97.1243°W) in south-central Texas, the Oklahoma State University Caddo Research Station near Ft. Cobb (35.0610°N, −98.2745°W) in southwestern Oklahoma, and in 2017 in the Texas High Plains at the Texas Tech Fiber and Biopolymer Research Institute (FBRI) near Lubbock (33.3527°N; −101.4657°W).

The test locations at Yoakum and Ft. Cobb were in the same general areas but different parts of the field in the 3 years. Other details of the test locations, peanut variety, and spray information are given in **Table 2**. The experimental design was a randomized complete block with three to four replications depending on location. An untreated check was included each year at all locations.

2.2 Plot size and weed populations

Each plot at Yoakum consisted of two rows spaced 97 cm apart and 7.6 m long, plots at Ft. Cobb were two rows spaced 91 cm apart and 9.1 m long, while plots at Lubbock were four rows spaced 102 cm apart and 9.5 m long. Plots were infested with naturally occurring weed populations at all locations. At the south Texas location, all field plots were naturally infested with populations of Texas millet, smellmelon, and Palmer amaranth with populations ranging from 6 to 14 plants/m^2. Pitted morningglory was only present in sufficient numbers (3–6 plants/m^2) in 2017. At the Oklahoma locations, Texas millet, ivyleaf morningglory, and yellow nutsedge (*Cyperus esculentus* L.) populations (present in 2017) ranged from 9 to 45 plant/m^2 while Palmer amaranth populations in all 3 years were similar to those mentioned above. Wild poinsettia (*Euphorbia heterophylla* L.) populations in 2018 were 3 to 4 plants/m^2. At the Lubbock location in 2017, golden crownbeard [*Verbesnia encelioides* (Cav.) Benth. & Hook. F. ex A. Gray] populations ranged from 8 to 10 plants/m^2, devil's-claw [*Proboscidea louisianica* (Mill.) Thell] populations ranged from 1 to 2 plants/m^2, and Palmer amaranth populations ranged from 6 to 8 plants/m^2.

2.3 Herbicide treatments and application

In 2017 at Ft. Cobb and in 2018 and 2019 at both Yoakum and Ft. Cobb, pendimethalin at 1.06 kg ha^{-1} was applied PRE to all treatments with the exception of the untreated check. The EPOST herbicide applications (also referred to as peanut cracking) were applied when the peanut plants had begun to emerge or were no bigger than saucer size (approximately 10–12 cm). In 2017, this was 21 days after planting (DAP) at Ft. Cobb. No EPOST treatments were applied at either Yoakum or Lubbock. In 2018, the EPOST treatments were applied 14 DAP at Yoakum and 22 DAP at Ft. Cobb while in 2019 EPOST treatments were applied 7 DAP at Yoakum and 27 DAP at Ft. Cobb. Postemergence (POST) treatments were applied 22–36 DAP at Yoakum, 35–41 DAP at Ft. Cobb, and 43 DAP at Lubbock. All EPOST and POST treatments included either a crop oil concentrate (Agridex®) at 1.25% v/v or a non-ionic surfactant (Induce®) at 0.25% v/v.

	Location						
Variables		**Yoakum**			**Ft. Cobb**		**Lubbock**
	2017	**2018**	**2019**	**2017**	**2018**	**2019**	**2017**
Soil name	Tremona	Tremona	Tremona	Binger	Binger	Binger	Estacado
Soil type	Loamy fine sand	Loamy fine sand	Loamy fine sand	Fine sandy loam	Fine sandy loam	Fine sandy loam	Clay loam
Sand (%)	65	65	65	50	50	50	50
Silt (%)	35	35	35	30	30	30	25
Clay (%)	10	10	10	20	20	20	25
pH	7.4	7.5	7.5	7.0	7.0	7.0	8.2
OM (%)	0.7	0.7	0.7	0.5	0.5	0.5	1.1
CEC	31	30	30	25	25	25	15
Planting date	June 14	June 26	June 24	May 5	May 8	May 15	April 26
Application							
PRE	June 15	June 28	June 24	May 5	May 8	May 15	April 28
EPOST	—	July 10	July 1	May 26	May 30	June 11	June 7
POST	July 6	July 31	July 30	June 15	June 12	June 25	—
Weed ht. at EPOST (cm)							
AMAPA	8–15	10–15	10–15	1.5–7.5	1.5–8.0	1.5–8	6–8
CUMME	4–8	5–8	5–8	—	—	—	—
CYPES	—	2–6	—	5–20	—	—	—
EPPHL	—	—	—	—	1.5–15	—	—
IPOHE	—	—	—	1.5–15	1.5–15	—	—
IPOLA	4–8	—	—	—	—	—	—
PROLO	—	—	—	—	—	—	6–8
UROTE	12–16	10–15	10–15	1.5–5	1.5–6	—	—
VEEN	—	—	—	—	—	—	10–15
Weed ht. at POST (cm)							
AMAPA	15–25	30–36	18–35	1.5–10	1.5–8	1.5–30	12–20
CUMME	10–30	5–25	5–25	—	—	—	—
CYPES	—	10–16	20–30	5–20	2.5–20	—	—
EPPHL	—	—	—	—	1.5–20	—	—
IPOHE	—	—	—	1.5–20	1.5–15	—	—
IPOLA	10–20	—	—	—	—	—	—
PROLO	—	—	—	—	—	—	12–16
UROTE	15–20	18–20	18–20	6–10	1.5–41	—	—
VEEN	—	—	—	—	—	—	10–16
Sprayer				CO_2 backpack			

	Location						
Variables		**Yoakum**			**Ft. Cobb**		**Lubbock**
	2017	**2018**	**2019**	**2017**	**2018**	**2019**	**2017**
Operating pressure (kPa)	207	207	207	161	165	165	220
Spray volume (L ha^{-1})	187	187	187	94	94	94	112
Spray nozzles	DG 11002	DG 11002	DG 11002	TT 110015	TTI 110015	TT 110015	TT1 10,015
Peanut variety	Georgia 09B	Georgia 13 M	Georgia 09B	Florida Fancy	Florida Fancy	Florida Fancy	Georgia 09B

[a]*Abbreviations: EPOST, early postemergence; POST, postemergence; PRE, preemergence.*
[b]*Bayer code for weeds: AMAPA, Palmer amaranth (Amaranthus palmeri L.); CUMME, smellmelon Cucumus melo L.); CYPES, yellow nutsedge (Cyperus esculentus L.); EPHHL, wild poinsettia (Euphorbia heterophylla L.); IPOHE, ivyleaf morningglory [Ipomoea hederacea (L.) Jacq.]; IPOLA, pitted morningglory (Ipomoea lacunose L.); PROLO, devil's-claw [Proboscidea louisianica (Mill.) Thell]; UROTE, Texas millet [Urochloa texana (Buckl.)]; VEEEN, Golden crownbeard [Verbesnia encelioides (Cav.) Benth. & Hook. f. ex A. Gray]* [22].

Table 2.
Variables associated with the study at each location.[a, b]

Herbicide standards varied between locations and years. In 2017 at both the Yoakum and Lubbock locations, pendimethalin at 1.06 kg ai ha^{-1} plus flumioxazin at 0.07 kg ai ha^{-1} applied PRE followed by *S*-metolachlor at 1.07 kg ai ha^{-1} applied POST was the standard comparison treatment. At Ft. Cobb in 2017, pendimethalin plus flumioxazin was applied PRE, and paraquat at 0.21 kg ai ha^{-1} was added to S-metolachlor at 1.42 kg ai ha^{-1} EPOST treatment. In 2018 and 2019 at both Yoakum and Ft. Cobb, flumioxazin at 0.11 kg ha^{-1} applied PRE followed by lactofen at 0.22 kg ha^{-1} plus *S*-metolachlor at 1.07 kg ha^{-1} plus 2,4-DB at 0.25 kg ha^{-1} applied POST was the standard.

2.4 Irrigation, weed control, and peanut harvest

Sprinkler irrigation was applied on a 2–3-wk schedule throughout the growing season as needed at all locations. Weed control and peanut leaf burn were visually estimated on a scale of 0–100 (0 indicating no control or plant death and 100 indicating complete control or plant death), relative to the untreated control [65]. Peanut yields were not obtained in any years due to the difficulty of digging plots with high weed pressure [23, 49, 50].

2.5 Data analysis

Weed control data were arcsine transformed prior to analysis of variance; however, because the transformation did not alter treatment means original data are presented. Means were compared with Fisher's protected LSD test at the 5% probability level. The untreated control was not included in the weed control analysis.

3. Results and discussion

Peanut injury (<15%) at all locations consisted of leaf burn, and this injury was visible for 7–21 days after a C + P application. Typically, the peanut leaf burn can be attributed to the carfentrazone in the premix. The peripheral leaves that were burned were replaced by new leaves that were void of any type of injury. This leaf burn has also been noted in another study with C + P [66].

Smellmelon data at Yoakum in 2018 and 2019 were combined over the 2 years because of a lack of treatment by year interaction. However, all other weed control data are presented separately by weed species, year, and location because there was a treatment by year and/or location interaction or herbicide treatments were different at the locations in some years.

3.1 Weed control

3.1.1 Palmer amaranth (AMAPA)

In 2017 at the Yoakum location when evaluated 16 weeks after planting (WAP), all herbicide systems that contained C + P at 0.007 kg ha^{-1} + at 0.11 kg ha^{-1} PRE controlled AMAPA at least 92% compared with the standard treatment of pendimethalin + flumioxazin PRE followed by *S*-metolachlor POST which provided 93% control (**Table 3**). Systems that included C + P at 0.005 kg ha^{-1} + 0.07 kg ha^{-1} PRE provided more variable AMAPA control which ranged from 76 to 100% while the system which included C + P POST provided 100% control. At the Lubbock location all C + P systems, with the exception of C + P at 0.005 kg ha^{-1} + 0.07 kg ha^{-1} PRE followed by imazapic POST, controlled AMAPA at least 90% compared with the standard treatment of pendimethalin + flumioxazin PRE followed by *S*-metolachlor POST which provided 67% control (**Table 4**). At the Ft. Cobb location, herbicide systems that included C + P EPOST controlled AMAPA 81–98% while systems that included *S*-metolachlor EPOST provided 81–96% control (**Table 5**).

In 2018 at Yoakum when evaluated 16 WAP, all C + P systems with the exception of C + P at 0.004 kg ha^{-1} + 0.05 kg ha^{-1} PRE, controlled AMAPA 97–100% (**Table 6**). At Ft. Cobb, herbicide systems that included pendimethalin + (C + P) PRE followed by imazapic + *S*-metolachlor POST or pendimethalin + flumioxazin PRE followed by imazapic + (C + P) POST provided 83–92% control of Palmer amaranth (**Table 7**). The three-way combination of bentazon + aciflurofen + *S*-metolachlor EPOST followed by imazapic + *S*-metolachlor POST provided 99% control while pendimethalin + flumioxazin PRE followed by imazapic + (C + P) POST controlled AMAPA 83–90%. No other herbicide systems provided better than 69% control.

In 2019 at Yoakum all herbicide systems provided almost perfect AMAPA control (**Table 6**). At Ft. Cobb, herbicide systems that included pendimethalin + flumioxazin PRE followed by imazapic + (C + P) POST controlled AMAPA 93–94%. Pendimethalin + flumioxazin PRE followed by imazapic + *S*-metolachlor POST provided 98% control and bentazon + aciflurofen + pyroxasulfone EPOST followed by imazapic + pyroxasulfone POST provided 97% control (**Table 7**). In previous work, Grichar et al. [66] reported that C + P provided at least 90% Palmer amaranth control while herbicide systems that included either pendimethalin or flumioxazin controlled this weed 97–99%. Steele et al. [39] reported, in corn, that pyroxasulfone at 0.125–0.5 kg ha^{-1}

Treatments[a,b]	Rate	Appl timing[c]	AMAPA	CUMME	IPOLA	UROTE[d]
			Weeks after planting			
				16		
	kg ai ha^{-1}			%		
Untreated	—	—	0	0	0	0
(C + P)	(0.005 + 0.07)	PRE	87	98	89	17
(C + P)	(0.007 + 0.11)	PRE	96	97	88	30
Flumioxazin + (C + P)	0.07 + (0.005 + 0.07)	PRE	100	100	79	27
Pendimethalin + (C + P)	1.06 + (0.005 + 0.07)	PRE	87	86	84	55
(C + P) Imazapic	(0.005 + 0.07) 0.07	PRE POST	79	99	99	53
(C + P) Imazapic	(0.007 + 0.11) 0.07	PRE POST	92	99	83	65
(C + P) + flumioxazin Imazapic	(0.005 + 0.07) + 0.07 0.07	PRE POST	100	100	92	58
(C + P) + pendimethalin imazapic	(0.005 + 0.07) + 1.06 0.07	PRE POST	76	100	99	75
Pendimethalin + flumioxazin (C + P)	1.06 + 0.07 (0.007 + 0.11)	PRE POST	100	99	95	37
Pendimethalin + flumioxazin *S*-metolachlor	1.06 + 0.07 1.07	PRE POST	93	88	82	70
LSD (0.05)			16	10	23	29

[a] *All POST treatments included Induce at 0.25% v/v.*
[b] *Abbreviations for herbicides: (C + P), premix of carfentrazone + pyroxasulfone.*
[c] *Application timing: PRE, preemergence; POST, postemergence.*
[d] *Bayer code for weeds: AMAPA, Palmer amaranth (Amaranthus palmeri S. Wats.); CUMME, smellmelon (Cucumis melo L.); IPOLA, pitted morningglory (Ipomoea lacunose L.); UROTE, Texasmillet [Urochloa texana (Buckl.)]* [22].

Table 3.
Late season weed control with carfentrazone + pyroxasulfone (C + P) during the 2017 growing season in south Texas (Yoakum).

provided comparable control of Palmer amaranth to *S*-metolachlor at rates of 1.1–4.3 kg ha^{-1}. Also, Knezevic et al. [44] found that pyroxasulfone at 0.16 kg ha^{-1} provided 90% control of tall waterhemp [*Amaranthus tuberculatus* (Moq.)] 28 days after treatment (DAT) and a higher rate was required to obtain the same control at 45 (0.2 kg ha^{-1}) and 65 DAT (0.27 kg ha^{-1}). Jha et al. [67] reported that the addition of pendimethalin to pyroxasulfone improved control of several weeds including common lambsquarters (*Chenopodium album* L.), kochia [*Kochia scoparia* (L.) Schard], and wild buckwheat (*Polygonum convolvulus* L.) over pyroxasulfone alone.

3.1.2 Texas millet (UROTE)

In 2017 at Yoakum, no herbicide system provided better than 75% UROTE control (**Table 3**). Herbicide systems that included C + P but not imazapic controlled UROTE

Treatments[a,b]	**Rate**	**Appl timing[c]**	**AMAPA**	**PROLO**	**VEEN**
			Weeks after planting		
				13	
	kg ai ha^{-1}			%	
Untreated	—	—	0	0	0
(C + P)	(0.005 + 0.07)	PRE	90	60	73
(C + P)	(0.007 + 0.11)	PRE	92	23	60
Flumioxazin + (C + P)	0.07 + (0.005 + 0.07)	PRE	100	63	62
Pendimethalin + (C + P)	1.06 (0.005 + 0.07)	PRE	100	17	67
(C + P) Imazapic	(0.005 + 0.07) 0.07	PRE POST	73	100	80
(C + P) Imazapic	(0.007 + 0.11) 0.07	PRE POST	92	100	85
(C + P) + flumioxazin Imazapic	(0.005 + 0.07) + 0.07 0.07	PRE POST	95	98	93
(C + P) + pendimethalin imazapic	(0.005 + 0.07) + 1.06 0.07	PRE POST	100	100	87
Pendimethalin + flumioxazin (C + P)	1.06 + 0.07 (0.007 + 0.11)	PRE POST	100	100	100
Pendimethalin + flumioxazin *S*-metolachlor	1.06 + 0.07 1.06	PRE POST	67	60	88
LSD (0.05)			32	54	20

[a]*All POST treatments included Induce at 0.25% v/v.*
[b]*Abbreviations for herbicides: (C + P), premix of carfentrazone + pyroxasulfone.*
[c]*Application timing: PRE, preemergence; POST, postemergence.*
[d]*Bayer code for weeds: AMAPA, Palmer amaranth (Amaranthus palmeri S. Wats.); PROLO, devil's-claw [Proboscidea louisianica (Mill.) Thell].; VEEN, Golden crownbeard [Verbesnia encelioides (Cav.) Benth. & Hook. f. ex A. Gray]* [22].

Table 4.
Late season weed control with carfentrazone + pyroxasulfone (C + P) during the 2017 growing season at Lubbock in the High Plains of Texas.

17–55% while systems containing imazapic provided 53–75% control. The standard of pendimethalin + flumioxazin PRE followed by *S*-metolachlor POST provided 70% control. At Ft. Cobb all herbicide systems that included C + P without imazapic provided 44–56% control while all systems that included imazapic POST controlled UROTE 88–96% (**Table 5**). The standard of pendimethalin + flumioxazin PRE followed by paraquat + *S*-metolachlor EPOST provided 33% control.

In 2018 at Yoakum, all herbicide systems that included C + P without imazapic controlled UROTE 88–92% while systems that included imazapic controlled UROTE 96–100% (**Table 6**). At Ft. Cobb, only herbicide systems that included pendimethalin + (C + P) PRE at the two rates of either 0.004 kg ha^{-1} + 0.05 kg ha^{-1} or 0.005 kg ha^{-1} + 0.07 kg ha^{-1} followed by imazapic + *S*-metolachlor

Treatments[a,b]	Rate	Appl timing[c]	14 weeks after planting			
			AMAPA[d]	CYPES	IPOHE	UROTE
	kg ai ha^{-1}		%			
Untreated	—	—	0	0	0	0
Paraquat + (C + P) Imazapic	0.21 + (0.005 + 0.07) 0.07	EPOST POST	85	85	95	94
Paraquat + (C + P) Imazapic + 2,4-DB	0.21 + (0.005 + 0.07) 0.07 + 0.25	EPOST POST	81	86	90	94
Paraquat + (C + P) Lactofen + 2,4-DB	0.21 + (0.005 + 0.07) 0.15 + 0.25	EPOST POST	95	30	74	56
Paraquat + *S*-metolachlor Imazapic	0.21 + 1.42 0.07	EPOST POST	86	84	89	95
Paraquat + *S*-metolachlor Imazapic + 2,4-DB	0.21 + 1.42 0.07 + 0.25	EPOST POST	81	76	89	88
Paraquat + *S*-metolachlor Lactofen + 2,4-DB	0.21 + 1.42 0.22 + 0.25	EPOST POST	94	28	43	33
Flumioxazin Paraquat + (C + P) Imazapic	0.07 0.21 + (0.005 + 0.07) 0.07	PRE EPOST POST	85	85	90	90
Flumioxazin Paraquat + (C + P) Imazapic + 2,4-DB	0.07 0.21 + (0.005 + 0.07) 0.07 + 0.25	PRE EPOST POST	83	85	93	94
Flumioxazin Paraquat + (C + P) Lactofen + 2,4-DB	0.07 0.25 + (0.005 + 0.07) 0.22 + 0.25	PRE EPOST POST	98	30	53	44
Flumioxazin Paraquat + *S*-metolachlor Imazapic	0.07 0.21 + 1.42 0.07	PRE EPOST POST	85	83	97	95
Flumioxazin Paraquat + *S*-metolachlor Imazapic + 2,4-DB	0.07 0.21 + 1.42 0.07 + 0.25	PRE EPOST POST	89	83	90	96
Flumioxazin Paraquat + *S*-metolachlor	0.07 0.21 + 1.42	PRE EPOST	96	13	61	33
LSD (0.05)			7	13	18	16

[a]*All herbicide treatments, except the untreated check, included pendimethalin at 1.06 kg* ha^{-1} *applied preemergence.*
[b]*All POST carfentrazone + pyroxasulfone treatments included Induce at 0.25% v/v while all other POST herbicide treatments included Agridex at 1.25% v/v.*
[c]*Abbreviations: (C + P), a premix of carfentrazone + pyroxasulfone; PRE, preemergence; EPOST, early postemergence (also referred to as peanut cracking); POST, postemergence.*
[d]*Bayer code for weeds: AMAPA, Palmer amaranth (Amaranthus palmeri S. Wats.); CYPES, yellow nutsedge (Cyperus esculentus L.); IPOHE, ivyleaf morningglory [Ipomoea hederacea (L.) Jacq.]; UROTE, Texas millet [Urochloa texana (Buckl.)]* [22].

Table 5.
Late season weed control with carfentrazone plus pyroxasulfone (C + P) during the 2017 growing season in Oklahoma (Ft. Cobb).

	16 weeks after planting						
Treatments[a,b]	Rate	Appl timing	AMAPA[d]			UROTE	
			2018	2019	CUMME[e]	2018	2019
	kg ai ha^{-1}				%		
Untreated	—	—	0	0	0	0	0
Flumioxazin Imazapic + *S*-metolachlor	0.11 0.07 + 1.07	PRE POST	97	99	98	99	94
(C + P) Imazapic + *S*-metolachlor	(0.004 + 0.05) 0.07 + 1.07	PRE POST	71	100	98	100	93
(C + P) Imazapic + *S*-metolachlor	(0.005 + 0.07) 0.07 + 1.07	PRE POST	99	99	99	97	90
Flumioxain Imazapic + (C + P)	0.11 0.07 + (0.007 + 0.11)	PRE POST	100	99	100	99	86
Flumioxazin Imazapic + C+ P	0.11 0.07 + (0.009 + 0.12)	PRE POST	99	100	99	96	83
Bentazon + acifluorfen + *S*-metolachlor Imazapic + S-metolachlor	0.37 + 0.19 + 1.07 0.07 + 1.07	EPOST POST	99	99	98	96	87
Bentazon + acifluorfen + (C + P) Imazapic + (C + P)	0.37 + 0.19 + (0.007 + 0.11) 0.07 + (0.007 + 0.11)	EPOST POST	99	100	95	99	91
Bentazon + acifluorfen + (C + P) Imazapic + (C + P)	0.37 + 0.19 + (0.008 + 0.13) 0.07 + (0.008 + 0.13)	EPOST POST	100	99	98	98	92
Bentazon + acifluorfen + pyroxasulfone Imazapic + pyroxasulfone	0.37 + 0.19 + 0.06 0.07 + 0.06	EPOST POST	100	100	99	98	92
Flumioxazin Lactofen + *S*-metolachlor + 2,4-DB	0.11 0.22 + 1.07 + 0.25	PRE POST	100	100	99	99	72
Flumioxazin Lactofen + (C + P) + 2,4-DB	0.11 0.22 + (0.009 + 0.12) + 0.25	PRE POST	99	100	99	92	66

Treatments[a,b]	16 weeks after planting Rate	Appl timing	AMAPA[d] 2018	2019	CUMME[e]	UROTE 2018	2019
	kg ai ha^{-1}				%		
Flumioxazin Lactofen + (C + P) + 2,4-DB	0.11 0.22 + (0.007 + 0.11) + 0.25	PRE POST	100	100	99	88	83
LSD (0.05)			9	2	4	6	18

[a]All herbicide treatments, except the untreated check, included pendimethalin at 1.06 kg ha^{-1} applied preemergence.
[b]All POST treatments included Induce at 0.25% v/v.
[c]Abbreviations: (C + P), a premix of carfentrazone + pyroxasulfone; PRE, preemergence; EPOST, early postemergence (also referred to as peanut cracking); POST, postemergence.
[d]Bayer code for weeds: AMAPA, Palmer amaranth (Amaranthus palmeri S. Wats.); CUMME, smellmelon (Cucumis melo L.); UROTE, Texas millet [Urochloa texana (Buckl.)] [22].
[e]Data combined over years because there was no treatment by year interaction.

Table 6.
Late season weed control with carfentrazone plus pyroxasulfone (C + P) during the 2018 and 2019 growing season in south Texas (Yoakum).

POST, pendimethalin + flumioxazin PRE followed by imazapic + (C + P) at 0.009 kg ha^{-1} + 0.12 kg ha^{-1} POST, or pendimethalin PRE followed by the 3-way mix of bentazon + acifluorfen + *S*-metolachlor EPOST followed by imazapic + *S*-metolachlor POST provided 90% or greater control (**Table 7**).

In 2019 at Yoakum, systems that did not include imazapic provided 66–83% UROTE control while systems that included imazapic controlled UROTE 83–94% (**Table 6**). Grichar et al. [66] reported that in 1 year at two locations, herbicide treatments that included either imazapic or imazethapyr provided 98% UROTE control while treatments that included the premix of C + P without imazapic controlled UROTE 88–98%. The herbicide systems that included the premix of C + P PRE and POST provided comparable control to those systems that included a POST application of imazapic. At another location, C + P PRE controlled UROTE 47–75%, EPOST provided 46–85% control, and POST provided 48–80% control while the herbicide standard of pendimethalin + *S*-metolachlor provided 66% control. Baughman et al. [50] and Grichar et al. [23] had previously reported that UROTE control with pyroxasulfone PPI or PRE was inconsistent when used alone; however, when used in a systems approach with either pendimethalin, *S*-metolachlor, or dimethenamid control was ≥90%.

3.1.3 Smellmelon (CUMME)

CUMME was only present at the Yoakum location. In 2017, all herbicide systems with the exception of pendimethalin + (C + P) at 0.005 kg ha^{-1} + 0.07 kg ha^{-1} PRE or the standard of pendimethalin + flumioxazin PRE followed by *S*-metolachlor POST controlled CUMME ≥97% (**Table 3**).

Data from 2018 and 2019 were combined since there was no treatment by year interaction. All herbicide systems provided ≥95% CUMME control (**Table 6**). Imazapic systems controlled CUMME 95–100% while systems containing lactofen provided 99% control. In earlier work, Grichar et al. [66] reported that C + P PRE and

			16 weeks after planting				
Treatments[a,b,c]	Rate	Appl timing	AMAPA[d]		EPPHL	UROTE	IPOHE
			2018	2019	2018	2018	2018
	kg ai ha^{-1}		%				
Untreated	—	—	0	0	0	0	0
Flumioxazin Imazapic + *S*-metolachlor	0.11 0.07 + 1.07	PRE POST	69	98	50	69	50
(C + P) Imazapic + *S*-metolachlor	(0.004 + 0.05) 0.07 + 1.07	PRE POST	91	73	80	91	73
(C + P) Imazapic + *S*-metolachlor	(0.005 + 0.07) 0.07 + 1.07	PRE POST	92	87	96	92	90
Flumioxain Imazapic + (C + P)	0.11 0.07 + (0.007 + 0.11)	PRE POST	83	94	98	83	94
Flumioxazin Imazapic + (C+ P)	0.11 0.07 + (0.009 + 0.12)	PRE POST	90	93	85	90	84
Bentazon + acifluorfen + S-metolachlor Imazapic + *S*-metolachlor	0.37 + 0.19 + 1.07 0.07 + 1.07	EPOST POST	99	73	100	99	97
Bentazon + acifluorfen + (C + P) Imazapic + (C + P)	0.37 + 0.19 + (0.007 + 0.11) 0.07 + (0.007 + 0.11)	EPOST POST	62	82	68	62	89
Bentazon + acifluorfen + (C + P) Imazapic + (C + P)	0.37 + 0.19 + (0.008 + 0.13) 0.07 + (0.008 + 0.13)	EPOST POST	58	87	71	58	96
Bentazon + acifluorfen + pyroxasulfone Imazapic + pyroxasulfone	0.37 + 0.19 + 0.06 0.07 + 0.06	EPOST POST	56	97	35	56	76
Flumioxazin Lactofen + *S*-metolachlor + 2,4-DB	0.11 0.22 + 1.07 + 0.25	PRE POST	10	81	21	10	98
Flumioxazin Lactofen + (C + P) + 2,4-DB	0.11 0.22 + (0.009 + 0.12) + 0.25	PRE POST	28	83	76	28	87
Flumioxazin Lactofen + (C + P) + 2,4-DB	0.11 0.22 + (0.007 + 0.11) + 0.25	PRE POST	25	81	79	25	98

Treatments[a,b,c]	Rate	Appl timing	16 weeks after planting				
			AMAPA[d]		EPPHL	UROTE	IPOHE
			2018	2019	2018	2018	2018
	kg ai ha^{-1}		%				
LSD (0.05)			24	12	25	24	18

[a]*All herbicide treatments, except the untreated check, included pendimethalin at 1.06 kg ha^{-1} applied preemergence.*
[b]*All POST treatments included Induce at 0.25% v/v.*
[c]*Abbreviations: (C + P), a premix of carfentrazone + pyroxasulfone; PRE, preemergence; EPOST, early postemergence (also referred to as peanut cracking); POST, postemergence.*
[d]*Bayer code for weeds: AMAPA, Palmer amaranth (Amaranthus palmeri S. Wats.); EPHHL,wild poinsettia (Euphorbia heterophylla L.); IPOHE, ivyleaf morningglory [Ipomoea hederacea (L.) Jacq.]; UROTE, Texas millet [Urochloa texana (Buckl.)]* [22].

Table 7.
Late season weed control with carfentrazone plus pyroxasulfone during the 2018 and 2019 growing season in Oklahoma (Ft. Cobb).

POST controlled CUMME 95–100% while the combination of C + P PRE followed by imazapic + *S*-metolachlor POST failed to control CUMME (63%). However, other research indicated imazapic provides excellent control of smellmelon [21] similar to the results from these studies. While lactofen provided excellent control in this study, previous research by Grichar [21] stated that lactofen control of CUMME could be inconsistent.

3.1.4 Pitted morningglory (IPOLA)

IPOLA was present in sufficient numbers to evaluate only at Yoakum in 2017. Herbicide systems that included imazapic POST controlled IPOLA 83–99% while systems that included C + P, but not imazapic provided 79–95% control (**Table 3**). The herbicide standard of pendimethalin + flumioxazin PRE followed by *S*-metolachlor POST controlled IPOLA 82%. Imazapic has been reported to effectively control IPOLA in peanut [6, 68]. Jordan et al. [68] reported that imazapic alone or in combination with aciflurofen, diclosulam, or 2,4-DB controlled IPOLA 89–97%.

3.1.5 Ivyleaf morningglory (IPOHE)

IPOHE was present only at the Ft. Cobb location in 2017 and 2018. In 2017, herbicide systems that included imazapic controlled IPOHE 89–97% while systems that included C + P but not imazapic provided 53–74% control. Herbicide systems that contained lactofen controlled this weed 43–74% (**Table 5**). In 2018 all systems, with the exception of either pendimethalin + flumioxazin or pendimethalin + (C + P) PRE followed by imazapic + *S*-metolachlor POST or pendimethalin PRE followed by bentazon + aciflurofen + pyroxasulfone EPOST followed by imazapic + pyroxasulfone POST, controlled IPOHE ≥ 84% (**Table 7**).

Grichar et al. [66] reported in a two-year study that in 1 year of the study, systems that included C + P PRE provided 83–100% IPOHE control while the standard of flumioxazin plus *S*-metolachlor provided 80% control. In another year, all systems that included C + P alone PRE provided 30–46% control, C + P alone EPOST provided 29–43% control, and C + P alone POST controlled IPOHE 30–40%. No application timing of C + P improved IPOHE control.

3.1.6 Golden crownbeard (VEEEN)

VEEEN was present only in the High Plains of Texas (Lubbock) in 2017. Herbicide systems that did not include a POST treatment controlled VEEEN ≤73% while systems which included imazapic POST provided 80–93% control (**Table 4**). Pendimethalin + flumioxazin PRE followed by C + P POST provided 100% control. Imazapic POST following ethalfluralin PPI has provided complete VEEEN control in a previous study [69]. Previously, Grichar and Sestak [70] reported imazapic provided inconsistent VEEEN control, and they attributed that to the amount and frequency of rainfall soon after application and taller weed size at the time of herbicide application.

3.1.7 Devil's-claw (PROLO)

PROLO was only present at the Lubbock location. Herbicide systems that included PRE treatments only or *S*-metolachlor POST provided 63% or less PROLO control while all POST treatments that included imazapic controlled PROLO at least 98% (**Table 4**). Pendimethalin + flumioxazin PRE followed by C + P POST also provided complete control.

3.1.8 Yellow nutsedge (CYPES)

CYPES was present only in sufficient populations to evaluate at Ft. Cobb in 2017. Herbicide systems that included imazapic provided 76–86% CYPES control while systems that did not include imazapic provided 30% or less control (**Table 5**). The standard of pendimethalin + flumioxazin applied PRE followed by paraquat + *S*-metolachlor applied EPOST controlled this weed only 13%. Imazapic typically provides excellent CYPES control [6, 71–74]. Grichar and Nester [71] found that imazapic at 0.05 and 0.07 kg ha^{-1} controlled CYPES 88% in the early season but was inconsistent later in the season. Wilcut et al. [6] reported that imazapic controlled CYPES greater than 90% at rates as low as 0.04 kg ha^{-1}. Grichar and Sestak [72] also reported that imazapic provided at least 80% control of CYPES. Richburg et al. [73] found that applying imazapic below CYPES and purple (*Cyperus rotundus* L.) (CYPRO) tubers did not reduce CYPRO or CYPES shoot numbers, shoot dry weight, shoot regrowth dry weight, or root tuber dry weight. However, when applied above CYPRO tubers, shoot dry weight was reduced; and when applied above CYPES tubers, shoot dry weight and root dry weight were reduced. Imazapic applied 5 cm above +5 cm below CYPRO tubers reduced shoot numbers, shoot dry weight, shoot regrowth weight, and root dry weight to 9, 4, 10, and 16% of the control, respectively, and with CYPES to 23, 16, 9, and 15% of the control, respectively.

3.1.9 Wild poinsettia (EPPHL)

EPPHL was present only at Ft. Cobb in 2018. Only pendimethalin + (C + P) PRE followed by imazapic + *S*-metolachlor POST, pendimethalin + flumioxazin PRE followed by imazapic + (C + P) POST, or pendimethalin PRE followed by bentazon + aciflurofen + *S*-metolachlor EPOST followed by imazapic + *S*-metolachlor POST controlled EPPHL at least 96% (**Table 7**). Systems that included lactofen + (C + P) provided 76–79% control while systems that include imazapic but not C + P provided variable control which ranged from 35 to 100%.

4. Conclusion

When used in a peanut herbicide system, the premix of C + P (both are HRAC Group 9 herbicides) provided excellent season-long residual control of several broadleaf weeds including Palmer amaranth, smellmelon, and morningglory spp. This herbicide combination will control pigweed species including ALS- and glyphosate-resistant Palmer amaranth, which are becoming major problems in many areas of Texas and Oklahoma. It also can provide burndown activity on emerged broadleaf weeds. The premix of C + P does not effectively control large-seeded annual grasses such as Texas millet and requires the use of a graminicide such as fluazifop-P-butyl, clethodim, or sethoxydim (WSSA Group 1 herbicides) to provide season-long control. This premix of C + P offers growers another option to help provide season-long control of problem weeds commonly found in the southwestern U.S. peanut production areas.

The challenges offered by evolved resistance to herbicides used in peanut combined with the limited competitive ability of peanut with weeds create a potentially devastating scenario for high-input, mechanized production systems in the U.S. While continuing to use a diversity of herbicides with multiple mechanisms of action either sequentially or in tank-mixtures is critically important in managing evolved resistance to herbicides, developing effective crop rotations and managing herbicide resistance with emphasis on the soil seedbank in all crops in the rotation is also paramount. Deep tillage and the burial of weed seed enables more effective management when used in combination with other control practices including herbicides and cover cropping.

Acknowledgements

The FMC Corporation, Texas Peanut Producers Board, Oklahoma Peanut Commission, and National Peanut Board provided funding for this research.

Author details

William James Grichar[1*], Peter A. Dotray[2] and Todd A. Baughman[2,3]

1 Texas A&M AgriLife Research and Extension Center, Corpus Christi, TX, USA

2 Texas A&M AgriLife Research and Extension Center, Lubbock, TX, USA

3 Institute for Agricultural Biosciences, Oklahoma State University, Ardmore, OK, USA

*Address all correspondence to: james.grichar@agnet.tamu.edu

References

[1] Creel JM Jr, Hoveland CS, Buchanan GA. Germination, growth, and ecology of sicklepod. Weed Science. 1968;**16**:396-400

[2] Prasad PVV, Boote KJ, Thomas JMG, Allen LH Jr, Corbet DW. Influence of soil temperatures on seedling emergence and early growth of peanut cultivars in field conditions. Journal of Agronomy and Crop Science. 2006;**192**:168-177

[3] Johnson WC III, Mullinix BG Jr. Weed management in peanut using stale seedbed techniques. Weed Science. 1995;**43**:293-297

[4] Johnson WC III, Brenneman TB, Baker SH, Johnson AW, Sumner DR, Mullinix BG Jr. Tillage and pest management considerations in a peanut-cotton rotation in the southeastern coastal plain. Agronomy Journal. 2001;**93**:570-576

[5] Henning RJ, Allison AH, Tripp LD. Cultural practices. In: Pattee HE, Young CT, editors. Peanut Science and Technology. Yoakum, TX: American Peanut Research and Education Society; 1982. pp. 123-138

[6] Wilcut JW, York AC, Grichar WJ, Wehtje GR. The biology and management of weeds in peanut (*Arachis hypogaea*). In: Pattee HE, Stalker HT, editors. Advances in Peanut Science. Stillwater, OK: American Peanut Research and Education Society; 1995. pp. 207-244

[7] Belel MD, Belel RD. Allelopathic effect of leaf and seed extract of nutgrass (*Cyperus tuberosus*) on the germination of beans (*Vigna unguiculata* (L.) Walp). Cogent Food & Agriculture. 2015;**1**:1102036. DOI: 10.1080/23311932.2015.1102036

[8] Elmore CD. Assessment of the allelopathic effects of weeds on field crops in the humid midsouth. In: Thompson AC, editor. The Chemistry of Allelopathy. Washington, DC: American Chemical Society; 1985. pp. 21-32

[9] Sorecha EM, Bayissa B. Allelopathic effect of *Parthenium hysterophorus* L. on germination and growth of peanut and soybean in Ethiopia. Advances in Crop Science and Technology. 2017;**5**:285. DOI: 10.4172/2329-8863.10000285

[10] Lancaster SH, Jordan DL, York AC, Burke IC, Corbin FT, Sheldon YS, et al. Influence of selected fungicides on efficacy of clethodim and sethoxydim. Weed Technology. 2005;**19**:397-403

[11] Lancaster SH, Jordan DL, Spears JF, York AC, Wilcut JW, Monks DW, et al. Brandenburg RL Sicklepod (*Senna obtusifolia*) control and seed production following 2,4-DB applied alone and with fungicides or insecticides. Weed Technology. 2005;**19**:451-455

[12] Grichar WJ, Dotray PA, Woodward JE. Weed and disease control and peanut response following post-emergence herbicide and fungicide combinations. In: Price A, editor. Herbicides - Current Research and Case Studies in Use. London, UK: IntechOpen; 2013. ISBN: 978-953-51- 1112-2. Available from: http://www.intechopen.com/books/herbicides-currentresearch-and-case-studies-in-use/weed-and-disease

[13] Royal SS, Brecke BJ, Shokes FM, Colvin DL. Influence of broadleaf weeds on chlorotalonil deposition, foliar disease incidence, and peanut (*Arachis hypogaea*) yield. Weed Technology. 1997;**11**:51-58

[14] Van Wychen L. Survey of the most common and troublesome weeds in broadleaf crops. fruits, and vegetables in the United States and Canada. Weed Science Society of America National Weed Survey Dataset. 2022. Available from: http://wssa.net/wp-content/uploads/2022-Weed-Survey-Broadleaf-crops.xlsx

[15] Webster TM. Weed survey—Southern states 2013; broadleaf crop section. Proceeding Southern Weed Science Society. 2013;**66**:279-280

[16] Balyan RS, BhanVM. Emergence, growth, and reproduction of horse purslane (*Trianthema portulacastrum*) as influenced by environmental conditions. Weed Science. 1986;**34**:516-519

[17] Chandra B, Sahai R. Autecology of *T. portulacastrum* Linn. Indian Journal of Ecology. 1979;**6**:17-21

[18] Elmore CD. Weed survey–southern states. Proceeding Southern Weed Science Society. 1989;**42**:416

[19] Grichar WJ. Horse purslane (*Trianthema portulacastrum*) control in peanut (*Arachis hypogaea*). Weed Technology. 1993;**7**:570-572

[20] Balyan RS, Malik RK. Control of horse purslane (*Trianthema portulacastrum*) and barnyardgrass (*Echinochloa crus-galli*) in mung bean (*Vigna radiate*). Weed Science. 1989;**37**:695-699

[21] Grichar WJ. Herbicide systems for control of horse purslane (*Trianthema portulacastrum* L.), smellmelon (*Cucumis melo* L.), and Palmer amaranth (*Amaranthus palmeri* S. Wats) in peanut. Peanut Science. 1998;**35**:38-42

[22] [SWSS] Southern Weed Science Society. Weed Identification Guide. Champaign, IL: Proceeding Southern Weed Science Society; 1999. p. 588

[23] Grichar WJ, Dotray PA, Baughman TA. Evaluation of weed control efficacy and peanut tolerance to pyroxasulfone herbicide in the South Texas peanut production area. Journal of Experimental Agriculture International. 2019;**29**(2):1-10. Article No. JEAI. 45347. DOI: 10.9734/JEAI/2019/45347

[24] Thompson AM, Rosales-Robles E, Chandler JM, Nester PR, Tingle CH. Crop tolerance and weed management systems in imidazolinone-tolerant corn (*Zea mays* L.). Weed Technology. 2005;**19**:1037-1044

[25] Tingle CH, Chandler JM. The effect of herbicides and crop rotations on weed control in glyphosate-resistant crops. Weed Technology. 2004;**18**:940-946

[26] Livingston SD, Janak JD, Matthies AZ Jr. Early and late-season suppression and control of Texas smellmelon in cotton. Proceeding Southern Weed Science Society. 2004;**57**:27

[27] Livingston SD. Over-the-top and lay-by herbicide treatments to control silverleaf nightshade, desert horse purslane, and Texas smellmelon in RR, RR-Flex, and Liberty Link transgenic cotton. Proceeding Southern Weed Science Society. 2006;**59**:225

[28] Tingle CH, Chandler JM, Jones CA, Steele GL. Competition and control of smellmelon (*Cucumis melo* L. Var *dudaim* Naud.) in cotton. Proceeding Southern Weed Science Society. 2000;**53**:207-208

[29] Grichar WJ. Horse purslane (*Trianthema portulacastrum*), smellmelon (*Cucumis melo*), and palmer amaranth (*Amaranthus palmeri*) control in peanut with postemergence herbicides. Weed Technology. 2007;**21**:688-691

[30] Hensleigh P, Pokorny M. Palmer amaranth (*Amaranthus palmeri* S. Watson). Agronomy Technical Note MT-92. USDA-NRCS; 2017

[31] Correll DS, Johnston MC. Manual of the Vascular Plants of Texas. 2nd ed. Richardson, TX: University of Texas at Dallas; 1979. pp. 555-556

[32] Grichar WJ. Control of Palmer amaranth (*Amaranthus palmeri*) in peanut (*Arachis hypogaea*) with postemergence herbicides. Weed Technology. 1997;**11**:739-743

[33] Grichar WJ. Control of hophornbeam copperleaf (*Acalyphao stryifolia* Riddell) and ivyleaf morningglory (*Ipomoea hederacea* (L.) Jacq.) in peanut (*Arachis hypogaea* L.). Texas Journal of Agriculture and Natural Resources. 1997;**10**:55-63

[34] Mosier DG, Oliver LR. Common cocklebur (*Xanthium strumarium*) and entireleaf morningglory (*Ipomoea hederacea van integriuscula*) interference on soybeans (*Glycine max*). Weed Science. 1995;**43**:239-246

[35] Lancaster SH, Jordan DL, York AC, Wilcut JW, Brandenburg RL, Monks DW. Interactions of late-season morningglory (*ipomoea* spp.) management practices in peanut (*Arachis hypogaea*). Weed Technology. 2005;**19**:803-808

[36] Barbour JC, Bridges DC. A model of competition for light between peanut (*Arachis hypogaea*) and broadleaf weeds. Weed Science. 1995;**43**:247-257

[37] Hang AN, McCloud DE, Boote KJ, Duncan WG. Shade effects on growth, partitioning, and yield components of peanuts. Crop Science. 1984;**24**:109-115

[38] Tanetani Y, Kahu K, Kawai K, Fujioka T, Shiminizu T. Action mechanism of a novel herbicide, pyroxasulfone. Pesticide Biochemistry and Physiology. 2009;**95**:47-55

[39] Steele GL, Porpiglia PJ, Chandler JM. Efficacy of KIH-485 on Texas panicum and selected broadleaf weeds in corn. Weed Technology. 2005;**19**:866-869

[40] Curran W, Lingenfelter D. Pyroxasulfone: The new kid in the neighborhood. Penn State Extension. 2013. Available from: http://extension.psu.edu/plants/crops/news/2013/04/pyroxasulfone-the-new-kid-in-the-neighborhood

[41] Anonymous Speciman Label-Zidua® Herbicide. BASF Corporation, Research Triangle Park, NC. 2015. 17 p

[42] King SR, Garcia JO. Annual broadleaf control with KIH-485 in glyphosate-resistant furrow-irrigated corn. Weed Technology. 2008;**22**:420-424

[43] Koger CH, Bond R, Poston DH, Eubank TW, Blessitt JB, Nandula VK. Evaluation of new herbicide chemistry: Does KIH-485 have a fit in the southern cotton producing region? In: Proceeding of the 2008 Beltwide Cotton Conference. Cordova, TN: The National Cotton Council; 2008. p. 1738

[44] Knezevic SZ, Datta A, Scott J, Porpoglia P. Dose-response curves of KIH-485 for preemergence weed control in corn. Weed Technology. 2009;**23**:340-345

[45] Nurse RE, Sikkema PH, Robinson DE. Weed control and sweet maize (*Zea mays* L.) yield as affected by pyroxasulfone dose. Crop Protect. 2011;**30**:789-793

[46] Hulting AG, Dauer JT, Hinds-Cook B, Curtis D, Koepke-Hill RM, Mallory-Smith C. Management of

Italian ryegrass in western Oregon with preemergence applictions of pyroxasulfone in winter wheat. Weed Technology. 2012;**26**:230-235

[47] Grier PW, Stahlman PW, Frihauf JC. KIH-485 and *S*-metolachlor efficacy comparisons in conventional and no-tillage corn. Weed Technology. 2006;**20**:622-626

[48] Prostko EP. Weed control update. In: Beasley JP Jr, editor. 2013 Peanut Production Update. Cooperative Extension Service Series CSS-12-0110. Athens, GA: University of Georgia; 2013. pp. 47-65

[49] Dotray PA, Baughman TA, Grichar WJ, Woodward JE. Performance of pyroxasulfone to control *Amaranthus palmeri* and *Salsola kali* in peanut. Journal of Experimental Agriculture International. 2018;**23**(2):1-10. Article No. JEAI. 41505

[50] Baughman TA, Grichar WJ, Dotray PA. Weed control and peanut tolerance using pyroxasulfone in Oklahoma. Journal of Experimental Agriculture International. 2018;**21**(3): 1-11. Article No. JEAI. 39881

[51] Theodoridis G, Baum JS, Hotzman FW. Synthesis and herbicidal properties of aryltriazolinones. A new class of pre- and post-emergence herbicides. American Chemical Society Symposium Series. 1992;**504**:135-146

[52] Dayan FE, Duke SO, Weete JD, Hancock HG. Selectivity and mode of action of carfentrazone-ethyl, a novel phenyl triazolinone herbicide. Pesticide Science. 1997;**51**:65-73

[53] Dayan FE, Weete JD, Duke SO, Hancock HG. Soybean (*Glycine max*) cultivar differences in response to sulfentrazone. Weed Science. 1997;**45**:634-641

[54] Becerril JM, Duke SO. Protoporphyrin IX content correlates with activity of photobleaching herbicides. Plant Physiology. 1989;**90**:1175-1181

[55] Sherman TD, Becerril JM, Matsumoto H, Duke MV, Jacobs JM, Jacobs NJ, et al. Physiological basis for differential sensitivities of plant species to protoporphyrinogen oxidase inhibiting herbicides. Plant Physiology. 1991;**97**:280-287

[56] Devine MD, Duke SO, Fedtke C. Physiology of Herbicide Action. N. J. Prentice Hall: Englewood Cliffs; 1993. pp. 177-188

[57] Anonymous. Speciman Label-Aim® Herbicide. Philadelphia, PA: FMC Corp., Agricultural Products Group; 2017. 20 p

[58] Culpepper AS, Grey TL, Vencill WK, Kichler JM, Webster TM, Brown SM, et al. Glyphosate-resistant palmer amaranth (*Amaranthus palmeri*) confirmed in Georgia. Weed Science. 2006;**54**:620-626

[59] Peterson DE. The impact of herbicide-resistant weeds on Kansas agriculture. Weed Technology. 1999;**13**:632-635

[60] VanGessel MJ. Glyphosate-resistant horseweed from Delaware. Weed Science. 2001;**49**:453-459

[61] Lovell ST, Wax LM, Horak MJ, Peterson DE. Imidazolinone and sulfonylurea resistance in a biotype of common waterhemp (*Amaranthus rudis*). Weed Science. 1996;**44**:789-794

[62] Shaner DL, Feist DA, Retzinger EJ. SAMOA: One company's approach to herbicide-resistant weed management. Pesticide Science. 1997;**51**:367-370

[63] Anonymous. Speciman Label-Anthem Flex® Herbicide. Philadelphia, PA: FMC Corp. Agricultural Products Group; 2020. 19 p

[64] Anonymous. FMC Offers New Postemergence Herbicide Option for Peanut Growers. Philadelphia, PA: FMC Corp. Agricultural Products Group; 2020. Available from: ag.fmc.com/us/en/fmc-news/news-release-fmc-offers-new-postemergence-herbicide-option-peanut-growers

[65] Frans R, Talbert R, Marx D, Crowley H. Experimental design and techniques for measuring and analyzing plant responses to weed control practices. In: Camper ND, editor. Research Methods in Weed Science. 3rd ed. Champaign, IL: Southern Weed Science Society; 1986. pp. 29-46

[66] Grichar WJ, Dotray PA, Baughman TA. Carfentrazone plus pyroxasulfone combinations for weed control in peanut (*Arachis hypogaea* L.). Journal of Experimental Agriculture International. 2021;**43**(10):52-63. Article no. JEAI. 75865

[67] Jha P, Kumar V, Garcia J, Reichard N. Tank mixing pendimethalin with pyroxasulfone and chloroacetamide herbicides enhances in-season residual weed control in corn. Weed Technology. 2015;**29**:198-206

[68] Jordan DL, Lancaster SH, Lanier JE, Lassiter BR, Johnson PD. Weed management in peanut with herbicide combinations containing imazapic and other pesticides. Weed Technology. 2009;**23**:6-10

[69] Grichar WJ, Sestak DC. Herbicide systems for golden crownbeard (*Verbesina encelioides*) control in peanut. Peanut Science. 2000;**27**:23-26

[70] Grichar WJ, Sestak DC. Control of golden crownbeard (Verbesnia enceloides) in peanut (*Arachis hypogaea*) with postemergence herbicides. Peanut Science. 1988;**25**:39-43

[71] Grichar WJ, Nester PR. Nutsedge (*Cyperus spp.*) control in peanut (*Arachis hypogaea*) with AC 263,222 and imazethapyr. Weed Technology. 1997;**11**:714-719

[72] Grichar WJ, Sestak DC. Effect of adjuvants on control of nutsedge (*Cyperus esculentus* and *C. rotundus*) by imazapic and imazethapyr. Crop Protect. 2000;**19**:461-465

[73] Richburg JS III, Wilcut JW, Wehtje GR. Toxicity of AC 263,222 to purple (*Cyperus rotundus*) and yellow nutsedge (*C. esculentus*). Weed Science. 1994;**42**:398-402

[74] Richburg JS III, Wilcut JW, Wiley GL. AC 263,222 and imazethapyr rates and mixtures for weed management in peanut (*Arachis hypogaea*). Weed Technology. 1995;**9**:801-806

Chapter 14

Actual Aspects of Growing Leguminous Crops in the South of Ukraine under Different Moisture Conditions

Sergiy Lavrenko and Nataliia Lavrenko

Abstract

As a member of the UN, Ukraine has joined the global process of ensuring sustainable development and fulfilling 17 Sustainable Development Goals and 169 tasks. One of the main aspects of fulfilling these tasks is the increase in the production of food protein of animal and plant origin. Unfortunately, it is not possible to rapidly increase the amount of animal protein, but it is possible to create a model of constant growth due to the cultivation of legumes, which are similar to animal protein. In this sense, the most suitable crops are chickpeas, grasspea, lentils, and beans. In the south of Ukraine, the main limiting factor in increasing the productivity of agricultural plants is the amount of moisture, its uneven distribution during the growing season, and critically high air temperatures. The specified conditions make the constant search for technological solutions to level the unfavorable conditions of the external environment difficult. Also, a method of overcoming stressful situations is the use of plant growth and development stimulants on leguminous crops. The specified environmental factors and technological ways of solving them in the conditions of the Southern Steppe of Ukraine are extremely relevant and timely, and they will be presented in the scientific work.

Keywords: grasspea, chickpea, lentils, beans, growth stimulant, growing techniques, irrigation, feed mixture

1. Introduction

Currently, global and food security have become integral parts of the strategic direction of development of any country. They are interconnected and inseparable. Therefore, the food supply of the country in full is the most urgent and main task of today.

In the last 25 years, the agriculture and crop production of Ukraine has been marked by significant changes. The structure of crop rotation areas, the crop fertilization system, rational measures of soil cultivation, and other elements in the general technology of growing crops are not observed by the majority of farms. This situation affected the grain yield of field crops in the direction of decrease and increase of degradation-deflation processes.

One of the reasons for such a situation is the decrease in the area of grain and leguminous crops, both under irrigation and dry land conditions.

In Ukraine, peas have been the main leguminous crop for many years, but their share is steadily decreasing every year, which is occupied by other less common crops but with a greater demand in the crop market, namely chickpeas, lentils, sorghum, mung bean and others.

The main reasons for this are the spread of diseases, pests, and weeds; significant losses during the two-phase harvesting of peas; difficulties in storage; high-cost price for low-cost grain, etc.

Leguminous crops are the basis of high-protein resources in animal feed and human nutrition, but in the process of agricultural intensification, they began to occupy smaller areas than the population needs. At the same time, the demand for peas, chickpeas, chickpeas, lentils, and other crops for food and fodder purposes, at the expense of own production, is far from being fully satisfied in most countries of the world [1].

Crops that are capable of providing stable grain harvests with a high content of food protein in the conditions of global climate change are grasspea, chickpeas, lentils, beans, and other leguminous crops.

Favorable soil and climatic conditions, favorable geopolitical and territorial-economic location of Ukraine, and hard-working people caused significant development of grain production in our country. Ukraine ranks 6–7 in terms of grain and leguminous crops, and 5–6 in terms of grain production per person [2].

Currently, issues related to protein supply should be resolved not only by increasing the area and volumes of grain and leguminous crops grown but also by taking into account the concepts of rational nature management, which aim to optimize land use, biologize agriculture, improve cultivation technologies, use of new varieties and hybrids of crops and land reclamation.

1.1 Grasspea (*Lathyrus sativus*)

Grasspea (*Lathyrus sativus*) is a leguminous crop known since prehistoric times. It is widespread in India, the countries of Southeast Asia, Algeria, Egypt, and in a number of countries in Western Europe (Spain, Italy, Albania, the former Yugoslavia, Portugal, Romania, Austria, and France). In Ukraine, grass pea (*Lathyrus sativus*) was produced in the nineteenth century. It began to be grown both in forest-steppe and steppe semi-arid and arid areas; the local name—"zubok"and "corner pea."

The genus Grasspea (*Lathyrus*) is an annual and perennial herb. The leaves are pinnate, ending in a tendril or a point. Grass pea belongs to the butterfly family, the legume-flowering order (Leguminosales, Fabales), the dicotyledons class (Dicotyledones), and angiosperms.

The genus *Lathyrus* has more than 200 species. But the following are used in production: seed grass pea (*L. sativus*) is an annual plant, the seeds are trapezoidal or tooth-shaped, common only in the cultivated form; a very drought-resistant plant; Tangier grass pea (*L. tingitanus*) is an annual plant that gives exceptionally high yields of green mass for fodder and siderate; chickpea grass pea (*L. cicer*)—used as an improver of soil properties; Meadow grass pea (*Lathyrus pratensis*) is a perennial fodder plant that grows in meadows; pink grass pea (*Catharanthus roseus*)—is eaten in Georgia [3, 4].

The total duration of the growing season of grass pea (*Lathyrus sativus*) ranges from 60 to 125 days. In wet years, the growing season is extended; during the late sowing period, it is reduced, which is associated with higher air temperature and low soil moisture in this period.

Grasspea (*Lathyrus sativus*) is a fairly tall plant. The stem is four-sided with two wings on opposite edges, and it branches well. The leaves are lanceolate or linear, unipinnate, sit on a relatively long petiole, and end with three to five branched tendrils at the end. Bean 1–7-seeded, broadly linear or elliptical, straight, less often curved, with two bent wings along the upper seam. The flowers are large, single, rarely two on a peduncle, white, rarely blue. A self-pollinated plant, but cross-pollination often occurs (10% or more). Grasspea is a good honey crop, which provides honey collection in the amount of 40 kg/ha.

The seeds are wedge-shaped, flat-wedge-shaped, rounded, or flat. The weight of 1000 seeds varies from 50 to 600 g. Plants with a seed weight of 50 to 100 g belong to the small-seeded group, from 150 to 250 g to medium-seeded and over 250 g to large-seeded. Varieties zoned in Ukraine have a mass of 1000 seeds from 120 to 350 g.

Grasspea (*Lathyrus sativus*) is a medium-demanding to warm, drought-resistant plant. The minimum temperature at which it can germinate, albeit slowly (for 10–20 or more days), is 1–2°C. The optimal temperature at which seeds begin to germinate is 3–4°C, but with a slight increase in temperature to 5–7°C, grass pea develops slowly, which contributes to the development of weeds [5].

From sowing to commercial maturity, grasspea need for heat, expressed as the sum of active average daily temperatures, is 1600–1700°C. When growing grasspea, this value is equal to 1650°C for grain, and 1100°C for green mass.

Grasspea is characterized by high drought resistance and is second only to chickpeas in this respect. Therefore, its advantage is especially noticeable in arid soil-climatic areas or in dry years. Drought resistance of grass pea is due to the biological and morphological features of the plant, primarily a well-developed root system that penetrates the soil to a depth of up to 2 m.

On the roots of grasspea, there are nodules that contain bacteria (*Rhizobium*) that assimilate nitrogen from the air. The mass of nitrogen fixation from the air on unirrigated lands reaches 200, and on irrigated lands—up to 300 kg/ha. After harvesting the crop, 60–70 kg of nitrogen is added to the soil with root residues, which is equivalent to applying 20–25 tons of manure.

1.2 Lentil (*Lens culinaris*)

Lentils are one of the first domesticated leguminous crops [6]. It formed the basis of nutrition of many prehistoric civilizations. Currently, lentils are one of the most widespread leguminous crops in the world, which are grown in 52 countries of the world on a total area of more than 4.0 million hectares [7].

The homeland of large-seeded lentils is considered to be the Mediterranean, and the homeland of small-seeded lentils is Iran and Afghanistan. It became most widespread in the following countries: Egypt, Spain, Romania, the Czech Republic, Slovakia, and Chile. Currently, small-seeded lentils are grown in Europe, Asia, Africa, and Latin America.

Seed color is a very unstable feature and depends on the region of cultivation and meteorological conditions of the year. If it rains during the ripening period, the seeds turn brown quickly, and their food value and, accordingly, their commercial value decreases. With untimely harvesting, as well as with storage in the light, the color changes from green to pink, and then to brown; the quality of seeds decreases sharply. The browning of seeds is explained by two reasons: oxidation of chlorophyll and the presence of tannins. Currently, varieties with a "non-browning" seed coat are being created [8]. The color of the cotyledons in lentils, unlike the color of the seed coat, is a permanent feature and comes in three types: yellow, orange, and green. Seeds with yellow and green cotyledons are used as a whole and for canning, and orange seeds, they are cleaned of shells and used to make cereals.

Green-seeded samples are divided by size into large, medium, and small. The most traditional market product is large-seeded green lentils, but recently, there has been an increase in demand for red-seeded lentils, food products from that have a pleasant aroma and delicate texture; red-seeded lentils are most often used as a meat substitute.

The grain is delivered to the market in whole with shells (whole), whole without shells (footbol), and split (split). The demand for "football" and "split" types is constantly growing.

Large-seeded lentils are mainly used as a food crop, while small-seeded lentils can be used for fodder purposes for all farm animals. Ground seeds are considered a good concentrated feed, and the tender green mass with a high-protein content is similar in quality to good meadow hay. The content of digestible protein in hay is 15%, and in chaff up to 20% [9]. Lentils are also used as an organic fertilizer.

Lentil plants react differently to growing conditions, which is expressed in a wide variation in the level of their productivity and morphological and biological features. In years with favorable weather conditions, seed yield reaches 1.5–1.9 t/ha, decreasing to 0.45–0.67 t/ha in unfavorable years. The mathematical analysis of these regularities established a high dependence of plant growth rates, interphase, and vegetation periods of lentils on the temperature regime ($r = 0.158–0.662$, reverse direction) and the amount of precipitation ($r = 0.596–0.850$, forward direction).

Lentils belong to the group of high-protein grain legumes. In lentils, an important component of quality indicators is the biochemical composition of the grain, namely the content of protein (lentil proteins consist mostly of water- and salt-soluble globulins), fiber, and fat. The protein content ranges from 21 to 36%. Lentil seed protein is well balanced in amino acid content, limited only by methionine and cystine. In addition, the insignificant content of fat (1–2%) in the seeds allows the protein to be extracted without prior defatting [10].

In terms of the content of lysine, phenylalanine, threonine, and leucine, lentil protein is similar to chicken egg protein. However, methionine and tryptophan are in short supply. In terms of digestibility by the human body (86%), lentil proteins are only slightly inferior to proteins of animal origin.

According to many scientists, lentils are environmentally friendly products that do not accumulate radionuclides, toxic elements, and other dangerous elements.

The biochemical composition of lentil seeds puts it in the category of the most important dietary food crops, which are used both in the daily diet and in medical, children's, and vegetarian food. In terms of taste and nutrition, lentils are one of the first among grain legumes. It boils well and has a delicate and pleasant taste.

Lentils, like other leguminous crops, accumulate atmospheric nitrogen in symbiosis with nodule bacteria and leave up to 90 kg/ha of biological nitrogen in the soil after harvesting [11].

1.3 Chickpeas (*Cicer arietinum*)

In modern economic conditions, where agricultural enterprises do not observe crop rotations, abandon the cultivation of leguminous crops, and use outdated equipment and technologies, there is a sharp decrease in the productivity of arable land. One of the ways to improve this situation is the introduction of chickpeas into production, which is one of the ways to increase soil fertility, solve the problems of fodder and food protein production, and establish the economic stability of farms.

The use of chickpeas makes it possible to rationally build a crop rotation, especially in the steppe zone of Ukraine, where, under the conditions of limited natural moisture supply, the choice of crops is small and there is an oversaturation of crop rotations with cereals, which means the accumulation of pests and diseases with the corresponding consequences [12, 13].

Chickpea is one of the most drought-resistant leguminous crops, the sown area of which is increasing every year. Culture is grown in more than 30 countries of the world, occupying the third place among legumes, second only to soybeans and beans. Its main cultivated areas are concentrated in India, China, and Pakistan [14, 15].

Chickpea, the most nutritionally valuable leguminous crop, contains a large number of vitamins and other biologically valuable substances. The grain contains 20–30% protein, which is close to egg protein in terms of quality; 4–7% fat; 2–7% fiber; 50–60% carbohydrates; 2–5% of mineral substances; a large amount of B vitamins (B_1, B_2, B_3, B_6), A, C (2.2 to 20 mg per 100 g of biomass), and PP. A significant amount of essential amino acids is important. The amino acid composition of grain protein is close to animal protein. Its digestibility ranges from 80 to 83%. The biological value of chickpea protein ranges from 52 to 78% [12, 14, 16].

In the south of Ukraine, chickpeas are a rare and understudied crop, and most of the recommended technologies for their cultivation are based on general approaches to soil cultivation and crop care. Ignorance of commodity producers in the peculiarities of its cultivation leads to obtaining low yields, high weediness of crops, and low quality of grain, and, as a result, makes it impossible to form sales trading lots. In the domestic market, the demand for chickpeas is insignificant, but the export potential of the culture is great.

Chickpea is a valuable crop in agrotechnical terms, as it is one of the best predecessors. After it is collected with post-harvest residues, as many nutrients as are contained in 15–20 t of manure enter the soil. Compared to other grain legumes, chickpeas are less demanding on moisture and have high drought resistance. The presence of such a biological feature of chickpea plants is explained by the fact that their cells contain less free and more bound water. As a result, their evaporation is lower than that of other leguminous crops. During a period of drought, chickpeas suspend growth and development, and when favorable conditions occur, they resume [17]. Irrigation of chickpeas makes it possible to almost double the yield in a dry year.

1.4 Beans (*Phaseolus vulgaris*)

Bean grains contain an average of 23–25% protein, which is characterized by high digestibility—up to 86–90%, which is higher compared to peas and chickpeas. High energy value of beans, so 100 g of grain contains 309 kcal (1293 kJ), which is two times higher than beef meat and seven times higher than fish [18–20]. In world agriculture, among leguminous crops, beans rank second after soybeans and are in high demand, especially in food quality [21–23].

Legumes of the so-called niche cluster are currently particularly popular, namely chickpeas, lentils, mung beans, and beans, which fully satisfy human needs for food protein [24]. Common bean is able not only to produce high grain yields in the conditions of non-irrigated and irrigated agriculture but also to improve the energy and economic balance of the economy by accumulating nitrogen in the soil [25, 26].

The bean genus *Phaseolus L.* includes up to 230 species. All known species are divided into two main groups—American and Asian. Beans of the American group have large flat beans with a long beak and large seeds; Asian, on the contrary—narrow beans without a beak with small seeds [27, 28]. In Ukraine, the most popular type is the bean of the American group—the common bean [29, 30].

According to the color of the seed coat, consumers in our country prefer white, but with the development of world culinary traditions, a new color—black—enters everyday life. Such a bean is called Preto—(Preto, Black Turtle, Frijol negro, Feijao preto)—a small bean, silky black on the outside. It has a delicate and at the same time dense texture, a sweetish taste with a slight bitterness, and a pleasant berry aroma.

Preto beans belong to the common bean species (*Phaseolus vulgaris*). It is very common in Latin American cuisine—it is used in meat and vegetable dishes, in soups and soup mixes, in stews, salads, and also in fried form. Beans have a dense, meaty texture that makes them popular in vegetarian dishes. This bean is the basis of Brazilian cuisine, Brazilians prepare "feijoada" from it—the main national dish. It is grown in industrial quantities in Canada, the United States, and China [23, 31].

Beans are capable of symbiosis with rhizobia. Under favorable conditions (optimal pH and soil moisture, air temperature, sufficient supply of macro- and microelements, and the presence of an active strain of bacteria), it absorbs 150–200 kg/ha of nitrogen from the air per vegetation.

Beans are a more drought-resistant crop than peas and lentils. It is especially picky about moisture during the germination period (100–120% of water from the weight of the seeds is needed), as well as during flowering and setting of beans. When overmoistened during grain pouring, the growth of beans stops, and ripening is delayed, which is the cause of the spread of fungal diseases and a decrease in yield. Drought during the period of flowering and ripening leads to the dropping of flowers and young beans and the formation of shallow seeds [32].

Despite the high export potential and natural and climatic conditions, the cultivation of common beans in Ukraine has not acquired large volumes, and in the future, it will remain a homestead crop. This trend is based on imperfect methods of cultivation, especially in the irrigated conditions of the Southern Steppe of Ukraine; at the same time, revealing the genetic potential of plants will allow agricultural enterprises to simultaneously improve the physical properties of the soil, their nutritional regime, increase the specific weight of leguminous crops in the structure of sown areas, and increase financial income from sales.

2. The influence of technological methods for growing leguminous crops on the yield level

2.1 Formation of productivity grasspea depending on technological methods of cultivation in different moisture conditions

Studying the characteristics of growth, the process of crop formation and grain quality depending on the predecessors, the depth of soil cultivation, the dose of mineral fertilizers, sowing dates, the width of the rows, the thickening of plants, and the pre-irrigation soil moisture when growing high-yielding varieties of grasspea is an unsolved scientific problem for the south of Ukraine and need theoretical justification and development of practical recommendations. The conducted studies showed that there is specificity in the formation of a high yield of grasspea when grown in the south of Ukraine under irrigation [33].

The formation of the grass pea crop took place under different environmental conditions, but the same patterns were observed during all years of research. Analyzing the data obtained in field experiments, it can be stated that the yield of grasspea varied widely from 0.99 to 2.94 t/ha depending on the combination of the studied factors (**Table 1**).

Seed sowing rates had the greatest influence on the formation of grass pea grain yield. Over the years of research, their share was from 34.92 to 46.37% in the formation of the grasspea grain crop. Also, the timing of sowing played a significant role in the formation of the crop, from 21.33 to 22.28%, and mineral fertilizers—10.94

Soil temperature during sowing, °C (factor B)	Row width, cm (factor C)	Seed sowing rate, million/ha (factor D)	Nutrition background (factor E)			
			Without fertilizers	$N_{30}P_{45}$	$N_{60}P_{90}$	$N_{90}P_{135}$
Tillage to a depth of 20–22 cm (factor A)						
4–5	15	1.0	1.85	2.02	225	2.00
		1.5	2.28	2.47	291	2.66
		2.0	1.97	2.13	242	2.17
	30	1.0	1.66	1.82	202	1.81
		1.5	2.07	2.31	258	2.45
		2.0	1.34	1.52	177	1.54
8–10	15	1.0	1.48	1.65	190	1.65
		1.5	1.85	2.03	242	2.14
		2.0	1.41	1.54	188	1.62
	30	1.0	1.40	1.53	177	1.55
		1.5	1.65	1.88	220	2.03
		2.0	0.99	1.17	140	1.22
Tillage to a depth of 28–30 cm (factor A)						
4–5	15	1.0	1.87	2.04	2.29	2.03
		1.5	2.32	2.51	2.94	2.70
		2.0	2.00	2.16	2.46	2.21
	30	1.0	1.69	1.85	2.06	1.84
		1.5	2.11	2.35	2.62	2.49
		2.0	1.36	1.54	1.80	1.57
8–10	15	1.0	1.52	1.69	1.95	1.69
		1.5	1.88	2.07	2.46	2.18
		2.0	1.44	1.58	1.92	1.66
	30	1.0	1.43	1.57	1.82	1.59
		1.5	1.67	1.92	2.25	2.08
		2.0	1.03	1.21	1.44	1.26

Soil temperature during sowing, °C (factor B)	Row width, cm (factor C)	Seed sowing rate, million/ha (factor D)	Nutrition background (factor E)			
			Without fertilizers	$N_{30}P_{45}$	$N_{60}P_{90}$	$N_{90}P_{135}$
LSD_{05}, t/ha (for factors)	A, B, C	0.03–0.04	AD, BD, CD	0.05–0.08	ADE, BDE, CDE, ABCD	0.11–0.15
	D	0.04–0.05	AE, BE, CE, ABC	0.63–0.87	ABCE	0.13–0.17
	E	0.04–0.06	DE, ABD, ACD, BCD	0.08–0.11	ABDE, ACDE, BCDE	0.15–0.21
	AB, AC, BC	0.04–0.06	ABE, ACE, BCE	0.09–0.12	ABCDE	0.22–0.30

Table 1.
Grain yield of grasspea depending on the studied factors during irrigation, t/ha.

to 22.55%. The share of participation of the sowing method in the formation of the crop was, over the years of research, from 3.20 to 16.63%. The depth of plowing had a slight effect on crop formation—it was only 0.16–0.22%.

The yield of grass pea grain, depending on the depth of tillage in the experiment, changed insignificantly—from 0.02 to 0.05 t/ha. Mathematical processing of these data proved that the yield increase was within the limits of experimental error. For plowing to a depth of 20–22 cm, the average yield, regardless of other investigated factors, was 1.88 t/ha. With an increase in the depth of cultivation to 28–30 cm, the yield of grasspea grain was 1.92 t/ha, which is 2.1% more.

One of the leading factors in crop formation is the time of crop sowing. It determines the further growth and development of the plant, the beginning of the main phases of development, and, in general, the main elements of grass pea productivity. The best time for sowing grass pea seeds in the conditions of the experiments was early, when the seeds were sown at a soil temperature of 4–5°C at the depth of wrapping. The grasspea harvest in these options, regardless of other investigated factors, significantly increased compared to the later period. The average yield of grass pea grain in the areas where it was sown in the soil warmed to 4–5°C is 2.10 t/ha, and with later sowing, it decreased by 0.40 t/ha or by 19.0%.

In the areas where grasspea was sown in wide rows, the yield was 1.75 t/ha. The continuous method of sowing, compared to the wide-row method, increases the productivity of grasspea plants by an average of 0.30 t/ha.

The yield of grasspea grain, depending on the sowing rate, also changed significantly in field experiments. It reached its maximum value in variants where grasspea was sown at the rate of 1.5 million seeds/ha and amounted to 2.26 t/ha. With reduced and increased sowing rates, the yield decreased by 0.47 and 0.61 t/ha, respectively.

Among the investigated factors, mineral fertilizers had the greatest impact on the change in grasspea grain yield. It is known that fertilizers are a leading factor in the intensification of irrigated agriculture. They are powerful regulators of the nutrient regime of the soil; in this regard, they significantly affect the productivity of plants, but in order to use fertilizers most productively, it is necessary to know under which agrotechnical conditions they are most effective.

In our experiments, regardless of the time of sowing, the width of the rows, and the rate of sowing, the applied doses of fertilizers significantly increased the yield of grasspea compared to the control (without fertilizers). At the rate of $N_{60}P_{90}$, the yield of grasspea grain was the maximum and was on average 2.15 t/ha, which is more than the control by 0.47 t/ha, and with the dose of $N_{30}P_{45}$ by 0.29 t/ha. A further increase in the dose of nitrogen-phosphorus fertilizers ($N_{90}P_{135}$) led to a decrease in the yield of grasspea grain by 0.23 t/ha.

The research conducted on the irrigated massifs of the southern region of Ukraine showed a significant influence of the studied factors: precursor, background nutrition, tillage, and pre-irrigation soil moisture on the amount of grasspea grain yield. The yield of grasspea grain over the years of research ranged from 2.06 to 3.34 t/ha, depending on the studied factors (**Table 2**).

The greatest influence on the grain yield of grasspea was the nutritional background. The share of participation of the nutritional background in the formation

Predecessor (factor A)	**Tillage (factor B)**	**Nutrition background (factor C)**				
		Without fertilizers	**$N_{30}P_{45}$**	**$N_{60}P_{90}$**	**$N_{90}P_{135}$**	
Pre-irrigation soil moisture is 60–65% of the lowest moisture content (factor D)						
Winter wheat	Disking to a depth of 10–12 cm	2.27	2.44	2.67		2.54
	Plow to a depth of 20–22 cm	2.51	2.71	2.99		2.82
Corn of milk-wax maturity	Disking to a depth of 10–12 cm	2.14	2.28	2.49		2.38
	Plow to a depth of 20–22 cm	2.34	2.51	2.76		2.62
Fodder beet	Disking to a depth of 10–12 cm	2.06	2.18	2.36		2.27
	Plow to a depth of 20–22 cm	2.23	2.37	2.59	2.47	
Pre-irrigation soil moisture 70–75% of the lowest moisture content (factor D)						
Winter wheat	Disking to a depth of 10–12 cm	2.47	2.67	2.95		2.78
	Plow to a depth of 20–22 cm	2.76	3.01	3.34		3.14
Corn of milk-wax maturity	Disking to a depth of 10–12 cm	2.31	2.47	2.72		2.58
	Plow to a depth of 20–22 cm	2.53	2.75	3.06		2.87
Fodder beet	Disking to a depth of 10–12 cm	2.18	2.33	2.56		2.43
	Plow to a depth of 20–22 cm	2.37	2.56	2.84		2.67
LSD_{05}, t/ha (for factors)	A	0.05–0.07	AB, AD	0.07–0.09	ABC, ACD	0.14–0.19
	B, D	0.04–0.06	AC, ABD	0.10–1,35	BCD	0.12–0.16
	C, BD	0.06–0.08	BC, CD	0.08–0.11	ABCD	0.20–0.27

Table 2.
Grasspea grain yield depending on the studied factors, t/ha.

of the grain yield by years of research ranged from 30.01 to 34.98%. Predecessors also had a significant impact on the change in grain yield, almost a quarter. Over the years of research, the share of changes was from 24.23 to 28.99%. Tillage during the research period influenced the amount of grasspea grain yield in the range from 19.26 to 24.60%. Among the studied factors, pre-irrigation soil moisture had the least influence on the change in grasspea grain yield, where the share of participation was from 14.82 to 17.06%.

Elements of cultivation technology, as well as crop yield, depend on the predecessor. The largest grain yield of grasspea was determined by the predecessor of winter wheat per grain, where its value was, regardless of other studied factors, an average of 2.75 t/ha. The use of corn of milk-wax maturity as a precursor reduced the grain yield of grasspea by 7.3% and amounted to 2.55 t/ha, and fodder beet—by 18.5% (2.24 t/ha).

The application of mineral fertilizers increased the yield of grasspea grain from 0.12 to 0.58 t/ha compared to the control option without fertilizers. The lowest yield of grasspea grain was formed in the variants of the experiment, where mineral fertilizers were not applied, and was, on average, 2.35 t/ha. Application of mineral fertilizers at the rate of $N_{30}P_{45}$ provided an increase in yield by 0.17 t/ha (7.2%), and $N_{60}P_{90}$–0.43 t/ha (18.3%).

A further increase in the rate of mineral fertilizers led to a decrease in the grain yield compared to the previous rate. Thus, the use of mineral fertilizers with the $N_{90}P_{135}$ norm led to the formation of grasspea grain yield, on average, according to the experiment, at the level of 2.63 t/ra, which is 5.4% less compared to the $N_{60}P_{90}$ and 4.4% more compared to the $N_{30}P_{45}$ and by 11.9%—with the control option without fertilizers.

When replacing shallow tillage with plowing in the conditions of experiments, the grasspea grain yield increased from 0.17 to 0.39 t/ra, regardless of other studied factors. Thus, when grasspea was planted, after the main tillage, disking to a depth of 10–12 cm ensured the formation of grain yield, on average, at the level of 2.44 t/ha. The use of face plowing to a depth of 20–22 cm increased grain yield, on average, by 10.7%.

Grasspea plants responded positively to the increase in irrigation intensity by increasing the yield from 0.12 to 0.35 t/ra. Maintenance of pre-irrigation soil moisture 60–65% of the lowest moisture content during the growing season of grasspea plants, regardless of other studied factors, formed the grain yield, on average, according to the experiment, at the level of 2.46 t/ra. An increase in pre-irrigation soil moisture to 70–75% of the lowest moisture content increased crop yield by 2.2 t/ha, or 8.9%.

2.2 Formation of productivity lentils depending on technological methods of cultivation in different moisture conditions

Traditionally, lentils are grown under conditions of natural moisture supply and less extreme conditions in terms of temperature gradient, in which they form a fairly high productivity [34, 35].

In the conditions of the Southern Steppe of Ukraine, climatic conditions have a significant impact on crop productivity (**Table 3**).

The creation of a deep arable layer determines the accumulation of more moisture and, accordingly, creates better conditions for the growth and development of lentils. According to the obtained data, the average yield after plowing to a depth of 20–22 cm was 1.08 t/ha. Cultivation deepening to 28–30 cm increased the yield of lentil grain to 1.11 t/ha or by 2.8%.

Tillage (factor A)	Nutrition background (factor B)	Plant density, million/ha (factor C)		
		2.0	2.5	3.0
No irrigation (factor D)				
Plowing to a depth of 20–22 cm	Without fertilizers	1.06	0.99	0.85
	$N_{45}P_{45}$	1.30	1.22	1.03
	$N_{90}P_{90}$	1.20	1.15	0.94
Plowing to a depth of 28–30 cm	Without fertilizers	1.08	1.01	0.87
	$N_{45}P_{45}$	1.35	1.26	1.05
	$N_{90}P_{90}$	1.24	1.17	0.96
Irrigation (factor D)				
Plowing to a depth of 20–22 cm	Without fertilizers	1.50	1.86	1.70
	$N_{45}P_{45}$	1.94	2.44	2.24
	$N_{90}P_{90}$	1.82	2.29	2.08
Plowing to a depth of 28–30 cm	Without fertilizers	1.51	1.89	1.74
	$N_{45}P_{45}$	1.98	2.48	2.30
	$N_{90}P_{90}$	1.86	2.33	2.15
LSD_{05}, t/ha (for factors)	A, D	0.02–0.03	BC	0.06–0.07
	B, C	0.03–0.04	ABD, ACD	0.07–0.08
	AD	0.04–0.05	BCD, ABC	0.08–0.09
	BD, CD, AB, AC	0.05–0.06	ABCD	0.12–0.13

Table 3.
Yield of lentil grain depending on the studied factors, t/ha.

During irrigation, the dynamics of changes were similar, but the yield level was 1.99 t/ha for plowing to a depth of 20–22 cm, on average, according to the experiment. A further increase in tillage by 10 cm resulted in a yield of 2.03 t/ha. But it is impossible to consider this increase as reliable because the dispersion analysis showed that the error of the experiment by the years of research ranged from 0.028 to 0.031 t/ha and was more than the increase. As a result of these data, increasing the plowing depth is impractical.

The growth and development of lentil plants can take place only under the conditions of a favorable nutritional regime. On the basis of the conducted research, the best conditions were created when applying $N_{45}P_{45}$ under both conditions of moisture supply.

Cultivation of lentils on variants of the natural level of fertility ensured obtaining from 0.85 to 1.08 t/ha of lentil grain on rainfed land and 1.50–1.89 t/ha under irrigation. Application of $N_{45}P_{45}$ increased grain yield, on average, by 22.4%—under natural moisture and 31.2%—under irrigation. A further increase in the dose of mineral fertilizers to $N_{90}P_{90}$ led to a decrease in plant productivity. Thus, when growing in non-irrigated areas, the grain yield decreased by 8.1% and ranged from 0.94 to 1.24 t/ha, and when irrigated—by 6.7% and 1.82–2.33 t/ha, respectively.

One of the factors in increasing the gross harvest of grain is the thickening of crops, but in conditions of moisture deficit, this is a very limited factor. Thus, when growing lentils at a plant density of 2.0 million/ha, the average grain yield was 1.21 t/ha. The thickening of crops to 2.5 million/ha led to a decrease in yield by 7.1%; further compaction of crops reduced the level of plant productivity to 27.4%.

With irrigation, when the limiting factor is no longer moisture but nutrients, the optimal level of plant thickening was 2.5 million/ha, where the grain yield was formed from 1.86 to 2.48 t/ha. The formation of the specified density of plants on the field ensured a significant increase in yield compared to the thickening of 2.0 and 3.0 million/ha—by 25.4 and 8.8%, respectively.

In the conditions of a significant moisture deficit in the conditions of the Southern Steppe of Ukraine, the level of productivity primarily depends on this factor, which was additionally proven in the culture of lentils. In non-irrigated conditions, the level of lentil grain yield ranged from 0.85 to 1.35 t/ha. Cultivation of the crop under irrigation allowed an increase in the productivity of the crop from 1.8 to 2.0 times, which was 2.01 t/ha on average, according to the experiment.

The main factor in the producer's decision to introduce this or that new or improved element of technology into production is the level of yield growth compared to the one currently used.

The greatest influence on the formation of lentil grain yield was caused by the factor "Moisture conditions"—79.73%, which is understandable in the conditions of the Southern Steppe of Ukraine. Other researched technological methods of growing had an order of magnitude less impact on productivity. A total of 9.54% of the formation of the yield level depended on plant thickening. The share of participation of the nutrition background was 2.87%, and the smallest—soil cultivation was 0.11%, which indicates the impracticality of increasing the cultivation depth deeper than 20–22 cm under lentils.

2.3 Formation of productivity chickpeas depending on technological methods of cultivation in different moisture conditions

Irrigation is the main factor in the formation of a high and high-quality crop of chickpeas in the south of Ukraine. Its implementation ensured a significant increase in crop yield. Under natural humidification, the average yield of chickpea grain over the years of research was 1.62 t/ha [36–38]. Cultivation of this crop under irrigation increased the indicator by almost 1.8 times. The yield increase of chickpea grain, where the limiting factor was leveled by artificial humidification, ranged from 0.93 to 1.62 t/ha (**Table 4**).

Carrying out cultivation of the soil to a depth of 28–30 cm compared to 20–22 cm increased the yield of chickpea grain in the non-irrigated variants by 0.02–0.08, and with irrigation—by 0.04–0.08 t/ha. However, this increase cannot be considered reliable because it was within the margin of error of the experiment.

The use of fertilizers is one of the main factors in the intensification of agricultural production. The cultivation of crops currently without their use is unprofitable and impractical. But, unlike other crops, legumes are able to supply themselves, as well as the following crops in the crop rotation, with nutrients, namely nitrogen.

On the unfertilized options, the yield of chickpea grain was the lowest and fluctuated under natural moisture—from 1.26 to 1.60, while under irrigation—from 2.18 to 2.74 t/ha. Fertilizer application had different effects on crop productivity under different moisture conditions. Thus, the introduction of $N_{45}P_{45}$ on variants without irrigation increased the yield of chickpea grain by 13.1%, which was 0.15–0.23 t/ha. The highest indicators of increase in grain yield were obtained when the crop was irrigated—0.47–0.63 t/ha.

Tillage (factor A)	Nutrition background (factor B)	Plant density, million/ha (factor C)		
		0.5	1.0	1.5
No irrigation (factor D)				
Plowing to a depth of 20–22 cm	Without fertilizers	1.26	1.48	1.55
	$N_{45}P_{45}$	1.41	1.68	1.77
	$N_{90}P_{90}$	1.52	1.80	1.90
Plowing to a depth of 28–30 cm	Without fertilizers	1.28	1.50	1.60
	$N_{45}P_{45}$	1.44	1.72	1.83
	$N_{90}P_{90}$	1.56	1.85	1.98
Irrigation (factor D)				
Plowing to a depth of 20–22 cm	Without fertilizers	2.18	2.48	2.70
	$N_{45}P_{45}$	2.66	3.02	3.31
	$N_{90}P_{90}$	2.83	3.24	3.53
Plowing to a depth of 28–30 cm	Without fertilizers	2.22	2.53	2.74
	$N_{45}P_{45}$	2.71	3.10	3.38
	$N_{90}P_{90}$	2.89	3.33	3.60
LSD_{05}, t/ha (for factors)	A, D	0.04–0.05	BC	0.08–0.10
	B, C	0.04–0.06	ABD, ACD	0.09–0.12
	AD	0.05–0.07	BCD, ABC	0.11–0.14
	BD, CD, AB, AC	0.06–0.08	ABCD	0.15–0.20

Table 4.
The yield of chickpea grain depends on the technological methods of its cultivation, t/ha.

The introduction of nitrogen-phosphorus fertilizers in the amount of 90 kg/ha of the active substance significantly increased the crop yield. At this dose, the average yield of chickpea grain in the experiment on variants without irrigation was 1.77 t/ha, and on irrigation——1.83 times more—3.24 t/ha. Compared with non-fertilized options, the increase under these conditions was, respectively, 0.26–0.38 and 0.64–0.85 t/ha.

The increase in the yield of chickpea grain, depending on the increase in the rate of mineral fertilizers under natural moistening, ranged from 0.11 to 0.15 t/ha, and irrigation—0.17–0.23 t/ha, or more by 53.3–54.5%.

The formation of the optimal density of the grass stand in conditions of moisture deficiency, both soil and air, is an important agrotechnological element. Sowing chickpeas with a plant density of 0.5 million/ha produced the lowest grain yield, which ranged from 1.26 to 1.56 without irrigation and from 2.18 to 2.89 t/ha with irrigation. As can be seen from the given data, irrigation increased the indicator, on average, by 83.0%. The thickening of plants to 1.0 million/ha increased the total yield of products (grains) from the area by 0.22–0.30 t/ha under natural and 0.30–0.44 t/ha—artificial moistening compared to the previous norm. At the maximum density of standing chickpea plants (1.5 million/ha), the highest yield was obtained under both moisture conditions. Thus, on the variants without irrigation, this indicator ranged from 1.55 to 1.98 t/ha, while with artificial irrigation – from 2.70 to 3.60 t/ha. Compared to the least thickening of plants, it provided additional formation of 0.30–0.42 t/ha of grain under natural and 0.51–0.71 t/ha—artificial moistening.

2.4 Formation of productivity beans depending on technological methods of cultivation in different moisture conditions

The formation of the harvest of any culture consists of many factors and elements. Therefore, the study of structural elements enables the researcher to learn more deeply about the nature of productivity formation and its influence on change [39–41].

Studies conducted on the dark chestnut soils of the south of Ukraine under irrigation conditions proved that the depth of the main tillage, mineral fertilizers, and the width of the row spacing had a significant effect on the grain yield of common beans, which ranged from 1.47 to 3.37 t/ha over the years of research (**Table 5**).

In a detailed assessment of the influence of the investigated elements of growing beans, special attention should be paid to soil cultivation. Previous studies on crop cultivation technology almost always suggest deep tillage. At the same time, during the last decade, leading scientists and manufacturers, on the contrary, are switching to surface and shallow tillage, or abandoning it at all using zero technology (no-till). Each of the specified processing methods has its advantages and disadvantages. In the soil and climatic conditions of conducting research, the application of surface and shallow tillage is ineffective because the soils are moderately saline, the irrigation water is of II-class quality, and the system of chemical protection of beans against weeds is imperfect. Considering the above, shelf processing is more appropriate.

According to our data, plowing to a depth of 28–30 cm provided better conditions for the growth and development of bean plants, which affected the level of their yield.

Creating a deep loose layer of soil to a depth of 28–30 cm provided the best conditions for the growth and development of common bean plants compared to a plowing depth of 20–22 cm. Using the deepest tillage made it possible to obtain 0.02–0.08 t/ha more yield grain. But these indicators, according to the conducted dispersion analysis,

Tillage to a depth, cm (factor A)	Nutrition background (factor B)	Row width, cm (factor C)			
		15	30	45	60
20–22	Without fertilizers	1.47	1.82	2.51	2.01
	$N_{45}P_{45}$	1.88	2.29	3.09	2.51
	$N_{90}P_{90}$	2.01	2.41	3.30	2.62
28–30	Without fertilizers	1.49	1.86	2.58	2.06
	$N_{45}P_{45}$	1.91	2.34	3.17	2.58
	$N_{90}P_{90}$	2.04	2.46	3.37	2.67
LSD_{05}, t/ha (for factors)	A	0.04–0.05	AB	0.07–0.08	
	B	0.05–0.06	BC	0.09–0.11	
	C	0.05–0.07	ABC	0.13–0.16	
	AC	0.08–0.09			

Table 5.
Grain yield of common beans depending on the studied factors under irrigation, t/ha.

are within the error of the experiment (LSD_{05} was for factor A – 0.04-0.05 t/ha). In this case, deepening the tillage of the soil is inefficient and impractical; therefore, it is better to use tillage to a depth of 20–22 cm in production. Additional confirmation of the effectiveness of plowing to a depth of 20–22 cm is the share of the factor in the formation of the crop, which, according to years of research, amounted to 0.18–0.29%.

In the conditions of intensification of agricultural production, organic and mineral fertilizers have the greatest impact on the yield of any crop. Statistical analysis of experimental data shows that the share of mineral fertilizers in crop formation over the years of research was 14.6–39.8%.

Conducting deep tillage of the soil by 20–22 cm on the experimental variants without introducing mineral fertilizers ensured the formation of a yield level of 1.95 t/ha. Application of $N_{45}P_{45}$ increased grain yield per crop by 0.41–0.78 t/ha and $N_{90}P_{90}$ by 0.54–0.78 t/ha.

When plowing to a depth of 28–30 cm, the dynamics of crop change were similar to plowing at 20–22 cm. The highest productivity of common bean plants was formed when $N_{90}P_{90}$ was applied, which was, on average, 2.64 t/ha—reducing the dose of fertilizers led to a decrease in the increase in yield, which amounted to 0.42–0.59 t/ha compared to the control options (without fertilizers). The lowest level of yield was formed in areas where mineral fertilizers were not applied and ranged from 1.49 to 2.58 t/ha.

Of the researched elements of the growing technology, the width of the row spacing had the greatest influence on the grain yield of common beans, and its share was 59.1–83.1% over the years of research. The conducted dispersion analysis proved the significance of the obtained grain yield increase depending on the studied factors.

The obtained experimental results confirm that the highest productivity of common bean plants was formed when sowing with a row spacing of 45 cm. When sowing with a row spacing of 60 cm, the yield of the crop decreased.

A detailed analysis of the results we obtained shows that the increase in the distance between the rows and the simultaneous decrease in the distance between the plants in the row led to their competition for the main factors of life. The increase in grain yield of common beans at a row width of 30 cm compared to the usual row method of sowing ranged from 0.35 to 0.43 t/ha. With a row spacing of 45 cm, the increase in grain yield increased compared to a width of 15 cm to 1.04–1.33 t/ha, and compared to a row spacing of 30 cm, the increase increased almost three times. Sowing beans with a row width of 60 cm had a negative effect on the crop yield. Under these conditions, the yield increase ranged from 0.54 to 0.66 t/ha, which, compared to the row spacing of 45 cm was two times smaller.

3. Effect of immune growth modulators on legumes

Currently, the use of not only biological drugs based on bacteria but also synthetic ones on various bases is very popular on leguminous crops. Usually, farmers use succinic acid and derivatives of heteroauxin. The latter is the most effective on legumes. One of the stacks is new immunomodulators of plant growth (MIR and MIR ORGANIC), which increases the speed of intracellular reactions of energy exchange, photosynthesis, synthesis, and accumulation of polysaccharides, protein substances, own enzymes and phytohormones; strengthens the immunity and resistance of plants to bacterial and fungal diseases; contains a special component—an adjuvant, which increases the selective permeability of cell membranes and the ability of cells to fully absorb the components of the drug. The composition of the drug also includes synthetic phytohormone—polyglycoside with acetylated L-isomer functional groups; heterocyclic nitrogen-containing compound with reversible protonation function; heterocyclic coupling with cationic functional groups capable of changing the ξ-potential of cell membranes; complex of trace elements (Cu, Fe, Zn, Mn, Mo, Mg, and Co) in chelate form [42–44].

Conducted research in the irrigated conditions of the Southern Steppe of Ukraine showed that the smallest number of beans of common beans was formed on the variant without tillage—11.3 pcs (**Table 6**). Treatment of seeds with the drug led to an increase in the indicator by 1.1 units. The use of the drug in the phase of formation of side shoots compared to the previous option reduced the number of beans on the plant by 0.7 pcs. The complex application of the drug both before sowing seeds and sowing in the phase of formation of lateral shoots ensured the formation of 12.7 pcs, on one plant, beans. Similar dynamics of changes were observed in the number of seeds on one plant. According to the obtained data, the highest yield of common bean grain under irrigation was formed during seed treatment before sowing—3.32 t/ha. Subsequently, the use of the drug during the growing season of plants led to a decrease in yield from 7.8% for processing during the growing season to 13.7% for complex processing with a stimulator of seeds and growing plants. The yield increase in the seed treatment option compared to the absolute control (no treatment) was 19.4%, where the yield was 2.78 t/ha. The increase in yield is due to the formation of more beans and the weight of 1000 seeds.

The smallest number of beans in lentils was formed on the option without cultivation—17.6 pcs. Treatment of seeds with the drug led to an increase in the indicator by 5.9 pcs. The use of the drug in the branching phase compared to the previous option reduced the number of beans on the plant by 1.1 pcs. Complex use of the drug both before sowing seeds and sowing in the branching phase ensured the formation of 21.7 pcs, on one plant, beans.

A similar dynamic of changes was observed in the number of seeds on one plant. Yes, the maximum amount is 24.6 pcs was formed on the option of using a growth stimulator on seeds before sowing. The smallest amount was formed on the control version, where treatment with the drug was not carried out—19.1 pcs.

According to the obtained data, the highest yield of lentil grain under irrigation was formed by seed treatment before sowing—2.51 t/ha. Subsequently, the use of the drug during the growing season of plants led to a decrease in yield from 10.8% for processing during the growing season to 13.9% for complex processing with a stimulator of seeds and growing plants. The yield increase in the seed treatment option compared to the absolute control (no treatment) was 17.5%, where the yield was 2.07 t/ha. The increase in yield is due to the formation of more beans and the weight of 1000 seeds.

According to the obtained data, the highest yield of chickpeas was formed by processing chickpea seeds before sowing—2.04 t/ha. Subsequently, the use of the drug during the growing season of plants led to a decrease in yield from 2.5% for processing during the growing season to 9.8% for complex processing with a stimulator of seeds and plants. The yield increase compared to the absolute control (without cultivation) was 15.9% (0.28 t/ha), where the yield was 1.76 t/ha. The increase in yield is due to the formation of more beans and the weight of 1000 seeds.

Conclusion: to increase the yield of leguminous crops, it is advisable to treat seeds with an immunomodulator of plant growth immediately before sowing. If it is

Experiment option	Number of beans on one plant, pcs.			Number of seeds on one plant, pcs.			Weight of 1000 seeds, g			Grain yield, t/ha		
	beans	lentils	chickpeas	beans	lentils	chickpeas	beans	lentils	chickpeas	beans	lentils	chickpeas
No cultivation	11.3	17.6	19	59.8	19.1	21	114	52.3	245	2.78	2.07	1.76
Seed treatment before sowing	14.4	23.5	31	66.4	24.6	29	122	55.7	258	3.32	2.51	2.04
Processing crops in the branching phase	13.7	22.4	27	63.3	23.1	26	119	54.9	254	3.08	2.24	1.99
Treatment of seeds before sowing + treatment of crops in the branching phase	12.7	21.7	23	61.1	22.5	24	117	54.1	250	2.92	2.16	1.84

Table 6.
The structure of the harvest and its size depend on the method of application of the growth immunomodulator on leguminous crops.

impossible to treat the seeds with the drug at this time, it is most effective to apply it to vegetative plants in the "branching" phase once.

4. Conclusions

1. To obtain the yield of grasspea grain at the level of 3.0 t/ha with a high content of digestible protein within 30% and the most rational consumption of water per unit of the main product of 118 m^3, farms in the southern region of Ukraine on irrigated lands are recommended to use the following agrotechnical complex: apply mineral fertilizers at the rate $N_{60}P_{90}$, carry out the main tillage to a depth of 20–22 cm, sow at a soil temperature of 4–5°C at the seed sowing rate of 1.5 million/ha in the usual continuous method with a row spacing of 15 cm.

2. In order to obtain a yield of grasspea grain at the level of 3.5 t/ha, farms in the southern region of Ukraine on irrigated lands are recommended to use the following agrotechnical complex: sowing should be carried out after the predecessor—winter wheat per grain, mineral fertilizers should be applied at the rate of $N_{60}P_{90}$, as the main tillage, shelf plowing should be carried out to a depth of 20–22 cm, during the growing season of the culture, maintain the moisture of the active layer of the soil at the level of 70–75% of the lowest moisture content.

3. Cultivation of lentils in the conditions of the Southern Steppe of Ukraine is a highly profitable production. Conducted experimental studies under different conditions of moistening make it possible to recommend the following agro-technological complex: (1) under irrigation conditions—plowing to a depth of 20–22 cm, applying mineral fertilizers in a dose of $N_{45}P_{45}$, sowing with a seeding rate of 2.5 million/ha ensures a profitability level of 260–270% and the grain yield level—2.3-2.7 t/ha; (2) without irrigation—plowing to a depth of 20–22 cm, applying mineral fertilizers in a dose of $N_{45}P_{45}$, sowing with a seeding rate of 2.0 million/ha provides a profitability level of 200–215% and a grain yield level of 1.2–1.5 t/ha.

4. In the irrigated conditions of the Southern Steppe of Ukraine, to form highly productive chickpea crops, it is necessary to apply the following set of technological methods of cultivation: plow the soil to a depth of 20–22 cm, apply fertilizers at the rate of $N_{45}P_{45}$ and ensure the formation of plant density within the range of 1.0–1.5 million/ha. The specified complex makes it possible to form a harvest within the range of 3.0–3.5 t/ha. In non-irrigated conditions, it is recommended to reduce the plant stand density to 0.5–1.0 million/ha, which will ensure plant productivity at the level of 1.5–2.0 t/ha.

5. On the basis of experimental studies conducted to study the productivity of plants and the quality of the obtained grain on the dark chestnut soils of the Southern Steppe of Ukraine, when irrigated with water of the II class, we recommend farms that grow common beans (Preto) to apply mineral fertilizers at a dose of $N_{45}P_{45}$, to plow to a depth of 20–22 cm and perform sowing with a row spacing of 45 cm. These measures will ensure grain yield at the level of 2.5–3.1 t/ha.

6. The presented results show that regardless of the technical support of farms, financial resources of farmers, and climatic and soil conditions, modern technologies for the use of immunomodulators of plant growth are able to form a high yield of leguminous crops, thereby not only obtaining a valuable protein product but also improving the quality of the soil—increasing the amount of biological nitrogen, reduce the chemical load on the soil and stabilize and reduce the amount of salts in the soil solution. The use of new modern growth stimulants makes it possible to increase the germination and energy of seed germination; stimulate growth, accelerates the development of plants; stimulates root formation in plants; ensure the active formation of shoots; increase the yield of various crops by 10–25% or more; to improve the quality of the obtained products; the ripening period is shortened up to 2 weeks; the immunity of plants increases, increases resistance to adverse conditions of the abiotic and biotic environment; negatively affects the development and action of pathogenic microflora (fungal and bacterial diseases); reduces the consumption of fertilizers and pesticides by 20–30%.

Conflict of interest

The authors declare no conflict of interest.

Author details

Sergiy Lavrenko* and Nataliia Lavrenko
Kherson State Agrarian and Economic University, Kherson, Ukraine

*Address all correspondence to: lavrenko.sr@gmail.com

References

[1] Isaev AP, Platonov AM. Make the most of the advantages of legumes. Agriculture. 1996;**5**:15-17

[2] Zbarskyi VK, Matsybora VI, Chalyy AA. Economics of Agriculture. Kyiv: Publishing House "Caravela"; 2009. 264 p

[3] Hay GK, Eliott. Tangier pea. Journal of the Department of Agriculture of Western Australia. 1939;**16**(2):165-170

[4] Songin H. Mieszanki pastewne w poplonach scierniskowych. Nowe Rolnictwo. 1967;**14**:31-33

[5] Lipes VI. New leguminous crops for the arid steppe of the Ukrainian SSR. Agriculture of Ukraine. K.-Kharkiv. 1946;**6**:28-31

[6] Zohary D, Hopf M. Domestication of Plants in the Old World. New York: Oxford University Press; 2000. p. 10

[7] Shevchenko AM, Shevchenko IA. Lentils - a Valuable Food Crop. Luhansk: LLC "Knowledge"; 2003: 27 p

[8] Vaillancourt R. Seed coat darkening and the inheritance of tannin content in lentil [thesis]. Saskatoon (Canada): University of Saskatchewan; 1984. pp. 126-139

[9] Harun-or Rashid M, Gonzalez J, Peter J, Young W, Wink M. Rhizobium leguminosarum is the symbiotic of lentils in the Middle East and Europe but not in Bangladesh. FEMS Microbiology Ecology. 2014;**87**(1):64-77

[10] Saskatchewan pulse growers. Lentil Production Manual. Saskatoon: Saskatchewan pulse growers; 2011: 60 p

[11] Hoque M, Farlul HH. Rhizobial inoculation and fertilization of lentil in Bangladesh. Lens News Fett. 1994;**21**(2):29-30

[12] Sichkar VI. The role of legumes in solving the protein problem in Ukraine. Fodder and Fodder Production. 2004;**53**:110-115

[13] Tolkachev MZ. Optimizing the symbiosis of leguminous plants and nodule bacteria in modern fodder production. Fodder and Production. 2003;**51**:133-136

[14] Sichkar VI. The status and prospects of the breeding of leguminous crops at the breeding and genetics Institute of the Ukrainian Academy of sciences. Collection of science proceedings of the breeding and genetic institute - National Center for seed science and varietal research. Odesa. 2002;**3**(43):92-103

[15] Sichkar VI, Bushulyan OV. Botanical Characteristics, Biological Features, Agricultural Technology and New Varieties Chickpeas. Odesa: SGI-NAS NAIS; 2007: 24 p

[16] Sichkar V, Bushulyan O. Growing technology chickpeas in Ukraine. PRO. 2001;**10**:42-43

[17] Didovych SV. Increasing chickpea productivity by nitrification of seeds. Scientific works of the Krymsky state agrotechnological university. Simferopol. 2005;**91**:25-31

[18] Golovan LV. Peculiarities of the use of different types of marker systems in breeding studies of the *Genus phaseolus* L [thesis]. Kharkiv: Institute of plant breeding named after V.Ya. Yuryeva; 2012: 27 p

[19] Hryshchenko OM. Biological features and selection value of variety samples of green beans for the conditions of the northern forest-steppe of Ukraine [thesis]. Kyiv: National University of Bioresources and Nature Management of Ukraine; 2015: 23 p

[20] Zengin M. Fasulyenin gübrelenmesi [Internet]. Available from: www.kutahyaazot.com/…/Fasulyenin_Gubrelenmesi.pdf [Accessed: January 20, 2024]

[21] Biological and Botanical Features of Growing Common Beans in the Druzhba Arboretum [Internet]. Mode of Access to Materials. Available from: http://www.br.com.ua/diplom/Biology/40696-8.html [Accessed: January 15, 2024]

[22] Beans. [Internet]. Available from: http://ldni.sumy.ua/agroscience/poperednyky-obrobitok-gruntu-ta-udobrenn [Accessed: January 20, 2024]

[23] Preto Black Beans. [Internet]. Available from: http://fasol.tv/fasolepedia/detail.php?ID=13 [Accessed: January 15, 2024]

[24] Sadova I. Innovative Soybeans and Beans Handmade [Internet]. Agribusiness Today. Kiev: LLC "Agrarian Vidavnitstvo"; 2015. Available from: http://agro-business.com.ua/2017-09-29-05-56-43/item/2273-novatorska-soia-i-kvasolia-handmade.html [Accessed: January 15, 2024]

[25] Graham PH, Ranalli P. Common bean (*Phaseolus vulgaris L.*). Field Crops Research. 1997;**53**(1-3):131-146

[26] Hegde DM, Dwivedi BS. Integrated nutrient supply and management as a strategy to meet nutrient demand. Fertilizer News. 1993;**38**(12):49-59

[27] Babych AO. World Land, Food and Fodder Resources. Kyiv: Agrarna nauka; 1996. pp. 147-271

[28] Biological Features of Beans [Internet]. Mode of Access to Materials. Available from: http://ultraagro.blogspot.com/2014/09/blog-post_949.html [Accessed: January 20, 2024]

[29] Likhochvor VV. Practical Advice on Growing Grain and Leguminous Crops in the Conditions of Western Ukraine. Lviv: Ukrainian Technologies; 2001: 128 p

[30] Shlyakhturov DS. The Yield of Common Beans Depends on the Cultivation Technology and Weather Conditions. Vol. 3-4. Kyiv: Collection of scientific works of the National Scientific Center "Institute of Agriculture of the Ukrainian Academy of Sciences". VD "EKMO"; 2008. pp. 85-89

[31] Turuko M, Mohammed A. Effect of different phosphorus fertilizer rates on growth, dry matter yield and yield components of common bean (*Phaseolus vulgaris L.*). World Journal of Agricultural Research. 2014;**2**(3):88-92

[32] Kolotilov VV, Kobizeva LN, Silenko SI. The collection of beans is a source of valuable economic traits for promising areas of breeding. Genetic Resources of Plants. 2006;**3**:61-67

[33] Ushkarenko VO, Lazer PN, Minkin MV, Lavrenko SO. Development of elements of growing grasspea under irrigation in the conditions of southern Ukraine. Tavriysk Scientific Bulletin: Collection of Scientific Works. 2004;**30**:8-14

[34] Bazaluk O, Havrysh V, Nitsenko V, Mazur Y, Lavrenko S. Low-cost smart farm irrigation systems in Kherson Province: Feasibility study. Agronomy. 2022;**12**(5):1013. DOI: 10.3390/agronomy12051013

[35] Lavrenko SO, Maksimov MV. The influence of technological methods

for growing lentils on photosynthetic activity under various moisture conditions. Environmental Management: Scientific and Practical Journal. 2016;**4**:80-85. Available from: https://www.webofscience.com/wos/alldb/full-record/RSCI:27261845

[36] Lavrenko S, Lavrenko N, Kazanok O, Karashchuk G, Kozychar M, Ye P, et al. Chickpea yields and water use efficiency depending on cultivation technology elements and irrigation. AgroLife Scientific Journal. 2019;**8**(2):59-64. Available from: http://agrolifejournal.usamv.ro/index.php/scientific-papers/452-chickpea-yields-and-water-use-efficiency-depending-on-cultivation-technology-elements-and-irrigation-452

[37] Lavrenko SO, Lavrenko NM, Lykhovyd PV. Effect of degree of salinity on seed germination and initial growth of chickpea (*Cicer arietinum*). Biosystems Diversity. 2019;**27**(2):101-105. DOI: 10.15421/011914

[38] Lavrenko N, Lavrenko S, Revto O, Lykhovyd P. Effect of tillage and humidification conditions on desalination properties of chickpea (*Cicer arietinum L.*). Journal of Ecological Engineering. 2018;**19**(5):70-75. DOI: 10.12911/22998993/91265

[39] Lavrenko S, Lykhovyd P, Lavrenko N, Ushkarenko V, Maksymov M. Beans (*Phaseolus vulgaris L.*) yields forecast using normalized difference vegetation index. International Journal of Agricultural Technology. 2022;**18**(3):1033-1044

[40] Lavrenko SO, Lavrenko NM, Maksymov DO, Maksymov MV, Didenko NO, Islam KR. Variable tillage depth and chemical fertilization impact on irrigated common beans and soil physical properties. Soil and Tillage Research. 2021;**212**:105024. Available from: https://www.sciencedirect.com/science/article/pii/S0167198721000945

[41] Ushkarenko VO, Lavrenko SO, Lykhovyd PV, Lavrenko NM, Maksymov DO. Yield components of haricot beans (*Phaseolus vulgaris L.*) depending on cultivation technology elements at the irrigated lands of the steppe zone. Modern Phytomorphology. 2018;**12**:73-79. DOI: 10.5281/zenodo.1295697

[42] Lavrenko SO, Lavrenko NN, Ambrosov SS. Method of Obtaining a Multi-Purpose Immunoregulator of Plant Growth (MIR). Ukraine: applicant and patent holder of Kherson State Agrarian University; 2013, Bull. No. 22: Utility model patent No. 85436, 4 p

[43] Lavrenko SO, Lavrenko NM. Method of Using a Multi-Purpose Immunoregulator of Plant Growth (MIR) on Leguminous Crops. Ukraine: applicant and patent holder of Kherson State Agrarian University; 2014, Bull. No. 3: 4 p, Utility model patent No. 87311

[44] Lavrenko SO, Lavrenko NM, Pichura VI. Method of Using an Organic Growth Stimulator Containing a Synthetic Phytohormone - Polyglycoside with Acetylated L-Isomeric Functional Groups; Heterocyclic Nitrogen-Containing Compound with Reversible Protonation Function; Heterocyclic Coupling with Cationic Functional Groups, Capable of Changing the ξ-Potential of Cell Membranes and the Complex of Microelements (Cu, Fe, Zn, Mn, Mo, Mg, Co) in Chelated Form, on Common Bean Crops. Ukraine: Kherson State Agrarian and Economic University; 2023, Bul. #49: 4 p, Patent for a utility model according to application No. 154691 on the topic

Chapter 15

Unraveling the Dynamics of Electromagnetic Fields in Legume Crop Cultivation

Tania María Guzmán Armenteros

Abstract

This chapter explores how magnetic fields can revolutionize plant productivity, particularly in legumes, by improving germination, growth, and resilience. It demonstrates how controlled exposure to magnetic fields increases the robustness and vitality of plants and examines how these fields protect legumes from environmental stresses, such as drought and soil salinity, by triggering beneficial molecular and physiological changes. Furthermore, it highlights the potential of magnetic fields to improve the nutritional quality of crops and the efficiency of photosynthesis, advocating their use as key tools to promote organic farming. This comprehensive analysis underscores the transformative potential of magnetic fields in promoting a productive and sustainable agricultural future.

Keywords: magnetic field, legumes, seeds, growth stimulation, grains

1. Introduction

The intersection between agriculture and science has yielded revolutionary discoveries that challenge conventional limits of vegetal productivity. Within this framework, the influence of magnetic fields on the growth and development of legumes has emerged as a dynamic and promising field of study. This chapter delves into the fascinating potential of magnetic fields to enhance the dynamism of legume growth, presenting notable advancements in understanding how these fields can act as crucial allies in promoting the development of resilient crops.

The first section of this chapter, "Enhancing Dynamics of Legume Growth," highlights how magnetic fields have been shown to influence the growth dynamics of legumes. It explores recent studies revealing how controlled exposure to magnetic fields can catalyze key germination and growth processes, boosting the robustness and vitality of these plants. These investigations offer profound insights into how magnetic fields could transform agricultural landscapes by optimizing legume productivity.

The second section of the chapter, "Magnetic Fields as Allies Against Abiotic Stress Challenges," discusses the relationship between magnetic fields and legumes' capacity to face abiotic stress challenges. From drought to soil salinity, legumes are consistently subjected to environmental pressures. This section illustrates how exposure to magnetic fields can serve as a protective shield, strengthening the resilience of legumes against these adverse conditions. The studies presented provide a glimpse into the transformative potential of this technology in mitigating stress effects on crops.

The third section of the chapter, "Unraveling Mechanisms: Magnetic Fields and Their Molecular Effects," delves into studies unraveling the molecular and physiological mechanisms at play when seeds and seedlings are exposed to specific magnetic fields. It discusses hypotheses on how these magnetic forces might influence the structure and function of key molecules within plant cells, such as proteins and nucleic acids, driving changes in metabolic and developmental processes.

The final section, "Towards Sustainable Agriculture with Transformed Legumes," presents a futuristic and optimistic view, highlighting the potential impact of magnetic fields on sustainable agriculture. From enhancing crop nutritional quality to optimizing photosynthesis efficiency, it examines how integrating magnetic fields into legume production could lay the foundation for a more sustainable agricultural future. These advancements promise to revolutionize our interaction with the land and its resources to feed a constantly growing global population.

Collectively, this chapter provides a deep dive into the promising convergence between magnetobiology and legume agriculture. By exploring the latest scientific advancements, it aims to inspire researchers, farmers, and decision-makers to recognize the vast potential of magnetic fields as key tools in the agricultural revolution toward a sustainable and productive future.

2. Enhancing dynamics of legume growth

Over the years, scientific literature has progressively explored the impact of magnetic fields on seed germination and early plant growth. In the late 1990s, research identified optimal magnetic field intensities to improve germination of various seeds, including legumes [1]. In 2000, studies were expanded to examine the effects of electric and magnetic fields on seed germination, although they did not focus specifically on legumes [2]. From that date, research began to refine the understanding of the effects of the magnetic field, identifying specific intensities that significantly improved germination and growth rates in legume seeds such as soybeans and beans [3–5]. Subsequently, from 2017 [6, 7], there was a notable increase in detailed research, which revealed how controlled exposure to magnetic fields can improve not only germination but also stress resistance, biochemical activity, and the efficiency of photosynthesis in legumes (**Table 1**).

Recent studies have continued this trend, optimizing magnetic field parameters to further boost germination rates and early growth characteristics under various environmental conditions. This body of work demonstrates a clear trajectory of increasing sophistication and specificity in the application of magnetic fields to improve legume seed germination and early plant development.

Seeds of broad beans (*Vicia faba*) exposed to 100 μT magnetic fields have revealed noteworthy findings concerning mitotic activity and DNA replication in resulting seedlings. Exposure of these seeds to low-intensity magnetic fields resulted in an increased mitotic index in the seedlings [8]. This phenomenon suggests an acceleration in the cell cycle, which could possibly be associated with changes in the regulation of growth and cell division cycles induced by magnetic field exposure [9, 10].

The accelerated DNA replication suggests a potential activation of cellular pathways related to cell division and genetic material synthesis [9]. Magnetic fields might influence the cellular machinery responsible for DNA replication, leading to

Legume and seeds	Optimal magnetic field intensity	Observed effect	Authors
Soybean, cotton, and wheat seed	0.10 T optimal with exposure time varying from 13 to 27 minutes	The three crops saw yield increases of roughly 46%, 32%, and 35%	Phirke et al. [1]
Tomato seeds	3–1000 G 30–45 s optimal	Increases germination rates	Moon and Chung [2]
Wheat and bean seedlings	7 mT for 7 days	Germination and growth improved	Cakmak et al. [3]
Soybean seed	200 and 150 mT at 60 minutes optimal	The improvements over the untreated control: 5–42% in germination speed, 4–73% in seedling length, 9–53% in fresh weight, 5–16% in dry weight, and 3–88% and 4–27% in vigor	Shine et al. [4]
Gourd seeds	25, 50, at 30, and 45 minutes 75 mT at 15 and 30 minutes optimal	Accelerates germination parameters	Iqbal et al. [5]
Soybean and maize seeds	200 mT at 1 hour optimal	Increases the germination percentage and early growth parameters of seedlings under different salinity levels	Kataria et al. [6]
Peas and beans	0.065 T optimal with a fourfold reversal and a conveyor movement speed of 0.4 meters per second	Increase in the yield of peas by 42% and beans by 23%	Kozyrskiy et al. [7]

Table 1.
Optimal magnetic field intensities for enhancing seed germination and plant growth.

an acceleration of the process [11, 12]. However, the precise molecular mechanisms underlying this influence remain subjects of ongoing research and detailed analysis.

The implications of accelerated cell division could have significant implications for plant growth and development. A higher mitotic index and faster DNA replication could lead to an increase in daughter cell production and, thus faster growth of vegetal tissues and organs [13]. This could translate into more vigorous seedling growth, potentially facilitating better adaptation to challenging environments or accelerating their life cycle in favorable conditions.

However, it is crucial to consider that while accelerated growth may be beneficial in terms of adaptability and initial performance, it could also lead to negative consequences, such as reduced genomic stability or a decrease in the quality of produced cells [14]. Understanding the molecular mechanisms behind these effects is therefore fundamental to fully assessing the long-term impacts of magnetic field exposure on plant development.

Similar results of improved germination and plant growth are obtained by exposing chickpea (*Cicer arietinum*) seeds to a static magnetic field of 150 mT for a period of 2 hours [15]. Particularly in other studies, the effects of exposure have revealed highly significant and promising results regarding the activity of the enzyme α-amylase in these seeds [16].

α-Amylase is a key enzyme involved in the seed germination process, playing a fundamental role in the hydrolysis of starch, the main energy reserve in seeds, into simple sugars [13]. This process is essential to provide the necessary energy for the initial growth of the seedling until it can perform photosynthesis on its own [13, 16]. The discovery of a highly significant increase in α-amylase activity as a result of exposure to static magnetic fields at a specific intensity of 150 mT for a period of 2 hours raises intriguing questions about the underlying molecular and physiological mechanisms related to this response [16].

One possible explanation for this phenomenon could be related to the influence of magnetic fields on biochemical processes and the structure of molecules within the seeds. Magnetic fields can interact with protein structures and cell membranes, subtly modifying their conformation or physicochemical properties [17, 18]. These changes could have a direct impact on enzymes such as α-amylase, altering their catalytic activity or their ability to bind to substrates, resulting in an increased reaction rate and thus measured enzymatic activity.

It is also important to consider the possibility of cumulative or secondary effects derived from exposure to magnetic fields. While the main focus may be on the enzyme α-amylase, it is likely that other cellular components and metabolic processes are simultaneously or indirectly affected. In this regard, it is crucial to consider the inherent variability and complexity of biological systems.

Exposure of soybean seeds to magnetic fields has also emerged as a novel approach in optimizing the nutritional quality and physiological efficiency of these seeds [1, 4, 6, 19]. Recent studies have specifically evaluated the effects of exposure to a magnetic field of 30 mT for a period of 60 seconds on soybean seeds and have revealed promising results suggesting a significant improvement in various key aspects related to plant health and growth [20].

Soybean seeds are exposed to the magnetic field compared to unexposed seeds. Essential components such as proteins, lipids, carbohydrates, vitamins, and minerals have shown quantifiable increases in seeds subjected to the magnetic treatment [20–22]. This improvement in nutritional composition could have significant implications for the quality of products derived from soy, promoting its utility in human and animal nutrition.

Enzymatic activity was also influenced by exposure to the magnetic field. In particular, a notable increase was observed in the activity of several key enzymes related to nutrient synthesis and metabolism in magnetically treated soybean seeds [21, 22]. Enzymes such as amylase, responsible for starch hydrolysis into simple sugars, and lipase, involved in lipid degradation, showed significant increases in activity in seeds subjected to the magnetic field [22]. This change in enzymatic activity could be linked to the improvement in nutrient availability and absorption by plants, thus contributing to their optimal growth and development [6, 23].

Exposure to the magnetic field also resulted in a measurable increase in chlorophyll content in soybean seeds [23]. Chlorophyll plays a crucial role in photosynthesis by capturing light energy and converting it into chemical energy, and its increase suggests a possible enhancement in the photosynthetic efficiency of developing plants [1–7, 23]. This finding could have significant implications for biomass production and the ability of plants to synthesize nutrients through photosynthesis, ultimately leading to improved agricultural yield.

In another study, exposure of these seeds to a pulsating magnetic field led to a notable increase in the biomass of emerging seedlings. An increase in the photosynthetic efficiency of seedlings was also demonstrated [20–22]. This suggests a possible influence on chloroplast activity and light capture rate, which could contribute to increased biomass yield and improved productivity. This improvement in initial growth could be attributed to the acceleration of physiological processes such as protein synthesis and the activation of key enzymes involved in metabolism, fundamental aspects for early plant development [1–7, 18].

Since the germination process is essential for the initial development of plants, an increase in the germination rate indicates greater efficiency in metabolic activation and in breaking seed dormancy. This phenomenon may be related to the influence of magnetic fields on the permeability of cell membranes and the regulation of ionic flows, which facilitate the absorption of essential nutrients and the water necessary for germination.

Although research on this topic abounds, analyzing the effect of magnetic fields on plants is not a simple task; researchers must begin with an exhaustive review of the literature to identify gaps and define the objectives of the study. Next, they design the experiment by selecting plant species, determining the intensity and duration of the magnetic fields, and defining control conditions. They prepare the plant material by obtaining uniform seeds and guaranteeing consistent growing conditions. The seeds and seedlings are then exposed to magnetic fields calibrated according to the experimental design. Researchers monitor and collect data on germination, growth, and physiological conditions, and evaluate different response variables in plants, such as resistance to abiotic stress. Molecular and biochemical analyses are also performed to examine changes in gene expression, protein activity, and metabolite levels. The data are analyzed statistically to identify significant trends, after which the findings are documented practical applications in agriculture are proposed, and future research directions are suggested to further explore the findings (**Figure 1**).

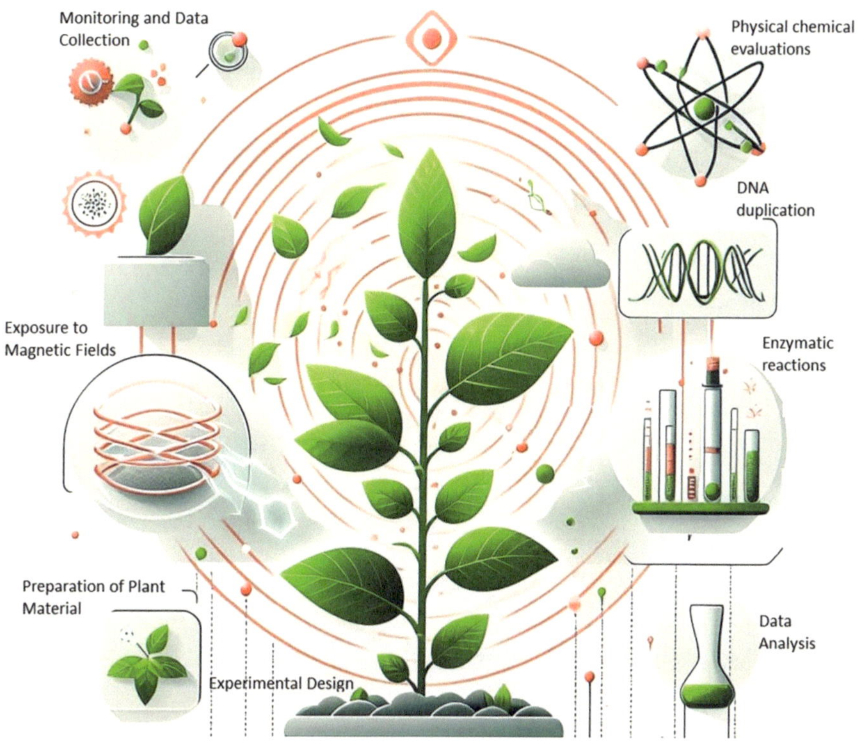

Figure 1.
Methodological steps for investigating the impact of magnetic fields on plant growth.

3. Magnetic fields as allies against abiotic stress challenges

Exposing *Phaseolus vulgaris* seeds to stationary electromagnetic fields before sowing has emerged as a potential technique to mitigate the negative impacts of drought on plant growth and development [24]. Specifically, applying electromagnetic fields of specific intensities, such as 100 and 150 mT for a short period of 10 minutes before sowing, has shown beneficial effects by enhancing several key physiological parameters in plants subjected to water stress conditions [24, 25].

One of the most notable effects has been the improvement in photosynthesis, a fundamental process for energy production in plants. Pre-sowing exposure to these electromagnetic fields has shown a positive influence on the photosynthetic efficiency of plants under drought stress [25]. This phenomenon suggests a greater capacity for capturing and utilizing solar light, resulting in an increase in the production of carbohydrates and energy compounds necessary for growth and development (**Figure 2**).

A significant impact has been observed on the transpiration rate and stomatal conductance of plants exposed to these electromagnetic fields before sowing [25]. Transpiration is a crucial process in the plant life cycle as it regulates water loss through evaporation from leaf stomata [24]. Under drought conditions, plants tend to reduce transpiration to conserve water, which can negatively affect their metabolic

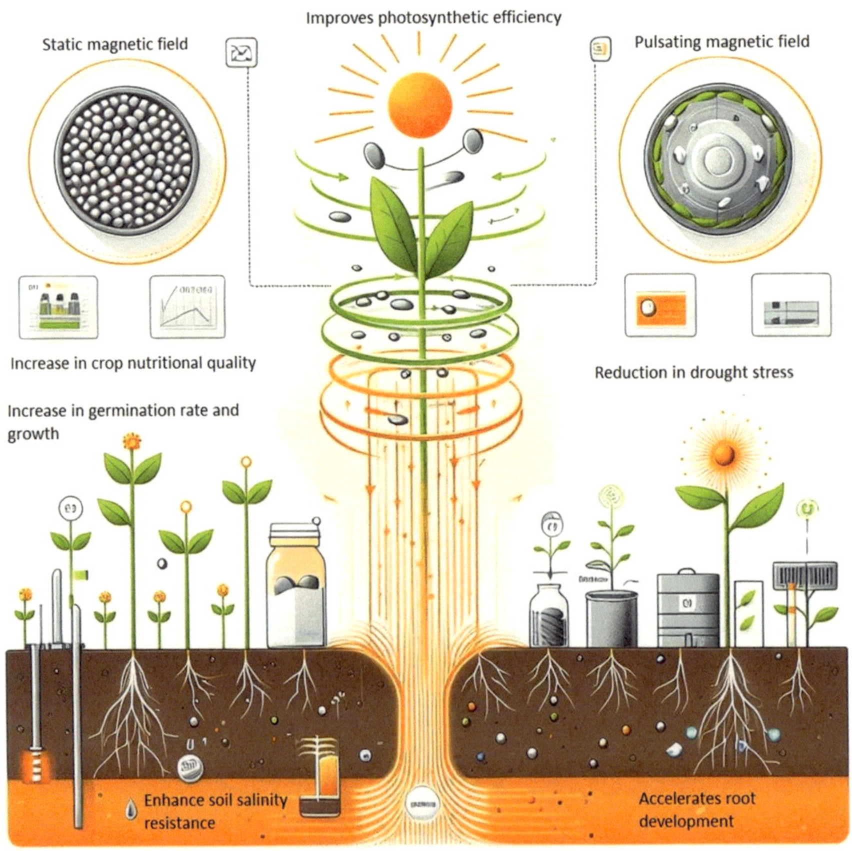

Figure 2.
Photosynthesis and stress resistance in plants under exposure to magnetic fields.

rate and gas exchange. However, exposure to these electromagnetic fields has shown the ability to maintain optimal levels of transpiration and stomatal conductance even under water stress conditions [24, 25].

This phenomenon suggests efficient regulation of stomatal opening and closure, enabling better plant adaptation to drought conditions. Adequate stomatal conductance facilitates gas exchange, such as the uptake of carbon dioxide (CO_2) needed for photosynthesis and the release of oxygen, contributing to the necessary gas balance for plant life [26, 27]. The underlying mechanisms behind these beneficial effects are not fully understood, but several hypotheses are postulated on how electromagnetic fields may interact with plant physiological systems. It is suggested that these fields could influence the activity of certain receptor proteins or ion channels in cell membranes, triggering specific responses that enhance the plant's ability to withstand water stress [24].

Exposure of soybean (*Glycine max*) seeds to pulsating magnetic fields has emerged as a promising area of research in improving germination and crop yield [28]. A recent study applied a pulsating magnetic field of 200 mT for 1 hour to soybean seeds before sowing, resulting in significant effects on multiple aspects of plant growth and productivity, particularly under saline stress conditions [29].

Soybean plants exposed to the pulsating magnetic field before sowing exhibited greater resistance to saline stress compared to nonexposed plants. This finding is significant, as saline stress is a global concern limiting agricultural production [30]. The improved ability of plants to mitigate the adverse effects of saline stress after exposure to the pulsating magnetic field indicates a possible adaptation or modulation of physiological responses that counteract the harmful effects of salinity on plants [31].

4. Unraveling mechanisms: Magnetic fields and their molecular effects

The effects of magnetic fields may depend on multiple factors, such as seed age, the specific phase of the plant's life cycle during which they are applied, as well as environmental and genetic conditions. The variability in responses to magnetic fields among different plant species and varieties underscores the need for a more detailed understanding of the specific interactions between these fields and biological processes [1–7, 11–13]. It is crucial to highlight that the duration and intensity of the magnetic treatment are variables that could influence the magnitude and nature of the observed effects.

The exact mechanisms underlying these observed effects are not yet fully understood. However, it has been speculated that exposure to magnetic fields may influence biophysical and biochemical processes at the cellular and molecular levels [6, 22]. The interaction between the magnetic field and biomolecules presents in seeds, such as proteins and nucleic acids, could trigger changes in their structure and conformation, potentially affecting their functionality and metabolic activities [11–15].

It has also been suggested that magnetic fields may affect ion flows across cell membranes. This could influence the ionic balance within seed cells, triggering physiological responses that eventually enhance enzymatic activity. Another possible explanation involves the stimulation of specific metabolic processes mediated by cellular signaling pathways. It has been postulated that magnetic fields might activate signaling pathways that, in turn, trigger gene expression and the synthesis of proteins related to germination, including those encoding enzymes (such as α-amylase) [11–13, 22].

The influence of electromagnetic fields could also relate to the organization and structure of water molecules within plant cells. In this regard, it could affect water availability and its movement within the plant [25, 32]. This modulation in water structure could potentially facilitate a more efficient distribution of moisture and nutrients, thereby contributing to the plant's resistance to drought.

At the molecular and cellular level, exposure to pulsating magnetic fields is presumed to trigger changes in gene expression and enzymatic activity, which may influence cellular homeostasis and stress adaptation [4, 16, 33]. The regulation of genes related to antioxidant defense, osmoprotectant solute accumulation, and modulation of signaling pathways could be involved in enhancing salt stress tolerance [11, 22, 34].

This phenomenon is not limited to legumes, as other species such as rice and maize also exhibit similar responses, indicating a generality in the influence of these fields on seeds [13, 15, 17, 31, 32]. For a more comprehensive understanding of this phenomenon, multidisciplinary approaches integrating molecular biology, biophysics, and plant physiology may be necessary.

Overall, the discoveries offer a fascinating insight into the effects of magnetic fields on fundamental cellular processes and their potential impact on plant development. However, it is clear that further research is needed to fully comprehend the involved mechanisms and their practical applications. Future studies could involve advanced molecular analysis techniques, such as spectroscopy, high-resolution microscopy, and more detailed enzymatic assays, to unravel specific biochemical changes induced by exposure to magnetic fields.

5. Toward sustainable agriculture with transformed legumes

The transformative potential of magnetic fields in legume agriculture lies not only in their observed effects but also in their practical applicability and sustainability. This promising technology offers an innovative perspective to enhance agricultural productivity and global food security. One of the most intriguing aspects of this technology is its ability to improve agricultural productivity. The controlled exposure of legume seeds to specific magnetic fields can trigger a series of beneficial effects on their growth and development. From accelerating germination to increasing enzymatic activity and improving seed nutritional quality, these effects have the potential to significantly boost crop productivity.

The prospect of increasing legume production with this technology is particularly relevant in the context of global food security. Legumes play a crucial role in human nutrition, being a valuable source of proteins, nutrients, and energy. Enhancing the productivity and quality of these crops through the application of magnetic fields could significantly contribute to mitigating food insecurity, especially in regions where these plants are essential in the local diet. Moreover, sustainability is a cornerstone of modern agriculture. Magnetic fields present an additional appeal by offering an alternative that might be more sustainable than some conventional crop improvement methods. As a noninvasive and nonchemical technique, it could reduce the dependency on fertilizers and agrochemicals, thus minimizing environmental impact and preserving soil health.

Agricultural sustainability is closely linked to the ability to conserve natural resources and maintain biodiversity. The application of magnetic fields in legume agriculture has the potential to promote more environmentally friendly agricultural practices by reducing the need for chemical inputs. This could contribute to conserving soil quality and preserving biological diversity in agricultural ecosystems.

The positive impact of magnetic fields in legume agriculture also relates to optimizing resources such as water. More vigorous plant growth facilitated by exposure to these fields could mean greater efficiency in water absorption from the soil. This is especially relevant in regions affected by water scarcity, where optimizing water use is crucial for maintaining agricultural production.

Additionally, the application of magnetic fields could help reduce crop losses due to adverse environmental conditions. By enhancing the resilience of legumes to factors like drought, high temperatures, or soil salinity, this technology could contribute to greater stability in agricultural production, thus reducing farmers' vulnerability to extreme weather events.

It is crucial to note that despite the promising effects observed in studies, a deeper understanding of the underlying mechanisms and comprehensive validation under real agricultural conditions is still required. Ongoing research is essential to establish optimal practices, identify effective doses, and fully comprehend the long-term impact of exposure to magnetic fields on crops and the environment.

In summary, the transformative potential of magnetic fields in legume agriculture extends beyond their observed effects and encompasses their practical applicability and contribution to agricultural sustainability. This technology offers a glimpse into a future where enhanced productivity, food security, and sustainability converge, providing innovative solutions to address global agricultural challenges.

6. Conclusions

The transformative potential of magnetic fields in legume agriculture extends to crucial realms:

Enhanced agricultural productivity: Magnetic fields exhibit the ability to positively influence the growth and development of legumes. Studies show their impact on accelerating germination, boosting enzymatic activity, and improving seed quality, potentially elevating overall crop productivity.

Global food security: By enhancing the yield and quality of legume crops, magnetic field applications hold promise in addressing global food insecurity. Legumes are vital in human nutrition, and augmenting their production through this technology could contribute significantly to food availability and nutrition.

Agricultural sustainability: Magnetic fields offer an eco-friendly alternative by potentially reducing dependency on chemical fertilizers and promoting environmentally conscious agricultural practices. This could aid in conserving soil quality and biodiversity while optimizing resource use, such as water.

Resilience to environmental stress: Legumes exposed to magnetic fields exhibit increased resilience to environmental stressors like drought or high salinity. This resilience could contribute to more stable agricultural production, reducing vulnerability to adverse weather conditions.

Despite these promising observations, a comprehensive understanding of underlying mechanisms and long-term impacts remains an ongoing challenge. Further research is crucial to validate optimal practices and understand the broader implications of magnetic field exposure on crops and ecosystems.

In essence, the transformative potential of magnetic fields in legume agriculture lies not only in observed effects but also in their capacity to offer practical, sustainable, and resilient solutions to address key global agricultural challenges. This technology presents an innovative pathway toward enhancing productivity, ensuring food security, and promoting sustainable agricultural practices.

Acknowledgements

Gratitude to the Litoral Polytechnic High School for their valuable collaboration in the preparation of the book chapter. Your contributions have been greatly appreciated and instrumental in enhancing the quality of the content. The author acknowledges the use of ChatGPT for language polishing and the use of DALL-E for figure creation in the manuscript.

Author details

Tania María Guzmán Armenteros
Litoral Polytechnic High School, Guayaquil, Ecuador

*Address all correspondence to: tamaguzm@espol.edu.ec

References

[1] Phirke PS, Patil MN, Umbarkar SP, Dudhe YH. The application of magnetic treatment to seeds: Methods and responses. Seed Science and Technology. 1996;**24**(2):365-373

[2] Moon JD, Chung HS. Acceleration of germination of tomato seed by applying AC electric and magnetic fields. Journal of Electrostatics. 2000;**48**(2):103-114. DOI: 10.1016/s0304-3886(99)00054-6

[3] Cakmak T, Dumlupinar R, Erdal S. Acceleration of germination and early growth of wheat and bean seedlings grown under various magnetic field and osmotic conditions. Bioelectromagnetics: Journal of the Bioelectromagnetics Society, The Society for Physical Regulation in Biology and Medicine, The European Bioelectromagnetics Association. 2010;**31**(2):120-129. DOI: 10.1002/bem.20537

[4] Shine MB, Guruprasad KN, Anand A. Enhancement of germination, growth, and photosynthesis in soybean by pre-treatment of seeds with magnetic field. Bioelectromagnetics. 2011;**32**(6):474-484. DOI: 10.1002/bem.20656

[5] Iqbal M, ul Haq Z, Malik A, Ayoub CM, Jamil Y, Nisar J. Pre-sowing seed magnetic field stimulation: A good option to enhance bitter gourd germination, seedling growth and yield characteristics. Biocatalysis and Agricultural Biotechnology. 2016;**5**:30-37. DOI: 10.1016/j.bcab.2015.12.002

[6] Kataria S, Baghel L, Guruprasad KN. Pre-treatment of seeds with static magnetic field improves germination and early growth characteristics under salt stress in maize and soybean. Biocatalysis and Agricultural Biotechnology. 2017;**10**:83-90. DOI: 10.1016/j.bcab.2017.02.010

[7] Kozyrskiy VV, Savchenko VV, Sinyavskiy AY. Pre-sowing treatment of leguminous crop seeds with a magnetic field. Agricultural Machinery and Technologies. 2019;**13**(1):21-26. DOI: 10.22314/2073-7599-2018-13-1-21-26

[8] Rajendra P, Sujatha Nayak H, Sashidhar RB, Subramanyam C, Devendranath D, Gunasekaran B, et al. Effects of power frequency electromagnetic fields on growth of germinating *Vicia faba* L., the broad bean. Electromagnetic Biology and Medicine;**24**(1):39-54. DOI: 10.1081/JBC-200055058

[9] Soultanas P, Janniere L. The metabolic control of DNA replication: Mechanism and function. Open Biology. 2023;**13**(8):230220. DOI: 10.1098/rsob.230220

[10] Saxena S, Zou L. Hallmarks of DNA replication stress. Molecular Cell. 2022;**82**(12):2298-2314. DOI: 10.1016/j.molcel.2022.05.004

[11] Stovbun SV, Zlenko DV, Bukhvostov AA, Vedenkin AS, Skoblin AA, Kuznetsov DA, et al. Magnetic field and nuclear spin influence on the DNA synthesis rate. Scientific Reports. 2023;**13**(1):465

[12] Rojeab AY. Magnetic properties govern the processes of DNA replication and the shortening of the telomere. International Journal of Biotechnology and Bioengineering. 2013;7(7):655-660

[13] Sablowski R, Carnier Dornelas M. Interplay between cell growth and cell cycle in plants. Journal of Experimental Botany. 2014;**65**(10):2703-2714. DOI: 10.1093/jxb/ert354

[14] Vashisth A, Nagarajan S. Exposure of seeds to static magnetic field enhances germination and early growth characteristics in chickpea (*Cicer arietinum* L.). Bioelectromagnetics. 2008;**29**:571-578

[15] Vashisth A, Nagarajan S. Effect on germination and early growth characteristics in sunflower (*Helianthus annuus*) seeds exposed to static magnetic field. Journal of Plant Physiology. 2010;**167**(2):149-156. DOI: 10.1016/j.jplph.2009.08.011

[16] Luo X, Li D, Tao Y, Wang P, Yang R, Han Y. Effect of static magnetic field treatment on the germination of brown rice: Changes in α-amylase activity and structural and functional properties in starch. Food Chemistry. 2022;**383**:132392. DOI: 10.1016/j.foodchem.2022.132392

[17] Maffei ME. Magnetic field effects on plant growth, development, and evolution. Frontiers in Plant Science. 2014;**5**:445. DOI: 10.3389/fpls.2014.00445

[18] Dziergowska K, Lewandowska S, Mech R, Pol M, Detyna J, Michalak I. Soybean germination response to algae extract and a static magnetic field treatment. Applied Sciences. 2021;**11**(18):8597. DOI: 10.3390/app11188597

[19] Kataria S, Jain M, Rastogi A, Brestic M. Static magnetic field treatment enhanced photosynthetic performance in soybean under supplemental ultraviolet-B radiation. Photosynthesis Research. 2021;**150**:263-278. DOI: 10.1007/s11120-021-00850-2

[20] Shaoyi JIA, Yong LIU, Songhai WU, Zhibin WANG. Effect of static magnetic field on α-amylase activity and enzymatic reaction. Transactions of Tianjin University. 2009;**15**:272-275. DOI: 10.1007/s12209-009-0048-8

[21] Michalak I, Lewandowska S, Niemczyk K, Detyna J, Bujak H, Arik P, et al. Germination of soybean seeds exposed to the static/alternating magnetic field and algal extract. Engineering in Life Sciences. 2019;**19**(12):986-999. DOI: 10.1002/elsc.201900039

[22] Kishore G, Singh RK, Saxena CK, Al-Ansari N, Vishwakarma DK, Heddam S. Magnetic treatment of irrigation water and its effect on French bean (*Phaseolus vulgaris*). Water Reuse. 2023;**13**(4):545-558. DOI: 10.2166/wrd.2023.064

[23] Sarraf M, Kataria S, Taimourya H, Santos LO, Menegatti RD, Jain M, et al. Magnetic field (MF) applications in plants: An overview. Plants (Basel). 2020;**9**(9):1139. DOI: 10.3390/plants9091139

[24] Selim AFH, El-Nady MF. Physio-anatomical responses of drought stressed tomato plants to magnetic field. Acta Astronautica. 2011;**69**(7-8):387-396. DOI: 10.1016/j.actaastro.2011.05.025

[25] Mshenskaya NS, Grinberg MA, Kalyasova EA, Vodeneev VA, Ilin NV, Slyunyaev NN, et al. The effect of an extremely low-frequency electromagnetic field on the drought sensitivity of wheat plants. Plants (Basel). 2023;**12**(4):826. DOI: 10.3390/plants12040826

[26] Saletnik B, Saletnik A, Słysz E, Zaguła G, Bajcar M, Puchalska-Sarna A, et al. The static magnetic field regulates the structure, biochemical activity, and gene expression of plants. Molecules. 2022;**27**:5823. DOI: 10.3390/molecules27185823

[27] Ayesha S, Abideen Z, Haider G, Zulfiqar F, El-Keblawy A, Rasheed A, et al. Enhancing sustainable plant production and food security: Understanding the mechanisms and

impacts of electromagnetic fields. Plant Stress. 2023;**9**:100198. DOI: 10.1016/j.stress.2023.100198

[28] Radhakrishnan R, Kumari BD. Influence of pulsed magnetic field on soybean (*Glycine max* L.) seed germination, seedling growth and soil microbial population. Indian Journal of Biochemistry & Biophysics. 2013;**50**(4):312-317

[29] Yalçin S. The effect of magnetic field on three different varieties of soybean seed. Doğu Fen Bilimleri Dergisi. 2018;**1**(1):1-8

[30] El Sabagh A, Sorour S, Ueda A, Saneoka H, Islam MS. Enhancing nitrogen fixation in soybean (*Glycine max* L.) by composting under saline stress condition. Agricultural Advances. 2016;**5**:345-348

[31] Baghel L, Kataria S, Guruprasad KN. Static magnetic field treatment of seeds improves carbon and nitrogen metabolism under salinity stress in soybean. Bioelectromagnetics. 2016;**37**(7):455-470. DOI: 10.1002/bem.21988

[32] Belyavskaya N. Biological effects due to weak magnetic field on plants. Advances in Space Research. 2004;**34**(7):1566-1574. DOI: 10.1016/j.asr.2004.01.021

[33] Yang X, Li Z, Polyakova T, Dejneka A, Zablotskii V, Zhang X. Effect of static magnetic field on DNA synthesis: The interplay between DNA chirality and magnetic field left-right asymmetry. FASEB Bioadvances. 2020;**2**(4):254. DOI: 10.1096/fba.2019-00045

[34] Dziwulska-Hunek A, Niemczynowicz A, Kycia RA, Matwijczuk A, Kornarzyński K, Stadnik J, et al. Stimulation of soy seeds using environmentally friendly magnetic and electric fields. Scientific Reports. 2023;**13**(1):18085